EMERGENCY RESPONSE

TO

CHEMICAL AND BIOLOGICAL AGENTS

EMERGENCY RESPONSE
TO
CHEMICAL AND BIOLOGICAL AGENTS

John R. Cashman

LEWIS PUBLISHERS
Boca Raton London New York Washington, D.C.

Library of Congress Cataloging-in-Publication Data

Emergency response to chemical and biological agents / John R. Cashman.
    p. cm.
    Includes bibliographical references (p.) and index.
    ISBN 1-56670-355-7 (alk. paper)
    1. Hazardous substances—Accidents—Handbooks, manuals, etc. I. Title.
T55.3.H3 C377 1999
628.9′2 21—dc21   99-043796
                                                               CIP

Visit the CRC Press Web site at www.crcpress.com

© 2000 by John R. Cashman
Lewis Publishers is an imprint of CRC Press LLC

No claim to original U.S. Government works
International Standard Book Number 1-56670-355-7
Library of Congress Card Number 99-043796
Printed in the United States of America   2 3 4 5 6 7 8 9 0
Printed on acid-free paper

Introduction

Emergency responders are trained primarily to respond to fires or hazardous materials accidents. However, with the increase of terrorism in the United States, it is increasingly important that emergency response teams are trained to handle incidents involving biological, chemical, and nuclear agents. Utilizing recent case studies and interviews, this book presents a framework for emergency response to terrorist and criminal acts. It provides the emergency responder with data on the safe handling and disposal of biological and chemical agents, information on hazardous materials teams' operations, and numerous resources.

A NEW BALLGAME FOR HAZARDOUS MATERIALS TEAMS

Chief John Eversole is the hazardous materials coordinator for the Chicago, IL Fire Department. He has been a member of the department for 30 years and worked on some of the busiest engines, trucks, hook-and-ladders, and squad companies in the western part of the city. Eversole is a member of the National Fire Protection Association standard committee that produced national hazardous materials NFPA-471, NFPA-472, and NFPA-473. He is chairman of the International Association of Fire Chiefs' hazardous materials committee.

A leader on the street, and in the committee rooms where the nitty-gritty of hazardous materials response in the United States is thrashed out, Chief John Eversole is approachable to all persons and eloquent in expressing his thoughts. His command presence on an incident scene comes from extensive experience leading Haz Mat teams at major incidents.

We asked Chief Eversole for his opinions about response to chemical, biological, and nuclear materials as they relate to domestic terrorism in the United States. "I believe we have to look at the problem from several levels. The biggest problem we face in the fire service today is not all the 'special' problems such as terrorism. The biggest problem we face today is keeping the Haz Mat program operational, keeping our fire departments moving along and producing. It doesn't matter how many times we go out the door; people expect us to be right every time. Nobody says, 'Well, that's okay; I know you messed up.' That's not acceptable to people. We have a number of problems. Obviously, the financial crunch continues. The constraints get tighter and tighter.

"The requirements that we have to fulfill with the federal government, national standards, and everything else become more and more burdensome. Our fire service emergency responders have to become more technically correct. They have to know more, be able to do more. Still, we are expected to do this with fewer people and fewer dollars. That is, the biggest challenge we face is just the ability to answer the call: every time the bell rings, we have to remain able to get out the door and still be able to do the job properly.

"I believe one of our new challenges is the question of how we are going to handle terrorism on a local basis. There is much question, much debate, and much unknown about possible methods. There are many opinions from people who are not well informed. What we need to do, and what the Hazardous Materials Committee of the International Association of Fire Chiefs is trying desperately to do, is to get a really good handle on how we are going to handle terrorism as an emergency responder. If we were to look at all terrorist incidents, it is obvious that the first responder is the critical factor in establishing control over the incident and thereby minimizing the disaster. If the first responders do not do a good job, the incident could become uncontrollable. We have to make sure we act appropriately in a safe manner and do not overreact.

"We can examine the recent anthrax hoax incident in Washington, D.C. Here is the seat of the federal government. Here's the place where every special 'whatever-they-are' and every whiz in the nation practices. If you look at the news clips, you could see everybody who was actually doing the work: the police were basically trying to control the area, and the firefighters were handling the problem. 'Was everything that was done in Washington right or wrong?' I don't think that is the question. What the incident shows us is that the local people are going to handle a terrorist incident. In Washington, I didn't see any soldiers, any Marines, or any special government agencies there. I saw police and fire departments protecting their communities, and I think we have to understand that's the way it's going to be.

"In this incident, a package marked 'anthrax,' it is my understanding that the name was actually misspelled, was delivered to a building. Of course, everyone became very concerned. Obviously, a terrorist act is intended to scare people. Well, anthrax did a pretty good job of scaring people, and the Washington, D.C. fire department was called as was the police department. What they did, had to be done. They took what they thought were basic, logical steps. They controlled the area, they backed away, they brought in people who had special protective clothing and decontamination equipment, and they handled the incident. The fire service stood proud in a fire department that is significantly stressed, yet they produced. If they did not produce, who would have been there to effectively and in a reasonable time frame protect their city? No one would be there. The point I am making again is that we have to be able to get out and do our everyday business every day. We'll handle all these special problems that come along, such as terrorist acts, as they come, but our operations have to be done in a reasonably intelligent manner.

"There are people going across the country making suggestions that are just unreasonable. We have to look at the whole picture and say what we can reasonably and intelligently do to respond to terrorist acts. Other than throwing millions of dollars at a program as our federal government often does, the problem does not have an answer. It's how the money is spent, how it is intelligently used that matters. We need to put money and effort into what will really solve problems. Oftentimes, I question the ability to deliver an effective program; the bottom line is that protection of communities has to be up to the local emergency responders. Even a small town has to be adequately prepared to handle their known hazards just like a big city does. How they are going to do that is critical. It does not matter if it's a big town or a little town; we must provide reasonable protection in our communities. When we cannot supply reasonable protection in our communities, there will be anarchy.

"The reality of the situation is that we cannot wait for somebody to come from some long distance to get into our town, even if assistance can get there in a few hours ... which is unlikely in this type of operation. They will come in simply to 'remediate,' which is kind of a fancy term for 'pick up the pieces.' They are not going to be the people who will make the life or death decisions that have to be made. It will be the local people who will make these decisions; it should be the responsibility of the federal and state governments to help them prepare to do that. If the local people are sitting in their communities thinking the federal government is going to totally bail them out of a terrorist Haz Mat situation, they are mistaken. I don't believe this is going to happen.

"In large metropolitan areas, we do have significant resources but we still need help. We are going to need some basic help that the federal government has not yet seen themselves ready to give. There are a lot of fancy words used, but the bottom line is that if we are going to deal with a terrorism situation, or a reported terrorism situation, we not only have to have trained people but we need to have basic specialized equipment that will allow us to determine if this incident is a terrorist incident or not.

"We've told firefighters for a long time that you can't stand there and sniff it. It takes some specialized equipment to determine the presence of a chemical warfare agent release. These instruments have to be made available to the emergency local responders. We again have to decide who is going to get them, what they are going to get, how they are going to be trained to use such equipment, what is practical and not practical. We've got to take these actions.

"The Hazardous Materials Committee within the International Association of Fire Chiefs met yesterday to make sure that our committee is actively pursuing decisions as to what should be done, how should we do it, who should do it, what we need to deal with the situation, and what kind of equipment is needed. We are going to have to make these recommendations like we've done many times in the past. People are aware we have not always taken popular stands, but the fire chiefs' committee studies varied issues and questions. There are a number of committee members who are very knowledgeable, not only in the fire service but in specialty areas. We have added a member to the committee who is perhaps one of the most knowledgeable persons in the fire service with regard to terrorism. We will make our recommendations to the board of directors of the International Association of Fire Chiefs. They will then take these measures and act on them, and hopefully agree with us. The Association will say this is what we think the management of the fire service feels is what we should be doing, how we should protect our people and our communities, and how we are going to react to terrorism.

"We are going to be worried and concerned about terrorist acts, but every day I have to be prepared respond to an anhydrous ammonia leak, a chlorine leak, or another type of spill. If we go back to the first part of our interview, we have still got to put capable teams on the street every day. That's going to be my number one priority for the next six months while I am chairman of the I.A.F.C. Hazardous

Materials Committee. Until the day I leave this job, when the bell rings, I will roll up the big door and put hazardous materials response teams on the street who will protect our communities. That's the biggest thing we have to worry about. We will deal with terrorism as we have dealt with communicable disease and everything else we have had to deal with. We will deal with terrorism just as another part of what we are already doing.

"Like everything else we do in the fire service, if we go to a house fire we need to bring engine companies, hose, water, ladders, and all associated tools. If we are going to a radiological incident, then we should know what the product is, how it reacts, how we control it, and the method required to handle the problem. It's just that simple. Let's not complicate this; don't let people play the part of the bogeyman and scare people. You cannot allow this to happen. This is an unbelievably great country and an advanced technological world we live in. The fire service has to maintain enough intelligence that they know they can do certain things. Like the poker player in the old Kenny Rogers song, firefighters have to know when to stand and when to fold. Well, we do that in fires. We say, 'Time to back out of this building that is not safe,' and we get out. We have to know what we can do with terrorist weapons of mass destruction, when to fold and when to hold. There is instrumentation and methodology that's readily available here so that we can establish meaningful operating procedures."

Contact: Chief John Eversole, Hazardous Materials Coordinator, City of Chicago Fire Department, 558 West DeKoven Street, Chicago, IL 60607; 312-747-6582.

BACKGROUND

A hazardous materials response team (HMRT) is an organized group of employees who are designated by the employer and who are expected to control actual or potential situations in which close approach to leaking or spilled hazardous substances may be required A Haz Mat team may be a separate component of a fire brigade or a fire department. While these teams are mainly concerned with handling hazardous materials accidents, more and more they are faced with intentional criminal and terrorists acts involving hazardous agents.

There is a lot of confusion about what terrorism is and what it is not. The Federal Bureau of Investigation defines terrorism as, "the unlawful use of force or violence against persons or property to intimidate or coerce a Government, the civilian population, or any segment thereof, in furtherance of political or social objectives. Domestic terrorism involves groups of individuals who are based and operate entirely within the United States and Puerto Rico without foreign direction and whose acts are directed at the elements of the U.S. Government or population. International terrorism is the unlawful use of force or violence committed by a group or individual who has some connection to a foreign power or whose activities transcend national boundaries, against persons or property to intimidate or coerce a government, the civilian population, or any subject thereof, in furtherance of political or social objectives."

According to the U.S. Department of Defense, terrorism is "the calculated use of violence or the threat of violence to inculcate fear; intended to coerce or to intimidate governments of societies in the pursuit of goals that are generally political, religious, or ideological."

Let's deal with some recent incidents that were **not** terrorist acts. A trailer truck carrying unirradiated nuclear fuel through downtown Springfield, MA was hit by a drunken driver going the wrong way on Interstate 91. The truck crashed and burned on an elevated section of highway in front of two major hotels at 3:18 a.m. The local fire department could not receive valid information from federal agencies, the General Electric Company who made and shipped the nuclear material, or the Vermont Yankee nuclear power facility where the cargo was headed until after the truck and sensitive cargo had been burned by flames estimated at 1200 degrees.

On April 2, 1997, Air Force Captain Craig D. Button flew his bomb-laden A-10 Thunderbolt jet, an aircraft costing $9 million, away from two other A-10s during a training mission at Davis-Monthan Air Force Base in Tucson, AZ and headed for Colorado. Button's disappearance sparked rumors of varied conspiracy theories coming as it did shortly before the anniversary of both the Waco, TX Branch Davidian killings and the killings at the Federal building in Oklahoma City, OK on April 19, 1995.

On Thursday, April 24, 1997, it became apparent that the F.B.I. and the Pentagon were looking for a tractor trailer carrying four training missiles, each valued at $150,000, from a Boeing plant in Duluth, GA to Cannon Air Force Base in Clovis, NM. The truck had been expected to arrive in New Mexico the previous Monday, and carried a satellite-monitored tracking beacon supplied by Defense Tracking System of Norfolk, VA. The truck vanished from computer screens on April 24, 1997. When a vehicle carrying munitions, weapons, or other sensitive equipment has disappeared for four hours, trackers call state police for assistance.

The F.B.I. eventually put out a detention alert on Ronald D. Coy, 42, of Middletown, OH. Coy was believed to be driving a black 1991 Kenworth with mauve and green pinstripes and the name, "Miss Honey Jean" on the bug shield. On April 26, 1997, Coy was found at the Flying J Truck Stop in Orange near the Louisiana state line, while the missiles were found 300 miles away at a fenced-in lumber yard at Ranger, TX. Reportedly, officials were expecting Coy to pull into that particular truck stop. There was an undercover official there who saw Coy pull in, watched the truck for about three minutes, and then gave a signal to about a dozen police officers from a number of different agencies who rushed in and surrounded the truck and arrested the driver. When taken into custody by the F.B.I., Coy was apparently alone, unarmed, and put up no resistance. Although he could face charges of theft of an interstate shipment and theft of government property, Coy's intentions were not immediately known. He was actually arrested on a charge of "wire fraud."

Another truck with a satellite tracking system was lost in Texas on the same day government agencies were searching for Coy. Carrying machine guns and mortars to the U.S. Marine Corps base at Camp Pendleton on the California coast, the truck was no longer responding to satellite tracking. The trucking company immediately canceled the driver's fuel credit card. When refused credit at a truck stop near El Paso, TX, the driver called the trucking company. The truck's tracing beacon had failed, and the driver did not know he was being sought.

No terrorist activity was involved in any of these incidents, but rather just run-of-the-mill, everyday problems in handling and security relative to transportation of hazardous materials. It is going to get worse.

High-level radioactive waste from Canada and other foreign countries is on the roads and rails of the United States and will undergo a massive increase in the near future. The U.S. Department of Energy (U.S. DOE) seeks to stem proliferation around the world of the sort of highly enriched uranium that can be used to make nuclear weapons.

U.S. DOE documents indicate more than 20 tons of such nuclear waste are to be transferred from 42 foreign countries ranging from Bangladesh to Romania to Brazil. Uranium containing more than 20% of the isotope uranium-235 is considered to be highly enriched. The more highly enriched the uranium, the more easily it can be used in weapons. Draft rules relative to this transportation of highly hazardous materials forbid notifying the general public, and state: "Motor Carrier Safety Coordinator shall forward notification of shipment to appropriate officials on a 'need to know' basis only. Otherwise information will be confidential."

Most of the foreign waste will come by ship to naval weapons stations in Charleston, SC and Concord, CA and will be shipped from these ports to Savannah River, SC and a similar facility in Idaho. About 15% of the total waste will come from Canada and will be shipped by road and rail through the United States. Shipping high-level radioactive waste around the United States has been called a "mobile Chernobyl bill" by some people.

Regarding another type of radioactive waste, the U.S. Senate passed a bill on April 16, 1997 that would allow shipments of nuclear waste from domestic commercial reactors to a "temporary" storage site in Nevada. At the time of this writing, such commercial sites have to store their radioactive waste "temporarily" on site at each local facility. The commercial facilities dearly want to get rid of what they now have to store since there is no permanent storage facility for high-level nuclear waste. President Clinton promised to veto the senate bill saying the temporary site in Nevada would relieve pressure to find a permanent waste site.

Hazardous materials response teams deal routinely with chemical and radioactive or nuclear threats, and some teams have handled biological hazards. Terrorists, however, tend to look at such materials as a land of opportunity for their respective causes.

Nuclear, biological, and chemical (NBC) materials are more readily available than ever before, and the total threat is growing day-by-day. Innocent people, including women and children, are major targets of terrorists because in threatening these, terrorists gain the most publicity for their cause. An infinite pool of potential victims is created when terrorists do not recognize the innocence of certain groups. Anyone, including you, can be a victim. The more common you are, the better it plays with the media, and the more publicity created. Terrorists demand publicity, and their particular targets are selected with an eye toward violence that will produce the most notoriety.

The world is changing to a point where horrendous violence has become possible by any person. You no longer need an army, a state, or a country behind you. You can do it alone. Rapid changes in technology have already stimulated broad scale terrorism in other countries. The question is, "Will the United States become engulfed by terrorists?"

In the past, most terrorist attacks were politically motivated. Now they can be religious, cult, ethnic, nationalistic or right-wing such as Neo-Nazi or anti-Semitic; or issue-specific movements like animals rights, anti-abortion, and environmentalism. Terrorist acts are cheap, can gain a lot of attention, and can be low-risk to the perpetrator. In both time and resources, an expenditure total can be small. With NBC materials, a weapon of mass destruction can be made, delivered, and set-off simply. The explosion that rent the federal building in Oklahoma City was not dynamite, a plastic explosive, nor any sophisticated weapon; it was fertilizer (ammonium nitrate) mixed with fuel oil. Such a product is available to anyone. One of the most horrendous Haz Mat incidents that ever occurred in the United States was a pier explosion of a cargo ship filled with ammonium nitrate that killed 561 Americans and injured thousands in Texas City, TX on April 16, 1947.

The lead-time required to mount a terrorist attack can be very short, the risk is currently minimal in the United States, and the potential for success is great. "Loners," like the one(s) who blasted the Oklahoma City federal building, are extremely difficult to identify before an attack. Security procedures are all-important to terrorists, so gaining information on them can be extremely difficult.

For foreign terrorists, the United States is thought of as the greatest power in the world, and we are no longer immune to terrorist acts. All the world's problems are laid at the feet of the United States, while at the same time the United States is blamed for not solving the world's problems. Our country could become the main target of international terrorism.

Anyone, legally or illegally, can enter the United States. It is a completely open country, and terrorists would be able to work safely here. By some accounts, immigration is out of control. In May, 1996, Border Patrol agents in the Swanton, VT sector covering 261 miles of the quiet U.S./Canadian border apprehended 123 deportable aliens from Canada, Mexico, El Salvador, Honduras, Cuba, Dominican Republic, India, Pakistan, Bangladesh, China, Philippines, South Korea, Lebanon, Sri Lanka, Togo, Spain, Romania, Somalia, Liberia, Jordan, Algeria, Sudan, Israel, Egypt, France, Turkey, Armenia, Bosnia, Bulgaria, and Afghanistan.

Recently in the United States, terrorism has been blamed for attacks on the World Trade center in New York City in 1993 where seven died and fourteen were injured. In 1995, a U.S. court convicted 21 people for involvement in terrorist-related activities, including Egyptian Shaykh Omar Abdel Rahman and nine followers found guilty of seditious conspiracy charges in plotting to bomb major New York City landmarks and assassinate prominent politicians. Other incidents included the U.S. Capitol Building in Washington, D.C., Mobil Oil headquarters in New York City, a derailment of Amtrak's Sunset Limited near Hyder, AZ on October 9, 1995, and the ammonium nitrate/fuel oil explosion at the Alfred P. Murrah Federal Building in Oklahoma City on April 19, 1995 which killed 168 American citizens and injured hundreds more. In 1995, a member of the El Rukns street gang in Chicago was charged with over 40 counts of conspiracy to conduct terrorist acts in the United States on behalf of a foreign country (Libya). In May of 1995, an American citizen obtained three vials of bubonic plague from a firm in Maryland, but it was unclear why he ordered the bacteria.

In July of 1995, a member of the Animal Liberation Front (ALF) pled guilty to arson at the Mink Research Facility at Michigan State University. In December of 1995, an Internal Revenue Service employee found a 30-gallon plastic drum loaded with 100 pounds of ammonium nitrate fertilizer and fuel oil in a parking lot behind the IRS building. Ramzi Ahmed Yowef went on trial May 13, 1996 for his alleged involvement in the World Trade Center bombing. Musa Abu Marzook of the HAMAS group was arrested at Kennedy International Airport while attempting to enter the United States.

Bombing is the most popular way of carrying out terrorists activities in the United States. Although not all were terrorist acts, during the period of 1990 to 1995, there were 12,512 total bombings and 5108 attempted bombings in the United States which killed 355 persons and injured 3176.

Most of the poison agent victims of World War I are now gone, but we could have a whole new generation of victims with us once again. It is important to realize how the victims would die. All eight of the following chemical agents — two nerve, two blood, two blister, and two choking agents — are all available to the U.S. military, to 26 other governments according to the F.B.I., and increasingly to terrorist organizations. The C.I.A. states there are at least ten countries that are believed to be conducting research on biological weapons.

A blister gas known as lewisite smells like geraniums to the victim. The median lethal dosage expressed as mg-min/m^3 is 1200 to 1500 by inhalation, or 100,000 by skin exposure. The action of lewisite is very rapid, and the best known decontamination method used by emergency responders to an incident would be to scrub victims down with calcium hypochlorite.

The blood agent hydrogen cyanide, known as AC to the military and HCN to commercial users, can be a colorless gas or liquid and smells like bitter almonds. Hydrogen cyanide is the only chemical in the United States which is required to be carried in a "candystriper," a specially colored railroad tank car that might hopefully warn away emergency responders who might be called to a railroad wreck. It has a median lethal dosage of 100 for resting persons, and its action as a chemical agent is extremely rapid. Decontamination would be by application and scrubbing with sodium hydroxide.

The nerve agent known as sarin, used by the Japanese cult Aum Shinrikyo to kill 12 people and injure 5500 in the Tokyo subway, has a rapid effect on victims and a median lethal dosage that varies widely. Another nerve agent that has a classified chemical formula and is known as VX has a median lethal dose of 100 and is also considered to have a rapid effect on victims. For the first, sodium hydroxide can be used for decon, while for the second, calcium hypochlorite could be used.

Phosgene, a choking agent that is common in commercial usage and has a median lethal dosage of 3200, can smell like newly mown hay or green corn, and is much slower in the rate of action, requiring up to three hours in some cases. Another choking gas, diphosgene, with the same median lethal dosage has the same smell and the same rate of action. The decon method used for either phosgene or diphosgene is simple aeration.

The blood agent, cyanogen chloride, smells like bitter almonds but less so than the other blood agent, hydrogen cyanide, yet it may have the fastest rate of action. The blister agent, distilled mustard or "HD," a colorless to pale liquid with a median lethal dosage of 15,000 by inhalation and 10,000 by skin exposure, may have the slowest rate of action which can be delayed for hours or days.

In addition to chemical materials like the above poison agents, terrorists also have access to biological materials such as anthrax, typhus, and cholera. Anthrax is an infectious, usually fatal disease of warm-blooded animals caused by *Bacillus anthracis* and transmission to man. The word itself comes from the Latin word which means "malignant boil," or the Greek word for "carbuncle" for a condition characterized by malignant ulcers on victims.

The name typhus is applied to various forms of infectious disease caused by microorganisms which in peace time can be flea-borne, louse-borne, or mite-borne. Symptoms include severe headache, sustained high fever, depression, delirium and red rashes.

Cholera is an acute, often fatal, infectious disease caused by the microorganism *Vibrio comma.* Symptoms are watery diarrhea, vomiting, cramps, suppression of urine, and collapse.

Ricin, which is made from castor beans, is one of the most toxic biological agents. Ricin first came to the attention of the public in 1978 when it was used to assassinate Georgi Markov who was stabbed with the point of an umbrella while waiting at a bus stop in London. On October 25, 1995, four Minnesota men were convicted under the Biological Weapons Anti-Terrorism Act of 1989 for manufacturing and intending to use ricin to kill a Deputy U.S. Marshal and a Sheriff.

Recently, the federal government has been quietly preparing for domestic terrorist attacks within the United States. The Federal Bureau of Investigation which is in charge of antiterrorist activities within U.S. borders operates a computer database containing information on 3000 suspected terrorist groups and 200,000 individuals. S.735, the "Antiterrorism and Effective Death Penalty Act of 1996," was passed by Congress and signed by President Clinton on April 24, 1996. The Federal Emergency Management Agency has issued an annex to its terrorist plan. The National Fire Protection Association has added a terrorist tentative interim amendment to its Standard 472, "Standards for Professional Competence of Responders to Hazardous Materials Incidents" which is used by both fire departments and industry. The U.S. Department of Defense has a newly created "Domestic Preparedness for Chemical and Biological Terrorism" unit stationed at the Aberdeen Proving Grounds in Maryland. This department will visit 120 cities to train local responders how to react to terrorist acts; that is, the U.S. military will be training civilians in peace time.

In addition, military units can now respond in military operations within the United States during peace time. Ten years ago, the Reagan Administration secured a loosening of the laws barring the use of military forces in domestic law enforcement.

Joint Task Force 6, a military unit based in El Paso, TX conducts anti-drug border operations for domestic Federal agents who work for the U.S. Border Patrol. On May 20, 1997, four camouflaged U.S. Marines from Camp Pendleton in California guarding the border in the little town of Redford, TX shot and killed an 18-year-old American goat herder.

The U.S. Marine Corps has a 400-person Chemical Biological Incident Force (CBIRF) stationed at Camp Lejeune, NC that has been established under Presidential Decision Directive 39. Part of this team recently conducted a training exercise at the Rayburn House Office Building in Washington, D.C. during a simulated sarin poison gas attack. In addition, there are 43 chemical response teams within the National Guard.

Every outdoor deposit box for the U.S. Postal Service now has the following notice attached: "Because of heightened security, the following types of mail may not be placed in this receptacle: *All domestic mail, weighing 16 ounces or over, that bears stamps. International mail and military APO/FPO mail weighing 16 ounces or over.* Please take this mail in person to a retail clerk in a post office."

For airline travel, a presidential commission had proposed that every passenger be matched with every bag on domestic flights. However, the Federal Aviation Administration (F.A.A.) directed that by December 31, 1997 domestic airlines match only about five percent of fliers with their bags (on international flights, all bags are matched). By the same date, domestic airlines must begin using a comprehensive profiling system that will identify those who might pose a security risk. As an example, if you pay for your airline ticket with cash, you will find your name on the list.

AUTHOR

John R. Cashman has studied, interviewed, or worked during the past 21 years with more than 700 hazardous materials response teams (HMRTs) that represented fire departments, police, health agencies, emergency management, civil defense, the military, state and federal governments, commercial response contractors, and private industry. He has provided consulting services in hazardous materials response to International Business Machines, Digital Equipment Company, Bombardier Rail Car Division, Dowty Electronics, Burlington Northern Railroad, Rossignol Ski Company, Cleveland State University, the State University of New York (S.U.N.Y.), fire service organizations, and emergency management agencies.

DEDICATION

This book is dedicated to the memory the 16 persons who died in Waverly, TN on February 24, 1978 and the 8 who died in Youngstown, FL two days later. These two rail incidents on the same weekend also seriously injured 184 persons and brought hazardous materials incidents to the attention of the media.

Table of Contents

1   Terrorists on U.S. Soil

"Did you really think we wanted these laws to be observed? We want them to be broken; there's no way to rule innocent men. The only power any government has is the power to crack down on criminals. When there aren't enough criminals, one makes them. One declares so many things to be a crime that it becomes impossible for men to live without breaking laws. Who wants a nation of law abiding citizens? What's there in that for anyone? But just pass the kind of laws that can neither be observed, nor enforced, nor objectively interpreted and you create a nation of lawbreakers and then you cash in on guilt. Now that's the system, Mr. Reardon, that's the game, and once you understand it you'll be much easier to deal with."

Ayn Rand, *Atlas Shrugged,* 1959

INTRODUCTION

President Bill Clinton has put 100,000 additional police officers on the streets of the United States during his terms in office. By December of 1997, Congress and the White House nearly tripled the F.B.I.'s counter-terrorism budget since 1994, allowing the bureau to add 350 new agents to domestic terrorism cases. The president's fiscal year 2000 budget request calls for $9.85 billion to support counter-terrorism efforts. The U.S. Department of Defense through the Pentagon decided in early 1999 to ask the president for the power to install a military leader for the continental United States due to a growing concern regarding domestic terrorist insurrection or foreign-led terrorist attacks.

For many years, violence has been as American as apple pie. Since 1991, the United States has been involved in six war-like acts: "Desert Storm" (Iraq One), Somalia, Haiti, Bosnia, "Desert Fox" (Iraq Two), and Yugoslavia/Kosovo. We have moved from the Manson Family in California to Eric Harris and Dylan Klebold in Littleton, CO.

THE WEATHER

In the last 30 years, insurrection against the U.S. government has moved from the radical left to the radical right. They were not called "terrorists" in 1969; they were called the "Weather," more commonly known as the "Weathermen," although a large number of the 1000 to 1500 members were female. They came from wealth or upper middle class families, attended the best colleges, hated individualism, and believed in armed struggle in the United States for political purposes. They may have harbored a feeling that they were members of an upper class, and may have abhorred dealing with the lower classes whom they attempted to organize. Their favorite singer, or poet, was Bob Dylan, and they adopted the name of their insurrection group from the lyrics of one of his songs: "You don't need a weatherman to know which way the wind blows."

The Weather members were bomb makers who caused millions of dollars in property damage with such devices; but since they were among the upper class, they never held blue-collar jobs that might have taught them about the process of electrical wiring and current. Such inexperience would prove detrimental. In New York City's Greenwich Village, on March 6, 1970 at a townhouse on West 11th Street, five members of the Weather group were making bombs when one exploded killing three members and badly injuring two. The two survivors, Cathy Wilkerson and another female, climbed out of the wrecked Wilkerson house with their clothes blown off and went into hiding, as did all other Weathermen who became the Weather Underground.

The Weather Underground members did continue bombings to make the capitalist system collapse, but underground they were out of touch and totally alienated themselves from the "lower classes" they had hoped to organize. Many stayed hidden for years with cash, new birth records, false names and false social security numbers. After several years, most of the leaders came out of hiding and turned themselves in to the F.B.I. and were given lenient sentences or no jail time. They were a thing of the past, gone and forgotten.

THE ORDER

Thirty years ago, the Weathermen ruled the radical left in insurrection attacks on the federal government. Today, the radical right seems to have led the way into the year 2000 with such acts. Selecting the most successful of the radical right groups is not that difficult. In the material that follows, the author deals with "The Order," a militant group started by nine men that drew members from throughout the Northwest who believed in armed struggle in the United States for political purposes. The leader of The Order was Robert "Bob" Jay Mathews who lived in Metaline Falls located in Pend Oreille County in northeast Washington state about 10 miles from British Columbia, Canada.

In 1982 and 1983, farms and ranches in mid-America and farther west were being broadly repossessed by government agencies, banks, credit unions, and other money lenders because of lack of payment for taxes and loans. Many farmers and ranchers lost their means of making a living and their homes. During these poor economic times, many farmers who saw their lives falling apart began to feel rage and fury. In North Dakota during 1977, Gordon Kahl, a tax protester, was sentenced to a year in jail plus five years of probation. Once released from prison, he neglected to show up for probation and again failed to pay his taxes. Federal officers were very slow, or afraid, to arrest to Kahl because they recognized the country was in turmoil due to foreclosures. On February 13, 1983, federal officers set up roadblock on a hill blocking Kahl's vehicle. He and his son, Yorie, fully armed, got out of the vehicle and waited. Someone fired a shot, and all Hell broke loose. Yorie was hit in the stomach, and Gordon Kahl shot two federal marshals to death and wounded two more. Gordon spoke to the two marshals who were lying wounded in the road, drove his son to the hospital, then went underground.

Kahl tried to hide in the Ozark Mountains of Arkansas, but an informant told federal agents where he was located, and well-armed agents staged an early morning attack on June 3, 1983. Kahl, a trained marksman, shot and killed a county sheriff. Kahl fired at the agents, and the agents fired at Kahl in what turned into a stand-off. Eventually, the agents used flares to set the house afire, and Gordon Kahl died in the inferno, a hero to the radical right.

Bob Mathews of The Order in the Northwest regarded Gordon Kahl as an inspiration and soon after Kahl was killed, swore an oath to fight the government of the United States in an armed struggle for political purposes. This 30-year-old blue-collar working man and farmer began a group that became the most organized collection of terrorists to ever operate in North America. Their issue was anger, fury with their own government which had turned against them. Morals were a factor, but The Order, nine men at the start, was strictly a political movement designed to rob and kill in the name of revolution.

They planned to destroy electrical power grids, place poison in water supplies, and create sabotage and mayhem at the Los Angeles Olympic Summer Games according to later reports by the media. It was reported that every member was assigned a well-known person to assassinate. They considered their planned actions as war against the state to protect their own race and beliefs. How do you take up war with your own government? Only in the United States do people ask questions like this. It is done every day by people around the world without a second thought.

In 1983, Bob Mathews turned words into deeds to create chaos and disrupt the system in the Northwest in order to bring down the recognized government. He was a good looking man with a baby-face who became action-oriented to achieve his goal, but he was not a good public speaker. However, he had fully evident charisma; even without ability in public speaking, he had a special quality that gave him influence over large numbers of people. He could achieve action in members of The Order who would back Mathews with their very lives if required. Bob Mathews knew it would take money and started out to obtain a lot of cash.

The Order first tried to create a war chest by counterfeiting $50 bills to bankroll the radical right. Their printing process left much to be desired, but they still managed to pass a lot of fake money through Tom Martinez, a money launderer they brought in from the eastern part of the country. More money was needed to fund groups in the Northwest who had the same ideology as the members of The Order, also known as "The Bruders Schweigen," "The Underground," "The White American Revolutionary Army," and "The Aryan Resistance Movement."

On December 20, 1983, a lone gunman robbed the Innis Arden branch of the City Bank in Seattle, WA. The take was $25,952 and The Order had its first sizable bankroll. Bob Mathews, an action person if ever there was one, had robbed his first bank, and it was easy. It got even easier as he progressed.

Money in his pocket was like a goad to Bob Mathews; he wanted more to give away to like-minded groups throughout the country, and used bank and armored car robbery, counterfeiting, and interstate transportation of stolen property to get it. Along the way he threatened and murdered people viewed as hostile to the aims and purposes of The Order and destroyed by fire or explosives the property of persons who did not hold the same opinions. The members of the insurgents used aliases, false identification, codes and code names to confuse federal agents and local police. As an example, one member, Bruce Carroll Pierce, was also known as Brigham, Logan, Brigham Young, William Allen Rogers, Scott Adam Walker, Roger Martin, Roger Morton, Michael Schmidt, Charles Lee Austin, Lyle Dean Nash, and Larry Martin. Members utilized communication centers linked by telephones located in various states, rented under false names a number of "safe houses" in various states, purchased or leased land, and bought firearms and munitions. They purchased explosives, vehicles, military equipment, and survival gear to equip and train members of The Order.

The following members were later indicted for murder among other crimes. Bruce Pierce was a leader of The Order, operated a "cell" meant to obtain money through robbery and other crimes, and headed an assassination squad within the group. Randolph Duey was another leader; he screened new recruits within The Order to determine ideological beliefs. Richard Kemp trained new recruits within the group. Richard Scutari was chief of security for The Order and gave members and recruits voice stress analyzer tests to attempt to determine if they were undercover federal agents or possible informants. David Lane operated a communications center in Boise, ID through which members maintained communication by telephone. James Dye and Jean Margaret Craig also committed the crimes of murder. David Tate joined a "cell" to obtain money by robbery and other crimes, as did Thomas Bentley.

Other members of The Order were charged with committing a number of crimes less serious than murder, including arson, counterfeiting, dealing in stolen property, robbery, and conspiracy. A central purpose of the organization was to distribute money gained from robbery and counterfeiting to individuals and groups sympathetic to the beliefs of The Order, thus allowing these other groups to recruit new members. Most of the charges stated as "dealing in stolen property" dealt with money from robberies and counterfeiting being transferred to other groups by members of The Order.

On January 30, 1984, Bruce Pierce and Gary Yarbrough held up the Valley Branch of the Washington Mutual Savings Bank located in Spokane, WA and came away with $3,600. Both men were leaders of The Order. Yarbrough was in charge of recruiting new members from prisons, but only a few members had trouble with the law before they joined the order. Yarbrough on April 22, 1984 caused a fire and explosion which damaged a building housing the Embassy Theater.

On April 23, 1984, a Continental Armored Transport vehicle at Northgate Shopping Mall in Seattle, WA was relieved of $500,000. Bob Mathews and other members of The Order in mid-1984 had decided to steal from armored cars. Ronald King utilized his position as Operations Manager for Brinks Armored Car Service to help The Order in the planning and facilitating of armed robberies. As an example, on August 29, 1984, King drew a diagram of the Brinks Armored Service Company cash vault in San Francisco, CA and gave it to Bruce Pierce, a member of The Order.

Walter Edward West was a member of The Order who fell out of favor with Bob Mathews because of suspicions that West might become an informer. On June 1, 1984, Randolph Duey, Richard Kemp, James Dye, David Tate, Thomas Bentley, and Bob Mathews tried to kill West, first by hitting him on the head with a hammer, and then putting him out of his misery with a rifle bullet. The victim's body was never found. He was the first known person killed by members of The Order, and the first member killed by other members. Seventeen days later in Denver, CO, Bruce Pierce, Richard Scutari, David Lane, Jean Margaret Craig, and Bob Mathews reportedly killed Jewish radio talk show host, Alan Berg, after following his Volkswagen to his home. Bruce Pierce was reportedly the trigger man who put 13 bullets from a .45 caliber automatic machine pistol similar to a MAC into Berg's body.

Near Ukiah, CA on July 19, 1984, Bruce Pierce, Gary Yarbrough, Randoph Duey, Andred Barnhill, Denver Parmenter, II, Richard Kemp, Richard Scutari, Randall Evens, Robert Merki, James Dye, Bob Mathews, and others stole $3.6 million in currency from a Brinks armored car, the biggest haul from an armored car in the United States at that time. However, Mathews made a possibly fatal mistake by dropping a traceable gun in the Brinks truck. Police would eventually know his name, but not where to find him.

On October 18, 1984, an arrest warrant was issued for Gary Yarbrough charging him with assaulting federal officers by firing a weapon at two F.B.I. agents. In November 1984, Ronald King, Operations Manager for Brinks, informed Bruce Pierce that no security changes had been made in the Brinks armored car route at Ukiah, CA.

Bob Mathews robbed his first bank on December 20, 1983; slightly less than a year later, he would be dead. The police had arrested Tom Martinez, the eastern money launderer who had passed counterfeit funds for The Order, for similar activity. For consideration in sentencing, Martinez told the F.B.I. that he knew where Bob Mathews would be on November 24, 1984 because he had an appointment in Portland, OR on that date with Mathews. Tom Martinez became an F.B.I. informant. On his arrival in Portland, he was picked up and taken to a motel with Mathews driving and Gary Yarbrough holding a machine gun in the rear seat. Martinez assumed he would be followed from the pickup point to the motel by an undercover F.B.I. vehicle, and he became a bit unhinged when Mathews, on a hunch or by a standard security practice, pulled into a dead-end street and parked the vehicle. No car followed into the dead-end. After a few minutes, Mathews drove the vehicle to the motel. After a short meeting, Martinez went to his Room 14 and Mathews and Gary Yarbrough went up to their rooms on the second floor. At dawn, an F.B.I. agent called Room 14 and told Martinez to lie low; they were going to take down Mathews and Yarbrough. Within minutes, there were sounds of running and guns firing. Mathews wounded one F.B.I. officer and was wounded in one hand himself in a running gun battle, but ran down the street and got away. One F.B.I. agent who fired a shot at Mathews wounded the motel manager instead. Yarbrough jumped from a second floor window but was captured. F.B.I. agents at the motel who searched the car that Mathews had been driving found a machine gun equipped with a silencer, a hand grenade, and approximately $30,000 in cash.

Bob Mathews apparently went to one or more of five "safe houses" in the area around the Mount Hood ski areas in Oregon, and treated the bullet wound in his hand. Two days later, he took a ferry from Mukilteo to Whidbey Island in western Washington, seemingly for a meeting of the clan including wives and children. The U.S. government charged in a subsequent trial that a number of people did conspire to harbor and conceal Bob Mathews in one of three "safe houses" used by The Order on Whidbey Island. These houses were located at 3306 South Smugglers Cove Road in Greenbank, 1749 North Bluff Road in Coupeville, and 2359 South Hidden Beach Road in Greenbank which are all located on Whidbey Island.

Members, wives, and children at one or more of these three houses were well armed with a grenade, 2 Uzi machine guns having their serial numbers obliterated and removed, 12 ounces of C-4 explosives, a Ruger Mini-14 .223 caliber semi-automatic rifle, a Universal 30 caliber rifle, an Ithaca 12-gauge shotgun, a Heckler & Koch 9mm handgun, a Ruger Speed Six .357 magnum handgun, and a Harrington & Richardson handgun. Ammunition for these weapons included 36 boxes of 5.56mm shells (20 rounds per box), 3 clips of 30 caliber shells (30 plus rounds per clip), 6 clips of 9mm shells (35 rounds per clip), 3 clips of 30 caliber shells (1 clip of 30 rounds/2 clips of 20 rounds), 6 12-gauge shotgun shells, 7 rounds of .38 special ammunition, 2 clips of 9mm shells (8 rounds per clip), 1 clip of 9mm shells (9 rounds), 1 shotgun belt (17 .12 gauge/14 .44 caliber rounds/24 .357 caliber rounds), 50 rounds of .22 caliber shells, 100 rounds of 9mm shells, and 30 rounds of .30 caliber shells.

Federal government agents watched the three houses and noticed the occupants visited among the houses. The agents obtained search warrants for all three houses. In the morning of December 7, 1984, (Pearl Harbor Day), reportedly up to 300 law enforcement officers took up positions in the general area of the three "safe houses," stopped airplane and boat traffic from coming into the area, and evacuated all nearby residents. At the house at 1749 North Bluff Road in Coupeville at 7:45 a.m., Randolph Duey exited the back door with an Uzi machine gun in one hand and a semi-automatic pistol in the other. F.B.I. agents with automatic weapons stepped out from hiding and demanded Duey surrender which he did. Duey admitted that Bob Mathews was in the house at 3306 Smugglers Cove Road, but refused to say how many others were with him.

At 2359 Hidden Beach Road, Robert Merki and Sharon Merki would not come out of the building for several hours until they burned a considerable quantity of documents and other apparent evidence. Shortly after 11:00 a.m., they did surrender and informed police agents there was an elderly woman still in the house. Both admitted Bob Mathews was at the house on Smugglers Cove Road, but refused to admit how many others were with him.

F.B.I. agents surrounding the house at 3306 Smugglers Cove Road eventually established contact on a field telephone with a male occupant who identified himself as "Robert Mathews." Mathews was not ready to surrender as yet, but said another male in the house did wish to do so. A "John Doe" with a list of aliases came out carrying a bag with $40,000 inside; "John Doe" claimed it was his money, but refused to speak about how many other people were in the house. There was a burst of automatic gunfire within the house; whether it signaled Mathew's decision to fight is not known, but it did tell the agents Mathews was armed with at least one automatic weapon. Attempts at negotiations continued on through the day and into December 8, 1984. Mathews then refused to negotiate any more or to surrender. About 2:00 p.m., a single shot was heard from inside the house as if Mathews had committed suicide. Agents fired tear gas shells into the house since they did not know Bob Mathews had a gas mask available.

After waiting a half-hour or so, a SWAT team of four agents entered the house expecting to find Mathews' body, but he had barricaded himself on the second floor and fired bullets from an automatic weapon through the floor/ceiling at the SWAT team. They were able to leave the house safely. Shortly after dark, an agent fired three white phosphorus "illumination" flares into the house. Anyone who has served in the military knows what white phosphorus can do. The house burst into flames as both sides fired away. Eventually, the house exploded. Bob Mathews' body was found in the wreckage of the house. A number of people from the radical right have noted that Gordon Kahl in Arkansas, Bob Mathews on Whidbey Island, WA, and the Branch Davidians in Waco, TX were burned out by U.S. government agents.

Chapter 10 covers more recent incidents involving chemical agents, biological materials, or terrorist actions in the United States. Are terrorist actions in the United States fact or fantasy?

2 Chemical Agents

"The enemy had in their trenches huge cylinders with this gas compressed to a very high pressure. When the wind was from them, the cocks of the cylinders were opened, allowing the gas to flow out. It rose to a height of 40 feet, and being heavier than air, it advanced slowly with the wind, forming a large sheet or cloud. The French Algerian troops ...stood it until it came up to them, and then, taken by surprise, they were unable to withstand its effects. One breath of this is sufficient to daze a man to such an extent as to render him absolutely helpless. Those who got a very bad dose of it turned purple in the face; they coughed and gasped for breath in awful pain. Gradually among untold sufferings, the lungs fill up with a sort of white secretion, which finally chokes the poor soldier and puts an end to his misery. All this takes is three or four days, and there are no antidotes known. There is no hope; there is nothing but to let the men die amid the most awful agony a human being can behold."

Aimar Auzias de Turerre, 1915

INTRODUCTION

A chemical agent incident on March 20, 1995 in Tokyo, Japan may have greatly changed the playing field, and the level of danger, for all local emergency responders (firefighters, police officers, emergency medical services personnel, hazardous materials response teams, ambulance crews) in developed and urban nations. On that day, 11 Japanese rush-hour commuters in Tokyo's very busy subway system, one that carries about 2.7 billion passengers a day — about double the number handled by the New York City subway — were killed, and approximately 5500 more were treated at hospitals and clinics, even though many were actually uninjured. The chemical agent was sarin, developed by the Germans during World War II. Religious terrorists from a group known as Aum Shinri Kyo (Supreme Truth) sponsored and carried out the sarin attack.

In the United States, persons responding to a chemical agent or biological agent would likely be firefighters, police officers, and emergency medical service providers. Perhaps they should buy canaries to warn of poison gas as the Japanese had to do in combating their sarin poison. Local response personnel have adequate knowledge of chemical agents, a number of which are important industrial products that they have dealt with before. These personnel also have the meters and sniffers to detect such products. Biological materials can be a different matter altogether. If local emergency personnel are going to respond to chemical or biological weapons, they will need massive additional training and specialized equipment. The requisite funding must be there.

In real life, there is top-heavy spending on federal programs concerned with responding to weapons of mass destruction (WMD) along with trickle-down economics for vastly important local programs. In response to terrorist chemical or biological attack, the current plan is that federal programs, of which there are as many as 40, will "assist" the locals while local Haz Mat and other local response programs will be called upon to do the actual work of command and control, evacuation, body clearing, triage, medical care, urban search and rescue, and mortuary services, at least in the first 24-hours of a terrorist incident. We presently have to move from a very deep hole in preparation and experience as federal agencies move where the money is and try to duplicate response techniques and tactics that local, commercial, and industrial HMRTs have already developed.

Firefighters tend to respond to hazardous materials most often. The National Fire Protection Association based in Quincy, MA has said in the past that "all volunteer" and "mostly volunteer" fire departments accounted for 90% of all fire departments in the United States. That is, most of the fire departments who respond to hazardous materials incidents in this country are *volunteer*. They work regular jobs in the community, and respond to fire calls and Haz Mat incidents as necessary. Lack of funding, expensive training to meet imposed standards, costly special equipment, and response supplies and materials can be critical problems for firefighters who respond to Haz Mat, chemical agents, and biological materials during their "free time." The training provided by the military will train only those local responders in the 120 largest metropolitan areas of the country. Those 120 areas will each be given up to $300,000 worth of specialized equipment to deal with WMD on the basis of a five year loan.

Is the United States prepared for terrorist chemical or biological agent incidents? The answer is **no**. The General Accounting Office in a recent report ("Chemical and Biological Defense, Observations on DOD's Plans To Protect U.S. Forces:" GAO/T-NSIAD-98-83) noted efforts to protect U.S. military forces against chemical and biological weapons during the Gulf War evidenced a number of problems. These weaknesses included shortages in individual protective equipment, inadequate chemical and biological detection devices, inadequate command emphasis on chemical and biological capabilities, and deficiencies in medical personnel training and supplies. GAO concluded that units went to war without the chemical and biological detection, decontamination, and protective equipment needed to operate in a contaminated environment, and that weaknesses could be traced to senior military leadership.

GAO found that the Department of Defense still needs to decide on major policy and doctrine issues, improve and increase its capability to detect toxic agents, provide forces with sufficient amounts of individual protective equipment, and deal with problems of collective protection and decontamination.

On March 16, 1988, a chemical attack by Iraq on Kurdish civilians in the town of Halabja, Iraq using a mixture of mustard agent (bis-(2-chloroethyl) sulphide) plus tabun (O-ethyl imethylamidophporlcyanide), sarin (isopropyl methylphosphonofluoridate), and VX (O-ethyl S-diisropylaminoethyl methylphophonothiolate) gave some idea of the long term effects that could be caused by poison gas. Mustard agent can be manufactured very simply, even in an undeveloped country. Tabun, also known as GA, is also easy to manufacture and therefore attractive to a third world country; however, in the United States it would be viewed as out-of-date or old-fashioned. Sarin, known in the military as GB, since as with all "G" poison agents it was developed by the Germans, is a volatile substance that should be used by inhalation of an aerosol in order to do its ultimate damage (on March 20, 1995 in Tokyo, Japan the Supreme Truth religious cult used sarin based on its volatility by punching holes in bags that contained the chemical, rather than transforming it to an aerosol which would have killed many more victims). VX is very persistent, capable of staying on equipment, material, and terrain for long periods. It is effective through the skin as well as through inhalation as a liquid or aerosol.

News reports estimated that 5000 immediate deaths occurred in Iraq, but no one was really prepared for the lingering effects of chemical agents on a rural, civilian population that had limited access to medical care. Noted ten years after the attack were serious eye, respiratory, neurological, and skin problems. A number of victims went blind. The rates of cancer in the neck, head, skin, breast, respiratory system, and gastrointestinal tract in exposed people were at least three times that of unexposed persons in a nearby city (among the mustard agents, HD has been found to be a carcinogen and lewisite should be treated as a suspected carcinogen, while GA/tabun, GB/sarin, and VX are not listed as carcinogens). There are indications even years after the attack that the effects of chemical agents can be genetically transmitted to following generations.

Chemical agents may be frightening, but biological weapons can cause absolute panic. Early in World War II, Winston Churchill approved the use of anthrax biological material on cattle in Europe if it proved necessary for the war effort. During 1941 and 1942, testing was carried out on Gruinard Island off the coast of Scotland. Due to the success of the Normandy landings on June 6, 1944, the plan was shelved, but the anthrax used in the testing proved very hardy and long lasting. Until 1988, the island was declared out-of-bounds and unsafe for unprotected persons.

DUTY ON CHEMICAL UNITS IN THE U.S. ARMY

Master Sergeant Dwight E. Stevenson has spent 21^1/$_2$ years with the military and is currently a senior NBC advisor with the 8th Training Support Battalion based in Aurora, CO. "I wanted to specialize in something that would help me when I got out of the military, so I decided to get into the chemical field," says Stevenson. "I entered the chemical field in 1979 and had duty in Germany, came back to Fort Bliss, TX, and returned to Germany where I was in a chemical unit for a year and a half. I returned to Fort Bliss for four years where I ran the post chemical school working with two civilians and two military personnel.

"This was during the time of Operation Desert Storm in Iraq and I was fully involved with deploying people to Iraq. We developed a three-day training program that dealt with what to look for, the basic types of chemical weapons, what form they were likely to take, how to recognize them, their appearance, the smell, how the agent can be deployed, detection, identification, first aid, and how to protect personnel and equipment. We did a little bit with biological substances but mainly dealt with chemical weapons: nerve agents, blood agents, and blister agents. The post commander made the decision that anyone who was going to be deployed in Desert Storm had to go through our course. Civilians ran the classes, while I and two other military personnel went out to the units and started training units. During this period, we trained 18,000 personnel, both military and civilian, including both reporters and television crews, who were then deployed to Iraq.

"I did not go to the desert until after Desert Storm when I worked with the Saudi Arabian National Guard teaching nuclear/biological/chemical safety for four years. Dealing with chemical agents, first of all you have to learn what type of agent you have. M8 detector paper detects and identifies blood and nerve agents and comes in a little booklet of 25 sheets. You tear off a sheet and dab it on the liquid; most chemical warfare agents are in liquid form. The paper turns a certain color. Whatever color the papers turns, you compare it with a color chart in the packet and you learn what type of agent it is (M8 VGH are dye impregnated papers that change color when exposed to liquid chemical agent. It only provides a qualitative measurement of the nerve and blister agent's presence. M8 paper cannot detect chemical agents in vapor form. It is primarily used in a chemically contaminated situation on the battlefield to identify unknown liquid droplets).

"There is also M9 detector paper, which is similar to M8 but it only detects; it tells you that you have a chemical agent but it does not identify the type of agent. It is normally worn on the uniform of soldiers or attached to equipment that may enter contaminated areas. If the M9 detector paper turns a reddish color, there is a chemical agent present. Another detector kit, the M2, is used to detect chemical agents in the form of liquid, vapor, and aerosols. For aerial monitoring, there is an M8A1 chemical agent alarm detector which has a radioactive source and an audio alarm and gives deployed military personnel early warning of chemical agents coming into an area. The chemical specialist within a unit has a set formula that can be used, depending on wind speed, to calculate how long it will take the chemical agent to travel from the M8A1 — usually set upwind about 300 meters — to reach the area where the unit is located.

"In addition, we have a piece of equipment that looks like a hand-held vacuum cleaner which is called a CAM (chemical agent monitor) which is used to monitor personal equipment and food from the effects of a chemical agent. (Its capability extends to nerve agents, G-series and V-series, and H-series blister agents. The CAM is capable of detecting, identifying, and providing relative vapor concentration levels for chemical agents. The CAM sensitivity levels are below the IDLH (Immediately Dangerous to Life or Health) concentrations for nerve agents, G series only, and above the IDLH for V series nerve and blister agents.)

"There is an M272 water test kit with little plastic bottles and tablets. Drop a certain tablet you're testing for into a water sample; whatever color it turns will tell you what type of agent is present. A M34 soil sampling kit is just a sample kit; you pick up a sample, label it, and send it back to a laboratory. The laboratory will inform you whether the soil is contaminated or not.

"As any U.S. chemical soldier, I participated in live agent training at the U.S. Army Chemical School located at Fort McClellan in Alabama," adds MSG Stevenson. "At that facility, you're actually working with mustard agent, and it's right beside you. You have a 6 to 8 hour practice session where you are outside the contaminated area. When you get proficient in a certain routine, they put you in a room where the actual agent will be placed. In what is normally a 6 to 12 hour operation, you don your protective equipment and clothing, the agent is brought in, and you are to identify the agent, tell the instructors the type of agent, do a complete decontamination of personnel and equipment, and ensure that the agent is completely gone."

What does a military chemical unit actually do? "Most people jokingly refer to a chemical unit as 'a car wash on wheels,'" according to Stevenson. "We actually do complete decon of personnel and equipment as required by Army Field Manual FM3-5. Complete decontamination is when you have set stations for people contaminated by chemicals to move through. They must drop certain pieces of clothing at each station, and by the time they get to the last station they have removed all of their clothing. They then go to a shower point and get scrubbed down, before they have all clothing reissued.

"An equipment decon would require about five stations to be established. As a vehicle enters, you remove mud or major decontamination at the first station. Another station might be application of decon solution to the vehicle where the driver would have to wait for awhile to give the decon solution a chance to remove contamination. The next station could be used for rinsing everything off and moving the vehicle out. This process would be a lot easier if we could come up with decon solution that could be universally applied. The problem is there are so many different chemicals out there; some of them react differently, so it is going to be difficult to come up with a single, set solution that could decon all the chemicals we work with. You may be able to minimize the number of decon solutions we use now, but there is no way you're going to come up with one solution that is going to decon everything.

"Some chemical units have a single mission; some have dual missions. Some may go out and provide decon support to an actual unit that is out in the field; also, you have units that are a combination decon/smoke unit. Normally, our mission is to do decon as far forward as possible. The decon unit goes into a battle area, sets up a decon area, the fighting unit comes back, goes through the decon site, gets its vehicles decontaminated, and goes back into battle. You cannot do a complete decon in a contaminated area, but you could remove gross contamination.

"In a complete decon, we use hot, soapy water or a mild decon solution to scrub down personnel. Supertropical bleach (STB), as well as decontamination solution number 2 (DS2), are very strong solutions used to break down chemical agents on a vehicle or on personnel when certain precautions are taken. (DS2 is effective against all known toxic chemical agents and biological materials, except bacterial spores, if sufficient contact time is allowed. Allow to remain in contact with contaminated surfaces for approximately 30 minutes, rinse off with water, and recheck for contamination. DS2 is most effective when accompanied by scrubbing action. DS2 is extremely irritating to the eyes and skin. Protective mask and rubber gloves must be worn. If DS2 contacts skin, wash the area with water. Do not inhale vapors. DS2 will cause green to black color change upon contact with M8 detector paper, cause a false/positive with M9 paper, and ignites spontaneously on contact with super tropical bleach (STB) and HTH. DS2 is a combustible liquid with a flash point of 160°F. STB is effective against lewisite, V and G agents, and biological agents. Allow to remain in contact with contaminated surface for at least 30 minutes, then wash off with clear water. STB ignites spontaneously with liquid blister agent or DS2, and gives off toxic vapors on contact with G agent. Protective mask or other respiratory protection device should be worn when preparing slurry; STB should not be inhaled or come in contact with the skin. Soap and detergents can be used for nuclear, biological, and chemical contamination; scrub or wipe contaminated surfaces with hot, soapy water solution or immerse item in the solution.)"

Currently, MSG Stevenson advises National Guard and military reserve units in five states on the current situation with NBC weapons. He provides assistance and makes sure such units do quality training to keep them current with active duty units.

Contact: MSG Dwight E. Stevenson, Senior NBC Advisor, 8th Training Support Battalion, Bldg. 421, USAG Fitzsimons, Aurora, CO 80045-5001; 303-361-8981; 303-361-8705 (Fax); stevensond@famc-rgdn.army.mil (E-mail).

CHEMICAL WEAPONS DEFINED

The Chemical Weapons Convention defines chemical weapons as any chemical which, through its chemical effect on living processes, may cause death, temporary loss of performance, or permanent injury to people and animals. Chemical agents are broadly defined as any chemical substance for use in military operations to kill, seriously injure, or incapacitate people through its physiological effects (excluded from consideration are riot control agents, and smoke and flame materials). The agent can be a vapor, aerosol, gas, solid, or liquid. A persistent agent upon its release retains its casualty-producing effects for an extended period of time, from 30 minutes to several days. Persistence of chemical agents can be influenced by the type of agent depending on consistency or viscosity, amount and dispersal, type of terrain as well as temperature, wind, humidity, and precipitation. A non-persistent agent would dissipate or vaporize rapidly after release and present a short-duration hazard.

GENERAL PRACTICES IN DEALING WITH CHEMICAL AGENTS

Some general practices, procedures, and methods can be applied to a number of different toxic chemical agents. Chemical weapons can work at maximum effectiveness when used against untrained people and unprotected targets. Chemical warfare attacks, by their very nature and history, can seriously effect warfare pressures through combat stress, poor morale, and general inefficiency. In this chapter, we deal only with chemical agents designed to kill or seriously injure; that is, we do not deal with riot control agents, incendiary agents, noxious chemicals, temporary incapacitating weapons, or smoke agents.

Chemical agents can be inhaled, absorbed through the skin/wounds/abrasions/eyes, or consumed through food and/or drink. They include blister agents, nerve agents, pulmonary (choking) agents, and blood agents. They can be distributed by spray devices, bombs, aircraft, rockets, missiles, mines, water supplies/reservoirs, and other methods. Some signs that a toxic chemical agent has been released might include an unexplained runny nose, an obvious attack of spray emitting from an aircraft, smoke/mist/fumes/clouds of unknown origin, laughter or strange behavior in other persons, slurred speech, difficulty in breathing, eyesight problems, unexplained vectors (hosts).

What are some of the indicators of a possible chemical incident when first responders are called to a site. A Chemical/Biological Incident Handbook published in 1995 by the Interagency Intelligence Committee on Terrorism represented by 38 U.S. Government agencies, listed these possibilities:

- Dead animals/birds/fish: Not just an occasional roadkill, but numerous animals (wild and domestic, small and large), birds, and fish in the same area.
- Lack of insect life: If normal insect activity (ground, air, and/or water) is missing, then check the ground/water surface/shore line for dead insects. If near water, check for dead fish/aquatic birds.
- Blisters/rashes: Numerous individuals experiencing unexplained water-like blisters, wheals (like bee stings), and other rashes.
- Mass casualties: Numerous individuals exhibiting unexplained serious health problems ranging from nausea to disorientation to difficulty in breathing to convulsions to death.
- Definite pattern of casualties: Casualties distributed in a pattern that may be associated with possible agent dissemination methods.
- Illness associated with confined geographic area: Lower attack rates for people working indoors versus outdoors, or outdoors versus indoors.
- Unusual liquid droplets: Numerous surfaces exhibit oily droplets/film; numerous water surfaces have an oily film (no recent rain).
- Areas that look different in appearance: Not just a patch of dead weeds, but trees, shrubs, bushes, food crops, and/or lawns that are dead, discolored, or withered (with no current drought).
- Unexplained odors: Smells may range from fruity to flowery to sharp/pungent to garlic/horseradish-like to bitter almonds/peach kernels to new-mown hay. It is important to note that the particular odor is completely out of character with its surroundings.
- Low-lying clouds: Low-lying cloud/fog-like condition that is not explained by its surroundings.
- Unusual metal debris: Unexplained bomb/munitions-like materials, especially if they contain a liquid (with no recent rain).

IMMEDIATE HELP FOR LOCAL RESPONDERS

DOMESTIC PREPAREDNESS CB HELPLINE

The Domestic Preparedness initiative formed under the FY 1997 Defense Authorization Bill (P.L. 104-201, September 23, 1996), commonly called the Nunn-Lugar-Domenici legislation, provided funding for the Department of Defense (DOD) to enhance the capability of federal, state and local emergency responders in incidents involving nuclear, biological and chemical terrorism.

The U.S. Army Chemical and Biological Defense Command (CBDCOM), Aberdeen Proving Ground, MD, the center of the DOD's chemical and biological expertise, is sharing their years of experience by establishing a Domestic Preparedness Chemical Biological (CB) Helpline to assist emergency responders and emergency planners in preparation for a chemical or biological incident.

The CB Helpline offers non-emergency, technical assistance to emergency responders. Potential subject matters may involve characterization of chemical agents, control of chemical agents, and defensive equipment or mitigation techniques. Trained experts staff the Helpline; all have experience working with CB agents and materials. Helpline staff are trained to listen and respond to questions from the field pertaining to CB preparedness issues and to quickly retrieve the most current information relevant to the specific emergency.

The Helpline is available for use by emergency responders across the United States; including firefighters, law enforcement officials, emergency medical personnel, and emergency management persons who have responsibilities for planning, training, and conducting exercises in domestic preparedness. CBDCOM has access to the following types of information that may assist Helpline staff respond to inquiries from emergency responders:

Physical properties of CB material
Toxicology information
Medical symptoms from exposure to CB material
Detection systems and methods
Hazard prediction methods
Applicable laws and regulations
Personal protective equipment

Additional information can be provided on specialized defense equipment, including masks, gloves, and other protective clothing, and emergency response activities and procedures.

The CB Helpline is staffed weekdays from 9 a.m. to 6 p.m. eastern standard time. On weekends, on holidays, and after normal business hours, callers can leave a voice mail message. Emergency responders can obtain information from the following sources: **CB Helpline 800-368-6498**, 410-612-0715 (Fax); cbhelp@cbdcom.apgea.army.mil (E-mail).

CHEM/BIO HOTLINE

The Chemical and Biological (CB) Hotline, an *emergency* resource for first responders to request technical assistance, is a joint effort shared by the Coast Guard, F.B.I., FEMA, EPA, DHHS, and DOD. The Nunn-Lugar II act demands "establishment of a designated telephone link to a designated source of relevant data and expert advice for the use of state and local officials responding to emergencies involving a weapon of mass destruction or related materials." The intended users of the 24-hour hotline include fire and police departments, EMS providers, state emergency operation centers, and hospitals that may treat victims of CB agent exposure.

Operators use extensive databases and references, and in certain cases immediate access to subject-matter experts in the field of CB agents. Frequently asked questions include CB agent identification, medical treatments, and information on military and civilian defense equipment. The National Response Center in Washington, D.C. is the entry point for the CB Hotline where basic incident information is assessed. The NRC may link the caller to DOD's and F.B.I.'s chemical, biological, and terrorism experts. If the situation warrants, a federal response action may be initiated. State and local officials can access the CB Hotline in emergency circumstances by calling **1-800-424-8802**, although such persons should first use local established policies and procedures for requesting federal assistance *before* contacting the CB Hotline.

ANTIDOTES, PRETREATMENT, AND DECON

Aerosolized Atropine (MANAA) is effective for nerve agent poisoning; including tabun (GA), sarin (GB), soman (GD), GF and VX. Atropine solution in a pressurized container with an inhaler, MANAA contains about 240 puffs for inhalation each holding 0.43 mg of atropine sulfate, equivalent to 0.36 mg of atropine. MANAA is designed to be used by ambulatory nerve agent casualties with respiratory symptoms as a supplemental treatment after adequate injectable atropine has been given. It is to be used under medical supervision and not for self/buddy aid. Limited side effects are usually deemed insignificant in a nerve agent casualty. MANAA is manufactured by 3M/Riker.

Convulsant Antidote for Nerve Agent (CANA) is used as a countermeasure to soman (GD). CANA is a single autoinjector holding 10 mg of diazepam to be used for the control of convulsions and to prevent brain and cardiac damage. CANA is to be used in conjunction with NAPP (Nerve Agent Pretreatment Pyridostigmine) and the Mark I kit (Nerve Agent Antidote Kit). Military personnel can be issued one CANA for self/buddy aid of nerve agent casualties. CANA may cause drowsiness, considered insignificant in a nerve agent casualty. CANA is manufactured by Survival Technology, Inc., in Rockville, MD.

Dimercaprol (British Anti-Lewisite or BAL) is a colorless, viscous oily compound with an offensive odor used in treating arsenic, mercury, and gold poisoning. It displaces the arsenic bound to enzymes. The enzymes are reactivated and can resume their normal biological activity. When given by injection, BAL can lead to alarming reactions which seem to pass in a few hours.

Nerve Agent Pretreatment Pyridostigmine (NAPP) provides a countermeasure to soman (GD) and/or tabun (GA). NAPP consists of 21 30-mg pyridostigmine bromide tablets in a blister pack within a sealed pouch. (In the military, pyridostigmine bromide tablets can only be used under what is called an "activated contingency protocol." If it is determined that soman or tabun are an actual threat, permission is given for pretreatment. It has been deemed that NAPP treatment substantially increases the effectiveness of the chemical components of the Mark I kit against soman and tabun. Tablets are taken every eight hours, and personnel are issued a total of 42 tablets each. Although a number of side effects could be apparent, they are still considered insignificant versus the tremendous improvement that NAPP provides to the Mark I kit's effectiveness against the nerve agent soman.

Nerve Agent Antidote Kit (NAAK or Mark I) consists of an atropine autoinjector (2 mg), a pralidoxime chloride autoinjector (2-PAM-Cl, 600 mg), the plastic clip joining the two injectors, and a foam case. The kit serves as a countermeasure to nerve agents, including tabun (GA), sarin (GB), soman (GD), GF, and VX. Military personnel can receive three Mark I kits for self/buddy aid. Possible side effects of atropine and/or 2-PAM-Cl are deemed insignificant in a nerve agent casualty. Intravenous atropine and 2-PAM-Cl can also be made available. The Mark I kit is manufactured by Survival Technology, Inc., in Rockville, MD.

Antidote Auto-injectors: Meridian Medical Techologies, Inc., a leader in the development of auto-injector drug delivery systems, currently responds to the domestic preparedness market. This company introduced the concept of auto-injector self-administration to the military medical community. Presently, their products include a range of auto-injector delivery systems and drug formulations for the domestic preparedness (civilian) market pertaining to nerve agent antidotes to be sold to first responders, firefighters, police officers, civil defense agencies, and cities and states developing anti-terrorism programs. In the event of a terrorist attack or criminal activity involving nerve agents such as VX, sarin, or soman, first responders must administer antidotes quickly and safely. Meridian Medical Technologies' Mark I kit (NAAK) allows first responders to conveniently and effectively treat not only contaminated victims but also themselves. A syringe can be slow and cumbersome, but Meridian's auto-injectors are compact, self-contained antidote delivery systems. The company is the only FDA-approved supplier of nerve agent antidotes to the U.S. Department of Defense and U.S. allies. Domestic preparedness nerve agent antidote auto-injectors come in the following applications:

AtroPen®: NSN 6505-00-926-9083. Contains: 2 mg atropine sulfate equivalent in 0.7 ml.
ComboPen®: (2-PAM-Cl) NSN 6505-01-125-3248. Contains: 600 mg pralidoxime chloride in 2 ml.
Nerve Agent Antidote Kit (NAAK)/MARK I: NSN 6505-01-174-9919. A kit containing the AtroPen and the ComboPen.
Diazepam (CANA): NSN 6505-01-274-0951. Contains: 10 mg diazepam in 2 ml.
Morphine: NSN 6505-01-302-5530. Contains: 10 mg/0.7 ml morphine sulfate.

Each auto-injector is a disposable, spring-loaded device prefilled with a prescribed drug. The concealed needle and speed of injection render it a quick, easy, and convenient application for self or buddy use. To activate, remove the safety cap found on the end of the unit, place the front end of the unit against the victim's outer thigh and push firmly. The pressure releases a spring which drives a concealed needle into the muscular tissue of the outer thigh. The auto-injector can be administered on bare skin, but it can also penetrate through heavy clothing.

When using the MARK I kit, the AtroPen would be administered first followed by the ComboPen. In addition to the classic symptoms, nerve agents may also induce convulsive seizures in some victims; the diazepam auto-injector would treat this indication. The morphine auto-injector is available for pain management needs.

To train personnel how to use auto-injectors for the domestric preparedness market, Meridian offers three training devices: the AtroPen Trainer, the ComboPen Trainer, and the Mark I Trainer. Each training device is the same size as the auto-injector that it simulates, works in the same manner, and can be recocked after activation and used again. These training devices contain no drug or needle.

Contact: Denise Nevins, Civil Defense Coordinator, Meridian Medical Technologies, Inc., 10240 Old Columbia Road, Columbia, MD 21046; 410-309-1477 or 800-638-8093; 410-309-1475 (Fax); http://www.meridianmeds.com (Website).

Skin Decontamination Kit (M291) is designed to work on liquid nerve agents including tabun (GA), sarin (GB), soman (GD), GF and VX, and vesicant agent sulfur mustard (HD). The kit contains six packets, each containing a pad filled with a mixture of activated resins that both absorb and neutralize liquid agents from a victim's skin. One pad will decontaminate both hands and the face, or an equivalent area of skin. Contact with open wounds, eyes, and mouth should be avoided, but the reactive and absorptive resins in the M291 kit are nonirritating and nontoxic, even after prolonged contact with skin. Manufactured by Rohm and Haas Co., Philadelphia, PA, the skin decontamination kit is FDA approved.

Decontamination Kit, Skin (258A1) carries six packets within a kit, three Decon 1 carrying a mixture of hydroxyethane and phenol which adsorbs and neutralizes the G-series nerve agents, and three Decon 2 with a mixture of chloramine B and hydroxyethane which are meant to absorb and neutralize the nerve agent VX and liquid mustard. For military troops, the 258A1 kit is currently the standard issue item for personal decontamination of liquid agents on the skin. A soldier would use a Decon 1 packet, fold on the solid line, tear open quickly at the notch, remove pad from envelope, unfold quickly, and wipe hands, neck and ears for one minute. Next, the soldier would take a Decon 2 pad, crush ampoules, fold packet at solid line, tear open quickly, remove pad letting screen fall away, and wipe exposed area for two to three minutes. If soldiers are certain they have an agent on their face, they would use a Decon 1 pad to wipe their hands. Then they would hold their breath, lift off the protective mask from their chin, decon the lower half of the face, then decon interior sections of the mask that contact the skin. They would then remask using the same Decon 1 pad to wipe neck and ears. They would repeat the same procedure using a Decon 2 pad on hands, face, mask, neck and ears. WARNING: The ingredients of the Decon 1 and Decon 2 packets of the M258A1 kit are poisonous and caustic and can permanently damage the eyes. The wipes must be kept out of the eyes, mouth, and open wounds. (The 258A1 kit will eventually be replaced with the M291 kit.)

BASIC BEHAVIOR OF CHEMICAL AGENTS

Attackers with a chemical agent may have more problems than the defenders have under chemical attack. If you chose to use a chemical agent to attack a military objective, or to stage a terrorist raid, your success can depend a great deal on the weather report and how you use such information. Success, or failure, of either the attackers or potential victims may depend as much on weather and terrain conditions as any other factor in determining where a chemical agent can be an effective weapon. The wind may carry vapors more or less than you might have intended. Choosing vapors, aerosols, liquids or other delivery systems will affect your intended outcome. Atmospheric stability, dispersion, temperature, humidity, and precipitation could be all involved. High wind speed decreases effective coverage of an area, and higher temperatures equal greater vaporization. Precipitation in the form in rain, snow, sleet and hail can wash chemical agents from vegetation, air, and material. Ground levels can create problems; higher concentrations of released agents can be obtained in narrow valleys. Trees, grass, and brush can cause a gas cloud to dissipate more rapidly.

Chemical agents can basically be described as "persistent" or "non-persistent." Persistent chemical agents can poison people and animals for a long time after delivery by remaining a contact hazard or vaporizing to form an inhalation hazard. Conversely, non-persistent chemical agents disperse quickly as airborne particles, gases or liquids to provide an inhalation threat of short duration. Generally, chemical agents are cumulative in their effects. The human body can detoxify them to a limited extent, but an individual who suffers a 30-minute exposure to distilled mustard or phosgene in two increments at two different times undergoes the same effect as in a one hour exposure. However, hydrogen chloride or cyanogen chloride can be detoxified by the human body to a significant degree rather quickly compared to other chemical agents. Therefore, it would take higher concentrations of these two agents to result in maximum casualties. Nerve agents such as tabun (GA), sarin (GB), and soman (GD) have a persistence that lasts from 10 minutes to 24 hours in summer and from 2 hours to 3 days during winter. Aerosol is most likely to enter the eyes or the lungs, while vapor may enter eyes, skin, or mouth. VX is more persistent than other nerve agents having a persistence of two days to one week in summer and two days to some weeks in winter. Possible points of entry for VX are the same as those listed for other nerve agents. Blood agents such as hydrogen cyanide (AC) or cyanogen chloride (CK) persist for one to ten minutes during warm weather and ten minutes to one hour during cold weather. Possible entry routes for AC and CK via vapor or aerosol are through the eyes and/or lungs, while liquid could enter through eyes, skin, and mouth. Blister agents distilled mustard (HD) and nitrogen mustard (HN) have a slow rate of action and might be persistent for three days to one week in warm weather and for some weeks in cold weather, while lewisite (L) and mustard/lewisite (HL) have a quick rate of action and would be persistent for 1 to 3 days during summer and for weeks during winter. Route of entry for these blister agents would be nearly the same: eyes, lungs, and skin for vapor or aerosol, but for liquid, the points of entry would be eyes and skin for HD and HN and eyes, skin, and mouth for L and HL.

TO DECONTAMINATE OR NOT TO DECONTAMINATE

Decontamination can be accomplished by neutralizing or removing a contaminant. Simple weathering could also remove a contaminant, but time is of the essence when dealing with chemical and biological agents. For example, if an individual is exposed to nerve and blister agents, that person should start decontaminating himself or herself within one minute of contamination. For some agents, you will want to decon contaminated people, equipment, and materials as quickly as possible. For other agents, such as arsine, cyanogen chloride, and hydrogen cyanide, there is little need to decon a victim exposed only to vapor; if liquid agent is present, remove clothing and wash the liquid off the victim's skin. For diphosgene and phosgene, no decon is likely to be needed in the field; however, provide aeration in closed spaces. The blister agent ethydichloroarsine (ED) does not usually require decontamination in the field; however, for enclosed areas, use HTH, STB, household bleach, caustic soda, or DS2 to decontaminate. For some agents, effective decon can be obtained with soap and water. Decontamination of most biologically contaminated patients and equipment can also be achieved with soap and water.

Ideally, skin decon by the individual person should start within one minute of contamination. Within two minutes, some chemical agents can cause serious injury to the victim. Phosgene oxime, a blister agent, can damage your skin within seconds. It is necessary to either neutralize or remove the contaminant as rapidly as possible following good practices and procedures. In many instances, chemical agents will undergo natural decon following the passage of sufficient time, but this weathering will not normally be acceptable at an incident scene, and certainly not when people have been contaminated. When chemical agents are used or abused, the basic of good response practice can be stated in four words: responders have to *detect*, *identify*, *control*, and *decontaminate* chemical agent(s). Controlling the scene is of the utmost importance using the techniques and tactics considered in Chapter 4, "Standard Operations of Haz Mat Teams" as well as basic police and security procedures. Anyone in a contaminated area should be detained, checked for contamination, and deconned if necessary. Contamination should be contained within the smallest possible area, 'and no one should be allowed to carry it into a wider area. Responders will have to deal with contamination that is solid (dust, powder, dirt), liquid (rain, mist, vapor), or gas (toxic gas clouds or residue). It is very difficult to stop the spread of contamination. You can walk through it, sit on it, wipe your nose, inhale it, or eat it to spread contamination from one surface to another. The more you spread it, the more you will have to decon people, equipment, and materials in formerly "clean" areas. Most chemical agents evaporate or disperse within a short period of time, but liquid agents on surfaces, equipment, and materials may release toxic gases for days.

The "enemy" may use two or more agents, chemical and/or biological, at the same time to confuse efforts to detect and identify exactly what the victims should be responding to. Responders should decon the chemical agent(s) first, since these agents often comprise the most lethal fastest-acting types of contamination. At the present time, no one in North America knows what will be the reaction of average citizens under an enemy attack or terrorist action involving chemical or biological weapons. We all have opinions, but practical knowledge is lacking. We may have people running amok, losing self control, and creating panic; or we may have trained and prepared citizens who have the necessary equipment, supplies, and knowledge needed to make a bad situation better than it might have been. History seems to indicate military organizations who are trained, prepared, and equipped to respond to chemical agents have better defense and survival rates. If civilian responders are also trained, prepared, and equipped to respond to chemical agents (many public safety agencies, commercial response contractors and industry response teams already meet these important requirements for survival), we will be much better off. However, the funding and resolve necessary to provide special training, preparation, and equipment to the many responders who have not yet benefited from such preparations will be expensive and time-consuming.

The military services use a number of standard decon solutions, and some not so standard. Even the standard ones have problems of which you should be aware and prepared to handle. Civilian fire responders often have to decon victims, but with chemical and biological agents they have to decon more equipment, materials, surfaces, buildings, and grounds than they might have done in the past. For the nerve agents sarin (GB) and tabun (GA), the following decon solutions are listed in order of preference: (1) caustic soda solution (sodium hydroxide), (2) DS2 (decontamination solution 2), (3) washing soda solution (sodium carbonate), (4) STB (supertropical bleach) slurry, and (5) hot soapy water.

With the following blister/vesicant agents sulfur mustard (H), distilled mustard (HD), nitrogen mustard (HN), nitrogen-T mixture (HT), mustard/lewisite mix (HL), and lewisite (L), the decon solutions to be used in preferential order would be (1) HTH-HTB calcium hypochlorite, (2) DS2, (3) STB slurry (supertropical bleach), (4) commercial or household bleach solution (sodium hypochlorite). For another blister agent, phosgene oxime, the solution of choice is DS2 used by the military.

The nerve agent VX should be handled by the following decon solutions in order of preference: (1) HTH-HTB solution (calcium hypochlorite), (2) DS2, (3) STB (supertropical bleach) slurry, (4) commercial or household bleach solution (sodium hypochlorite). For the choking agent phosgene (CG), and the blood agents, cyanogen chloride (CK) and hydrogen cyanide (AC), the first choice would be DS2 solution and the second choice caustic soda solution (sodium hydroxide).

In preparing smaller amounts of either STB (supertropical bleach) or HTH (calcium hypochlorite) solution, the following mixtures would apply. For a 5% solution, mix 0.6 pounds of STB or HTH with one gallon of water; or mix 3.6 pounds of STB or HTH with five gallons of water. For a 10% solution, mix 0.75 pounds of STB or HTH with one gallon of water; or mix 4.5 pounds of STB or HTH with five gallons of water. A 5% solution (in water) of either sodium hypochlorite (household bleach) or calcium hypochlorite (HTH) will provide an effective decon of most chemical and biological agents (such mixtures can be used to decon skin if a soap and water bath follows the application of the decon solution).

Decontaminating Solution No.2 (DS2) can be used against all known toxic chemical agents and biological materials (except for bacterial spores) when allowed to remain in contact with contaminated surfaces for approximately 30 minutes and rinsed with water. DS2 is most effecting when scrubbing action is used. However, a number of cautions need to be followed. Decon workers should wear protective masks and rubber gloves and vapors should not be inhaled. The solution is combustible and extremely irritating to the eyes and skin. When DS2 contacts the skin, it is necessary to wash the area with water. The solution ignites spontaneously with STB and HTH, and will cause a black color change with M8 detector paper and a false/positive with M9 paper.

Supertropical Bleach (STB) is appropriate for both chemical and biological agents, being most effective against lewisite, V and G nerve agents, and biological agents. STB must have contact with the contaminated surface for at least 30 minutes, then be washed off with water. Problem areas with STB include spontaneous ignition with liquid blister agent or DS2, emission of toxic vapors on contact with G nerve agents, and corrosion of most metals and damage to most fabrics so that after use, it is necessary to oil surfaces and rinse fabrics. Both dry and slurry mixtures of STB do not decon mustard agents in a satisfactory manner if such agents have solidified at low temperatures. STB should not be inhaled or come in contact with the skin, and a protective mask or other respiratory protection should be donned when preparing slurry. STB can be used as slurry paste (approximately equal parts by weight of STB and water), dry mix (two shovels of STB to three shovels of earth or inert materials), and slurry mix (for chemical decon, use 40 parts of STB to 60 parts of water; for biological decon use 7 parts of STB to 93 parts of water by weight).

Soaps and Detergents handle both chemical and biological decon when contaminated surfaces are scrubbed or wiped with hot, soapy water solution or when items are soaked. Soaps and detergents physically remove contamination, but the runoff water must be considered and handled as contaminated, as it may cause casualties. For smaller amounts of soap solution, use about one pound of powdered soap per gallon of water; for larger amounts, mix 75 pounds of powdered soap in 350 gallons of water. Laundry soap could also be used; cut 75 pounds of laundry soap into one-inch pieces and dissolve it in 350 gallons of hot water. Using detergent, the mixture would be about 1 pint of detergent to 225 gallons of water.

Sodium Hypochlorite Solution (household bleach) can be used against both chemical and biological agents. It is effective and fast acting against blister and V nerve agents, but requires a contact time of 10 to 15 minutes for all biological materials. It is available in food stores as a 5% solution under various brand names, and as a 14 to 19% solution at commercial laundries. Undiluted, household bleach is harmful to skin and clothing, and corrosive to metals unless the metal is rinsed, dried, and lubricated following decon. For decon spray, dilute half and half with water. No mixing is required for chemical decon; for biological decon, add two parts bleach to ten parts water.

Calcium Hypochlorite (HTH, HTB, or high test hypochlorite) can decon lewisite, V nerve agents, and all biological agents, including bacterial spores. It reacts within 5 minutes with mustard agents and lewisite, but must have 15 minutes contact time for biological materials. HTH is more corrosive than STB. Undiluted it will burn on contact with VX, HD, or DS2. It can be used as a slurry or dry mix, has a toxic vapor and will burn the skin. Therefore, a mask and rubber gloves are the minimum protective equipment required when dealing with this substance. For chemical decon, mix five pounds HTH to six gallons of water for a 10% solution; for biological decon, mix one pound HTH to six gallons of water for a 2% solution. A slurry with 3 parts HTH and 97 parts water can be used on flat surfaces by spreading one gallon per eight square yards.

Sodium Hydroxide (caustic soda or lye) decons G nerve agents, neutralizing them on contact. It also decons lewisite and all biological materials including bacterial spores. Contact time with the contaminated surface should be about 15 minutes. Sodium hydroxide can damage skin, eyes, and clothing on contact, can damage lungs or the respiratory system via inhalation of dust or concentrated mist, and will cause a red color change with M8 detector paper. In dealing with this decon solution, full rubber protective clothing, gloves, boots, and mask are required. Flush with large volumes of water and rinse with acetic acid or vinegar if contact with skin occurs. The runoff is highly corrosive and toxic. Mixtures for a 10% solution are 10 pounds of lye to 12 gallons of water. Never use aluminum, zinc, or tin containers for mixing; use strictly iron or steel mixing containers, and add the lye to water since excessive heat will be formed. Do not handle the mixing utensil with bare hands. Lye should not be used as a decon solution if less caustic solutions are available.

Sodium Carbonate (washing soda, soda ash, or laundry soda) normally reacts within five minutes against G nerve agents. Sodium carbonate is unable to detoxify VX nerve agent and when mixed with it creates very toxic by-products. Sodium carbonate solution is also ineffective against distilled mustard (HD) blister agent as it does not dissolve the agent. For a 10% solution, mix 10 pounds of sodium carbonate with 12 gallons of water. This is a chemical decon solution and should not be used on biological agents.

Potassium Hydroxide (caustic potash) will work on certain chemical and biological agents. Remarks under sodium hydroxide apply across the board to potassium hydroxide.

Ammonia or Ammonium Hydroxide (household ammonia) can be used to decon G nerve agents but takes longer than sodium hydroxide or potassium hydroxide. SCBA or a special purpose mask is required when working with this product. Ammonium hydroxide needs no further mixing; it is a water solution of ammonia.

Perchloroethylene (tetrachloroethylene) is a nonflammable solvent of low toxicity that dissolves and removes H blister and V nerve agents but does not neutralize them. NIOSH has recommended that this substance be treated as a potential human carcinogen. It does not work with G nerve agents.

TABLE 2.1

Technical Decontamination Solutions Used by the U.S. Environmental Protection Agency

For unknown products and known products within 10 hazard classes

Solution	Composition
Solution A	Five percent (5%) sodium carbonate and 5% trisodium phosphate. Mix 4 pounds of commercial-grade trisodium phosphate with each 10 gallons of water.
Solution B	Solution containing 10% calcium hypochlorite. Mix 8 pounds with 10 gallons of water.
Rinse Solution C	A general purpose rinse to be used for both solutions. Five percent (5%) solution of trisodium phosphate with each 10 gallons of water.
Solution D	A dilute solution of hydrochloric acid (HC1). Mix one pint of concentrated HC1 into 10 gallons of water (acid to water only). Stir with wood or plastic stirrer.

Guideline for Selecting Degradation Chemicals for Specific Types of Hazards

Hazard	Solution
Inorganic acids, metal processing wastes	Solution A
Heavy metals: mercury, lead, cadmium, etc.	Solution B
Pesticides, chlorinated phenols, dioxins, PCPs	Solution B
Cyanides, ammonia, and other non-acidic inorganic wastes	Solution B
Solvents and organic compounds such as trichloroethylene chloroform, and toluene	Solution C or A
PBBs and PCBs	Solution C or A
Oily, greasy, unspecified wastes not suspected to be contaminated with pesticides	Solution C
Inorganic bases, alkali, and caustic wastes	Solution D

Nerve agents are very toxic organophosphorous esters, chemically similar to organophosphorous insecticides. Tabun, sarin, soman and GF are basically non-persistent, while V agents are more persistent. Immediate decontamination is the ultimate requirement when combating nerve agents, and decon workers must wear a respirator or self-contained breathing apparatus (SCBA) and full protective clothing. Nerve agents can readily penetrate street clothing and can be absorbed through any body surface. Nerve agents are known as anticholinesterase chemicals that inhibit the cholinesterase enzymes, thus interfering with the usual transmission of nerve impulses. Medical care to a nerve agent victim would include pretreatment if possible (pyridostigmine was given to large numbers of American military troops before action during the Gulf War of 1991), early treatment, assisted ventilation, bronchial suction, muscarinic cholinergic blockade (atropine), enzyme reactivating agents (oximes), and anticonvulsants (benzodiazepines).

The blister agent (vesicant), sulfur mustard (HD), was widely used in World War I, both to produce casualties and to reduce the efficient operation of the enemy force. There are a number of blister agents: distilled mustard (HD), ethyldichloroarsine (ED), lewisite (L), methyldichloroarsine (MD), mustard/lewisite mix (HL), three nitrogen mustards (HN-1, HN-2, and HN-3), phenyldichloroarsine (PD), and phosgene oximedichloroforoxime (CX). However, HN-3 is the only likely nitrogen mustard to be used in battle. Blister agents, in general, are able to penetrate cell tissue membranes, work well in cold and temperate climates, and are very persistent. Their persistence can be increased further by mixing mustard with non-volatile solvents which makes them more difficult to decontaminate. Street clothing provides little or no protection against mustard gas, and responders should wear a respirator or SCBA and full protective clothing. Mustard agents, particularly HD and HN-3, are stable, persistent, attack through skin/eyes/respiration, and have little or no medical options for their effects. Many soldiers were seriously injured by mustard agents during World War I but significantly fewer were killed by these agents.

Pulmonary or "choking" agents such as phosgene (CG), diphosgene (DP), chlorine (CL), and chloropicrin (PS) are also called lung damaging agents because they injure lung tissue by causing pulmonary edema. Since phosgene was responsible for 80% of all chemical fatalities during World War I, it is viewed as the most dangerous of the pulmonary agents.

So-called blood agents, properly called cyanogen agents, include hydrogen cyanide (AC) and cyanogen chloride (CK) among others. They injure a victim by disturbing the exchange of oxygen and carbon dioxide between blood and tissues. The central nervous system is affected, respiration ceases, and cardiovascular collapse ensues.

TYPES OF CHEMICAL AGENTS

The chemical agents most likely to be used as WMD in wartime or in a terrorist incident are listed in a standard format in the pages that follow. The U.S. Department of Transportation assigns the following hazard classes to chemical agents involved in transportation. Nerve agents such as tabun (GA), sarin (GB), soman (GD), and V agent (VX) are all classified as DOT hazard class 6.1 (Poison, Packing Groups I and II). The blister agents, such as mustard (H), distilled mustard (HD), nitrogen mustard (HN), and lewisite (L), are also assigned to DOT hazard class 6.1. Of the blood agents, hydrogen cyanide (AC) is classified as a 6.1 poison while cyanogen chloride (CK) is assigned a hazard class 2.3 as a Poison Gas. The choking agents, chlorine (CL) and phosgene (CG), are also classified as 2.3 hazard class. (For the purposes of comparison, biological agents and toxins such as anthrax, mycotoxin, plague, and tularemia are assigned the DOT Hazard Code of 6.2 — infectious substance or etiologic agent.)

For each chemical on the following pages a standard format is used. Please refer to the glossary for a definition of terms. Temperatures are provided in Celsius degrees. (If you wish to convert Celsius to Fahrenheit: degrees Fahrenheit = 1.8 (degrees Celsius) + 32 degrees.)

BLISTER AGENTS (VESICANTS)

Blister agents such as mustard (H), distilled mustard (HD), nitrogen mustard (HN), and lewisite (L) were used in gross amounts during World War I where they were intended to contaminate troops, force enemy troops to wear full protective equipment, and reduce combat efficiency. They can be very persistent. Many troops were put out of service for hours, days, weeks, and even years, but actual deaths were limited even though exposure to blister agents can be fatal. Both lewisite (L) and phosgene oxime (CX) are immediately painful, while the mustard agents may cause very little or no pain for as long as several hours after exposure. Victims contaminated with blister agents create a danger for those around them. Other staff having contact with contaminated persons need to wear full protective clothing, and decontamination should be done out-of-doors to prevent vapor accumulation indoors. Separate contaminated persons from those uncontaminated. Contaminated items of any type should be left outdoors, and decon should be done on equipment and vehicles used to treat or transport victims. Mustard (H and HD) is used primarily to bring about delayed casualties.

Persistence of mustard vapor or liquid is affected by a number of conditions: the level of contamination, nature of terrain and soil, type of mustard, and the weather. Persistence increases in terrain that is wooded rather than open, and in winter rather than summer. However, the threat from vapor increases under hot conditions and decreases under cool temperatures. Not all persons respond in the same manner to any chemical agent. All responders should understand the principles of toxicology, the scientific discipline of the effects, antidotes, detection, and other studies related to poisons. People respond in various ways to a given dose of chemical. If someone enters a chemically contaminated environment and happens to be susceptible to that chemical or particular class of chemicals, he or she could experience dire effects. Someone else who has developed a resistance or a protective mechanism could be working beside the first person yet not be as affected. Studies done on military troops who are fit, healthy, and young will yield different results from studies done on older persons who are unfit and have minor health problems. Young troops will have a better survival rate in a chemical agent attack, all other conditions being equal, than older civilians. Currently, when we talk about terrorism, WMD, and chemical and biological agents we have to think of casualties as civilians who are likely not going to survive as well as young, fit military troops. Age and health of a population are just some of the factors to consider in assessing reactions to mustard gases.

Distilled mustard (HD) and nitrogen mustard (HN-3) are feared because they stack the deck against their victims; they are persistent in the field, chemically stable, wage a three-pronged attack against the skin/eyes/respiratory tract, and are resistant to drugs that may counter such effects. Responders must rely on their personal protective clothing and respirator or self-contained breathing apparatus (SCBA) to deal with these poisons. Regarding sulfur mustard agents H and HD, HD is H that has been purified through washing and vacuum distillation to reduce sulfur impurities. The U.S. supply of sulfur mustard consists of one-ton containers, artillery shells, and other munitions stored at Aberdeen Proving Grounds, MD; Anniston Army Depot, AL; Blue Grass Army Depot, KY; Pine Bluff, AR; Pueblo Depot, CO; Tooele Army Depot, UT; and Umatilla Depot, OR. Sulfur mustard exposure also attacks the skin/eyes/respiratory tract, but a victim can remain symptom-free for some time after exposure. Method of exposure, temperature, and the physical condition of the victim before exposure can affect the length of time before symptoms appear and the severity of skin blisters. Either vapor or liquid can prompt blisters, and some persons are much more sensitive to mustard agents than others. Burns, blisters, and lesions, are somewhat different terms. Blisters, reportedly, are not very painful. Burns can lead to full-thickness skin loss, particularly in the penis and scrotum. Lesions tend to be painful, or extremely painful, and slow in healing.

The respiratory tract is subject to mustard attacks on the mucous membranes. Mustard gas also causes depletion of all elements of bone marrow and temporary blindness, but the majority of victims survive mustard agent contamination (about 30% die; in World War I, about 80% of victims contaminated with phosgene died). Victims may have lengthy psychological problems such as chronic depression, loss of libido, and anxiety. Sulfur mustard is a known carcinogen. No effective medical care exists for persons with mustard agent exposure and lesions. Care is directed toward relieving symptoms and preventing infections in order to promote healing.

Blister Agent Distilled Mustard (HD)

Vapor density (air = 1): 5.4
Vapor pressure (mm Hg): 0.072 at 20°C
Molecular weight: 159.08
Liquid density (g/cc): 1.27 at 20°C
Volatility (mg/m³): 610 at 20°C
Persistence: Persistent
Median lethal dosage (mg-min/m³): 1500 by inhalation; 10,000 by skin exposure
Physical state (at 20°C): Oily, colorless to pale yellow liquid
Odor: Similar to garlic or horseradish
Freezing/melting point: 14.45°C
Boiling point: 217°C
Action rate: Delayed, usually 4 to 6 hours until first symptoms appear
Physiological action: Blisters, destroys tissue, injures blood cells
Required level of protection: Protective mask and clothing. Firefighters, or other responders who have such equipment, should wear full firefighter protective clothing, or Level A (vapor protective suit) or Level B (liquid splash protective suit), plus positive pressure, full face piece, NIOSH-approved, self-contained breathing apparatus (SCBA), until they know what they are up against.

(A terrorist attack might use mixed chemical agents, mixed chemical and biological agents, or so-called "thickened" agents having polymers added to make the product highly-viscous or thickened. Thickening might also occur in colder weather, making chemical agents more persistent. The use of personal protective equipment along with positive pressure, self-contained breathing apparatus will greatly increase your safety in such situations.)

Decontamination: Bleach, fire, DS2, M258A1, M280, supertropical bleach (STB), wash with soap and water

Detection in the field: Draeger Thioether tube, M18A2, M256, M256A1, M8 and M9 paper

Carcinogen: Yes

Use: Delayed-action casualty agent

CAS registry number: 505-60-2

RTECS number: WQ0900000

LD_{50}: (oral) 0.7 mg/kg

WARNING: Do not breathe fumes. Avoid skin contact. Contact with liquid or vapor can be fatal.

Blister Agent Ethyldichloroarsine (ED)

Vapor density (air = 1): 6.0

Vapor pressure (mm Hg): 2.09 at 20°C

Molecular weight: 174.88

Liquid density (g/cc): 1.66 at 20°C

Volatility (mg/m^3): 20,000 at 20°C

Persistence: Persistent

Median lethal dosage ($mg-min/m^3$): 3000 to 5000 by inhalation; 100,000 by skin exposure

Physical state (at 20°C): Colorless liquid

Odor: Fruity, but biting; irritating

Freezing/melting point: –65°C

Boiling point: 156°C

Action rate: Irritating effect on nose and throat is intolerable after one minute at moderate concentrations; delayed blistering

Physiological action: Damages respiratory tract, affects eyes, blisters, can cause systemic poisoning

Required level of protection: Protective mask and clothing. Firefighters, or other responders who have such equipment, should wear full firefighter protective clothing, or Level A (vapor protective suit) or Level B (liquid splash protection suit), plus positive pressure, full face piece, NIOSH-approved, SCBA.

Decontamination: Not usually necessary in the field. If necessary for enclosed areas, use HTH, STB, household bleach, caustic soda, DS2, or soap and water. (Decon liquid agent on skin with M258A1, M258, or M259 skin decon kit. Decontaminate individual equipment with M280 decon kit.)

Detection in the field: Draeger Organic Arsenic Compounds and Arsine, M18A2

Carcinogen: Unknown

Use: Delayed-action casualty agent

CAS registry number: 598-14-1

RTECS number: CH3500000

LCt_{50}: (percutaneous) 3000 to 5000 mg-min/m^3, depending on the period of exposure

WARNING: Do not breathe fumes. Avoid skin contact. Contact with liquid or vapor can be fatal.

Blister Agent Lewisite (L)

Vapor density (air = 1): 7.1

Vapor pressure (mm Hg): 0.394 at 20°C

Molecular weight: 207.35

Liquid density (g/cc): 1.89 at 20°C

Volatility (mg/m^3): 4480 at 20°C

Persistence: Persistent

Median lethal dosage (mg-min/m^3): 1200 to 1500 by inhalation; 100,000 by skin exposure

Physical state (at 20°C): Colorless to brownish liquid

Odor: Like geraniums, but very little odor when pure

Freezing/melting point: −18°C

Boiling point: 190°C

Action rate: Rapid

Physiological action: Similar to HD, may cause systemic poisoning

Required level of protection: Protective mask and clothing. Firefighters, or other responders who have such equipment, should wear full firefighter protective clothing, or Level A (vapor protective suit) or Level B (liquid splash protection suit), plus positive pressure, full face piece, NIOSH-approved, SCBA.

Decontamination: Bleach, fire, DS2, caustic soda, M258A1, M280, soap and water

Detection in the field: Draeger Organic Arsenic Compounds and Arsine, M18A2, M256, M256A1, M8 and M9 paper

Carcinogen: Suspected carcinogen

Use: Moderately delayed-action casualty agent

CAS registry number: 541-25-3

RTECS number: CH2975000

LCt_{50}: (inhalation, man) 1200 to 1500 mg-min/m^3
(skin vapor exposure, man) 100,000 mg-min/m^3

WARNING: Do not breathe fumes. Avoid skin contact. Contact with liquid or vapor can be fatal.

Blister Agent Methyldichloroarsine (MD)

Vapor density (air = 1): 5.5
Vapor pressure (mm Hg): 7.76 at 20°C
Molecular weight: 160.86
Liquid density (g/cc): 1.836 at 20°C
Volatility (mg/m³): 74,900 at 20°C
Persistence: Persistent
Median lethal dosage (mg-min/m³): Estimated at 3000 to 5000
Physical state (at 20°C): Colorless liquid
Odor: None
Freezing/melting point: –55°C
Boiling point: 133°C
Action rate: Immediate irritation of eyes and nose, delayed blistering
Physiological action: Irritates respiratory tract, injures lungs and eyes, causes systemic poisoning
Required level of protection: Protective mask and clothing. Firefighters, or other responders who have such equipment, should wear full firefighter protective clothing, or Level A (vapor protective suit) or Level B (liquid splash protection suit), plus positive pressure, full face piece, NIOSH-approved, SCBA.
Decontamination: Bleach, DS2, caustic soda, M258A1, M280, soap and water
Detection in the field: M18A2
Carcinogen: Unknown
Use: Delayed-action casualty agent
CAS registry number: 593-89-5
RTECS number: CH4375000
LCt$_{50}$: No accurate data, probably similar to ED: 3000 to 5000 mg-min/m³
WARNING: Do not breathe fumes. Avoid skin contact. Contact with liquid or vapor can be fatal.

Blister Agent Mustard/Lewisite Mix (HL)

Vapor density (air = 1): 6.5
Vapor pressure (mm Hg): 0.248 at 20°C
Molecular weight: 186.4
Liquid density (g/cc): 1.66 at 20°C
Volatility (mg/m³): 2730 at 20°C
Persistence: Persistent
Median lethal dosage (mg-min/m³): 1500 by inhalation; over 10,000 by skin exposure
Physical state (at 20°C): Dark, oily liquid
Odor: Garlic
Freezing/melting point: –42°C for plant purity; –25.4°C when pure
Boiling point: Less than 190°C

Action rate: Immediate stinging of skin and redness within 30 minutes; blistering delayed about 13 hours

Physiological action: Similar to HD, may cause systemic poisoning

Required level of protection: Protective mask and clothing. Firefighters, or other responders who have such equipment, should wear full firefighter protective clothing, or Level A (vapor protective suit) or Level B (liquid splash protection suit), plus positive pressure, full face piece, NIOSH-approved, SCBA.

Decontamination: Bleach, fire, DS2, caustic soda, M258A1, M280, soap and water

Detection in the field: M18A2, M256, M256A1

Carcinogen: Should be treated as a suspected carcinogen

Use: Delayed-action casualty agent

CAS registry number: Mixture

RTECS number: Mixture

LCt_{50}: (respiratory) About 1500 mg-min/m^3

WARNING: Do not breath fumes. Avoid skin contact. Contact with liquid or vapor can be fatal.

Blister Agent Nitrogen Mustard (HN-1)

Vapor density (air = 1): 5.9

Vapor pressure (mm Hg): 0.24 at 25°C

Molecular weight: 170.08

Liquid density (g/cc): 1.09 at 25°C

Volatility (mg/m^3): 1520 at 20°C

Persistence: Persistent

Median lethal dosage (mg-min/m^3): 1500 by inhalation; 20,000 by skin exposure

Physical state (at 20°C): Oily, colorless to pale yellow liquid

Odor: Faint, fishy or musty

Freezing/melting point: −34°C

Boiling point: 194°C At atmospheric pressure, HN-1 decomposes below boiling point

Action rate: Delayed, 12 hours or longer

Physiological action: Blisters, affects respiratory tract, destroys tissue, injures blood cells

Required level of protection: Protective mask and clothing. Firefighters, or other responders who have such equipment, should wear full firefighter protective clothing, or Level A (vapor protective suit) or Level B (liquid splash protection suit), plus positive pressure, full face piece, NIOSH-approved, SCBA.

Decontamination: Bleach, fire, DS2, M258A1, M280, soap and water

Detection in the field: Draeger Basic Nitrogen Compounds, M18A2, M256, M256A1, M8 and M9 paper

Carcinogen: Should be treated as a suspected carcinogen

Use: Delayed-action casualty agent
CAS registry number: 538-07-8
RTECS number: YE1225000
LCt_{50}: (respiratory) 1500 mg-min/m³
WARNING: Do not breathe fumes. Avoid skin contact. Contact with liquid or vapor can be fatal.

Blister Agent Nitrogen Mustard (HN-2)

Vapor density (air = 1): 5.4
Vapor pressure (mm Hg): 0.29 at 20°C
Molecular weight: 156.07
Liquid density (g/cc): 1.15 at 20°C
Volatility (mg/m³): 3580 at 25°C
Persistence: Persistent
Median lethal dosage (mg-min/m³): 3000 by inhalation
Physical state (at 20°C): Dark liquid
Odor: Soapy in low concentrations, fruity in high concentrations
Freezing/melting point: –65° to –60°C
Boiling point: 75°C at 15 mm Hg
Action rate: On the skin, delayed 12 hours or more; on the eyes, faster than HD
Physiological action: Similar to HD, bronchopneumonia possible after 24 hours
Required level of protection: Similar to HD
Decontamination: Bleach, fire, DS2, M258A1, M280, soap and water
Detection in the field: Draeger Organic Basic Nitrogen Compounds, M18A2, M256, M256A1, M8 and M9 paper
Carcinogen: Should be treated as a suspected carcinogen
Use: Delayed-action casualty agent
CAS registry number: 51-75-2
RTECS number: 1A1750000
LCt_{50}: (respiratory) 3000 mg-min/m³
WARNING: Do not breathe fumes. Avoid skin contact. Contact with liquid of vapor can be fatal.

Blister Agent Nitrogen Mustard (HN-3)

Vapor density (air = 1): 204.54
Vapor pressure (mm Hg): 0.0109 at 25°C
Molecular weight: 204.54
Liquid density (g/cc): 1.24 at 25°C
Volatility (mg/m³): 121 at 25°C
Persistence: Persistent
Median lethal dosage (mg-min/m³): 1500 by inhalation; about 10,000 by skin exposure

Physical state (at 20°C): Dark, oily liquid

Odor: None if pure

Freezing/melting point: −3.7°C

Boiling point: 256°C

Action rate: Serious effects same as for HD (4–6 hours), minor effects sooner (such as eye irritation, tearing, and light sensitivity)

Physiological action: Similar to HN-2

Required level of protection: Protective mask and clothing. Firefighters, or other responders who have such equipment, should wear full firefighter protective clothing, or Level A (vapor protective suit) or Level B (liquid splash protection suit), plus positive pressure, full face piece, NIOSH-approved, SCBA.

Decontamination: Bleach, fire, DS2, M258A1, M280, soap and water

Detection in the field: Draeger Organic Basic Nitrogen Compounds, M18A2, M256, M256A1, M8 and M9 paper

Carcinogen: Should be treated as a suspected carcinogen

Use: Delayed-action casualty agent

CAS registry number: 555-77-1

RTECS number: YE2625000

LCt_{50}: (respiratory) 1500 mg-min/m^3

WARNING: Do not breathe fumes. Avoid skin contact. Contact with liquid or vapor can be fatal.

Blister Agent Phenyldichloroarsine (PD)

Vapor density (air = 1): 7.7

Vapor pressure (mm Hg): 0.033 at 25°C

Molecular weight: 222.91

Liquid density (g/cc): 1.65 at 20°C

Volatility (mg/m^3): 390 at 25°C

Persistence: Persistent

Median lethal dosage (mg-min/m^3): 2600 by inhalation

Physical state (at 20°C): Colorless liquid

Odor: None

Freezing/melting point: −20°C

Boiling point: 252° to 255°C

Action rate: Immediate eye effect, with skin effects appearing in $1/2$ to 1 hour

Physiological action: Irritates; causes nausea, vomiting, and blisters

Required level of protection: Protective mask and clothing. Firefighters, or other responders who have such equipment, should wear full firefighter protective clothing, or Level A (vapor protective suit) or Level B (liquid splash protection suit), plus positive pressure, full face piece, NIOSH-approved, SCBA.

Decontamination: Bleach, DS2, caustic soda, M258A1, M280, soap and water

Detection in the field: Draeger Organic Arsenic Compounds and Arsine, M18A2

Carcinogen: Should be treated as a suspected carcinogen
Use: Delayed-action casualty agent
CAS registry number: 696-28-6
RTECS number: CH5425000
LCt_{50}: (respiratory) 2600 mg-min/m^3
WARNING: Do not breathe fumes. Avoid skin contact. Contact with liquid of vapor can be fatal.

BLOOD AGENTS

Blood agents are substances that injure a person by interfering with cell respiration (the exchange of oxygen and carbon dioxide between blood and tissues). Cyanide, for example, blocks the use of oxygen in cells of the body causing asphyxiation in each cell. Arsine (SA) interferes with blood and damages the liver and kidneys. A severe exposure could cause anemia. Arsine is a carcinogen. Hydrogen cyanide (AC) is lighter than air, while cyanogen chloride (CK) is heavier than air and is rapidly metabolized to cyanide when it enters the body, thereby creating the same biological effects as hydrogen cyanide. CK differs from AC in that it has strong irritating and choking effects that slow breathing. Nitrites and thiosulfates can be effective antidotes if given in time. AC and CK are very volatile. If vapor-only is used, there is little need for decontamination; if liquid is present, wash skin promptly to remove contamination while removing contaminated clothing, including shoes. Cyanide liquid absorbed through the skin can continue to penetrate the bloodstream. Wash clothing and destroy contaminated shoes. Potential blood agents that might be used as WMD are listed in a standard format below.

Blood Agent Arsine (SA)

Vapor density (air = 1): 2.69
Vapor pressure (mm Hg): 11,100 at 20°C
Molecular weight: 77.93
Liquid density (g/cc): 1.34 at 20°C
Volatility (mg/m^3): 30,900,000 at 0°C
Persistence: Non-persistent
Median lethal dosage (mg-min/m^3): 5000
Physical state (at 20°C): Colorless gas
Odor: Mild, garlic-like
Freezing/melting point: –116°C
Boiling point: –62.5°C
Action rate: Delayed action, 2 hours to 11 days
Physiological action: Damages blood, liver, and kidneys
Required level of protection: Protective mask
Decontamination: None needed
Detection in the field: Draeger Arsine 0.05a tube
Carcinogen: Yes
Use: Delayed-action casualty agent

CAS registry number: 7784-42-1
RTECS number: CG6475000
LCt_{50}: 5000 mg-min/m^3

Blood Agent Cyanogen Chloride (CK)

Vapor density (air = 1): 2.1
Vapor pressure (mm Hg): 1000 at 25°C
Molecular weight: 61.48
Liquid density (g/cc): 1.18 at 20°C
Volatility (mg/m^3): 2,600,000 at 12.8°C; 6,132,000 at 25°C
Persistence: Non-persistent
Median lethal dosage (mg-min/m^3): 11,000
Physical state (at 20°C): Colorless liquid or gas
Odor: Pungent, biting; can go unnoticed
Freezing/melting point: –6.9°C
Boiling point: 12.8°C
Action rate: Very rapid
Physiological action: Chokes, irritates, causes slow breathing rate
Required level of protection: Protective mask
Decontamination: None needed in the field. There is little need for decon if
    victim is exposed to vapor alone. If any liquid agent is present, remove
    clothing and wash liquid off skin.
Detection in the field: Draeger Cyanogen Chloride 0.25a tube, M18A2, M256,
    M256A1, M8 alarm
Carcinogen: No
Use: Quick-action casualty agent
CAS registry number: 506-77-4
RTECS number: GT2275000
LCt_{50}: 11,000 mg-min/m^3

Blood Agent Hydrogen Cyanide (AC)

Vapor density (air = 1): 0.990 at 20°C
Vapor pressure (mm Hg): 742 at 25°C, 612 at 20°C
Molecular weight: 27.02
Liquid density (g/cc): 0.687 at 20°C
Volatility (mg/m^3): 1,080,000 at 25°C
Persistence: Non-persistent
Median lethal dosage (mg-min/m^3): Varies widely with concentration
Physical state (at 20°C): Colorless gas or liquid
Odor: Bitter almonds
Freezing/melting point: –13.3°C
Boiling point: 25.7°C

Action rate: Very rapid

Physiological action: Interferes with body tissues' oxygen use and rate of breathing

Required level of protection: Protective mask, protective clothing in unusual situations

Decontamination: None needed in field. There is little need for decon if victim is exposed to vapor alone. If liquid is present, remove clothing and wash liquid off skin.

Detection in the field: Draeger Hydrocyanic Acid 2a tube, M18A2, M256, M256A1, M8 alarm

Carginogen: No

Use: Quick-action casualty agent

CAS registry number: 74-90-8

RTECS number: MW6825000

LCt_{50}: Varies widely with concentration because of the rather high rate at which the human body detoxifies hydrogen cyanide

Nerve Agents

Nerve agents, known as cholinesterase inhibitors, interfere with the central nervous system by reacting with the enzyme acetylcholinesterase and creating an excess of acetylcholine which affects the transmission of nerve impulses. Nerve agents tend to be the most toxic of known chemical agents. Sample nerve agents include GB (sarin), GA (tabun), GD (soman), GF, and VX. The G agents tend to be non-persistent, while the V agents are persistent. However, the G agents can be "thickened" with other materials which will increase their persistence. Some are volatile and some are non-volatile at room temperature. As an example, GB is a comparatively volatile liquid, and thus, non-persistent. The near complete volatilization of sarin means it is mainly a vapor hazard. VX, on the other hand, is a relatively non-volatile liquid and is thus persistent. It is mainly a liquid contact hazard.

Nerve agents can be absorbed through any of a victim's body surfaces. A spray or aerosol can be absorbed through the eyes, skin, or respiratory tract; a vapor would be mainly absorbed through the respiratory tract. Liquid nerve agent could be absorbed through the skin, eyes, mouth, and membranes of the nose and through the gastrointestinal tract when eaten with food or water. Decon at once if you suspect you have been contaminated; speed is all important. If there is no treatment available for a nerve agent victim, the person will likely die of a diminished amount of oxygen in blood and tissues resulting from airway obstruction, weakness of respiration muscles, and depression of respiration. However, a patient may survive, even after several lethal doses of nerve agent, if assisted ventilation is provided, airway secretions are drained and/or suctioned, and if sufficient amounts of atropine are provided. If a very large dose of nerve agent has been acquired quickly, death can occur rapidly. The dose rate is critical. Daily exposure to small doses can be cumulative, resulting in symptoms after several days. If the small, daily exposures continue, the effects can be increasingly severe.

Regular clothing is penetrated by nerve agents whether they occur in a liquid or vapor form. Protective clothing is required, consisting of a respirator or positive-pressure self-contained breathing apparatus, an NBC suit or full protective clothing (designed to keep gases, vapor, liquids, and solids from any contact with the skin while preventing ingestion or inhalation), as well as gloves and overboots. The respirator or self-contained breathing apparatus will guard the mouth, respiratory tract, and eyes against spray vapor and aerosol. Nerve agent vapor is absorbed slowly through the skin if at all. Where only a vapor hazard exists, an NBC suit or full protective clothing may not be required since the respirator or SCBA might provide a sufficient level of protection. The respirator/SCBA should be donned immediately upon evidence of a nerve agent's presence.

Speed is essential in the decontamination of nerve agent victims. Skin decon must be done as soon as possible, and decon personnel should wear a respirator or SCBA and full protective clothing while decontaminating victims. Liquid agent may be chemically deactivated using decon solutions or removed using Fullers' earth. When an individual victim has been decontaminated, or has fully absorbed the agent, his or her body fluids, bowel movements, and urine do not present a contamination threat to other persons.

WARNING: Do not give nerve agent antidotes as preventive medicine before contemplated exposure to a nerve agent. Doing so may enhance respiratory absorption of nerve agents by inhibiting bronchoconstriction and bronchial secretion. Giving atropine before exposure to a nerve agent may degrade performance requiring maximal visual acuity or impede an individual's ability to work in high temperatures.

Atropine, 2-PAM-Cl, and Diazepam are antidotes for exposure to nerve agents. Atropine sulfate, familiar to most paramedics, is used to treat everything from irregular heartbeats to nervous disorders to nerve agent poisoning. Side effects might include chest pain, dry mouth, rapid heart beat, nausea, blurred vision, pain in the eyes, diarrhea, constipation, or skin rash. It limits a victim's ability to perspire which can be a serious problem when wearing chemical protective clothing. This drug can be incredibly important to victims of nerve agent poisoning since it blocks the effects of acetylcholine by muscarine receptors. For a nerve agent victim, atropine is several times more effective in saving lives when used in tandem with assisted ventilation than when either is used alone. Present supplies of atropine could be used quickly in any major incident.

2-PAM-Cl is an oxime that reverses the bonding of a nerve agent to the acetylcholinesterase. It works in a synergistic fashion with atropine so the total effect of both drugs is greater than the sum of individual effects.

Diazepam is a sedative and tranquilizer used to treat anxiety, nervous tension, muscle spasms, and convulsions. It can be given as a pretreatment as well, and tends to increase the effectiveness of other nerve agent antidotes. Recognize, however, that a single dose of 10 mg, the dose carried by one autoinjecter of CANA, could reduce a person's performance level for up to five hours causing poor vision, reduced decision-making ability, and decreased alertness.

Potential nerve agents that might be used as WMD are listed in a standard format below.

Nerve Agent Tabun (GA)

Vapor density (air = 1): 5.63
Vapor pressure (mm Hg): 0.037 at 20°C
Molecular weight: 162.3
Liquid density (g/cc): 1.073 at 25°C
Volatility (mg/m³): 610 at 25°C
Persistence: Non-persistent
Median lethal dosage (mg-min/m³): 400 for a resting person
Physical state (at 20°C): Colorless to brown liquid
Odor: Faintly fruity; none when pure
Freezing/melting point: –50°C
Boiling point: 240°C
Action rate: Very rapid
Physiological action: Cessation of breathing, death may follow
Required level of protection: Protective mask and clothing. Firefighters, or other responders who have such equipment, should wear full firefighter protective clothing, or Level A (vapor protective suit) or Level B (liquid splash protection suit), plus positive pressure, full face piece, NIOSH-approved, SCBA.
Decontamination: Bleach, slurry, diluted alkali, or DS2; steam and ammonia in confined areas; M258A1, M280, soap and water
Detection in the field: Draeger Phosphoric Acid Ester 0.05a tube, M18A2, M256A1, M8 and M8A1 alarms, M8 and M9 paper
Carcinogen: No
Use: Quick-action casualty agent
CAS registry number: 77-81-6
RTECS number: TB4550000
LCt_{50}: (respiratory) Approximately 400 mg-min/m³
WARNING: Do not breathe fumes. Avoid skin contact. Contact with liquid or vapor can be fatal. Fire may form hydrogen.

Nerve Agent Sarin (GB)

Vapor density (air = 1): 4.86
Vapor pressure (mm Hg): 2.9 at 25°C, 2.10 at 20°C
Molecular weight: 140.1
Liquid density (g/cc): 1.0887 at 25°C
Volatility (mg/m³): 22,000 at 25°C, 16,090 at 20°C
Persistence: Non-persistent
Median lethal dosage (mg-min/m³): 100 for a resting person
Physical state (at 20°C): Colorless liquid
Odor: Odorless
Freezing/melting point: –56°C
Boiling point: 158°C

Action rate: Very rapid

Physiological action: Cessation of breathing, death may follow

Required level of protection: Protective mask and clothing. Firefighters, or other responders who have such equipment, should wear full firefighter protective clothing, or Level A (vapor protective suit) or Level B (liquid splash protection suit), plus positive pressure, full face piece, NIOSH-approved, SCBA.

Decontamination: Steam and ammonia in confined area; hot soapy water; M258A1, M280

Detection in the field: Draeger Phosphoric Acid Ester 0.05a tube, M18A2, M256, M256A1, M8 and M8A1 alarms, M8 and M9 paper

Carcinogen: No

Use: Quick-action casualty agent

CAS registry number: 107-44-8

RTECS number: TA8400000

LCt_{50}: (inhalation) 70 mg-min/m^3

WARNING: Do not breathe fumes. Avoid skin contact. Contact with liquid or vapor may prove fatal. Hydrogen produced by corrosive vapors may be present.

Nerve Agent Soman (GD)

Vapor density (air = 1): 6.33

Vapor pressure (mm Hg): 0.4 at 25°C

Molecular weight: 182.178

Liquid density (g/cc): 1.0222 at 25°C

Volatility (mg/m^3): 3900 at 25°C

Persistence: Non-persistent

Median lethal dosage: (mg-min/m^3): Approximately GB, GA range

Physical state (at 20°C): Colorless liquid

Odor: Fruity, camphor when impure

Freezing/melting point: –42°C

Boiling point: 198°C

Action rate: Very rapid. Death usually occurs within 15 minutes after absorption of fatal dose

Physiological action: Cessation of breathing, death may follow

Required level of protection: Protective mask and clothing. Firefighters, or other responders who have such equipment, should wear full firefighter protective clothing, or Level A (vapor protective suit) or Level B (liquid splash protection suit), plus positive pressure, full face piece, NIOSH-approved, SCBA.

Decontamination: Bleach slurry, diluted alkali; hot soapy water in confined area; M285A1, M280, DS2, STB

Detection in the field: Draeger Phosphoric Acid Ester 0.05a tube, M18A2, M256, M256A1, M8 and M8A1 alarms, M8 and M9 paper

Carcinogen: No
Use: Quick-action casualty agent
CAS registry number: 96-64-0
RTECS number: TA8750000
LCt_{50}: (respiratory) 70 mg-min/m^3
WARNING: Do not breathe fumes. Avoid skin contact. Contact with liquid or vapor can prove fatal.

Nerve Agent GF (GF)

Vapor density (air = 1): 6.2
Vapor pressure (mm Hg): 0.044 at 20°C
Molecular weight: 180.2
Liquid density (g/cc): 1.1327 at 20°C
Volatility (mg/m^3): 438 at 20°C
Persistence: Non-persistent
Median lethal dosage (mg-min/m^3): 30 mg
Physical state (at 20°C): Liquid
Odor: Sweet, musty, peaches, shellac
Freezing/melting point: –30°C
Boiling point: 239°C
Action rate: Very rapid
Physiological action: Cessation of breathing, death may follow
Required level of protection: Protective mask and clothing. Firefighters, or other responders who have such equipment, should wear full firefighter protective clothing, or Level A (vapor protective suit) or Level B (liquid splash protection suit), plus positive pressure, full face piece, NIOSH-approved, SCBA.
Decontamination: Bleach slurry, diluted alkali, or DS2; steam or ammonia in confined area; M258A1, M280, soap and water
Detection in the field: M18A2, M256, M256A1, M8 and M8A1 alarms, M8 and M9 paper
Carcinogen: No
Use: Quick-action casualty agent
CAS registry number: 329-99-7
RTECS number: TA8225000
LD_{50}: (subcutaneous) Values are reported from 16 µg/kg to 400 µg/kg for mice

Nerve Agent VX (VX)

Vapor density (air = 1): 9.2
Vapor pressure (mm Hg): 0.0007 at 20°C
Molecular weight: 267.38
Liquid density (g/cc): 1.0083 at 20°C
Volatility (mg/m^3): 10.5 at 25°C

Persistence: Persistent
Median lethal dosage (mg-min/m^3): 100
Physical state (at 20°C): Colorless to amber, oily liquid
Odor: None
Freezing/melting point: Below −51°C
Boiling point: 298°C
Action rate: Very rapid
Physiological action: Produces casualties when inhaled or absorbed
Required level of protection: Protective mask and clothing. Firefighters, or
    other responders who have such equipment, should wear full firefighter
    protective clothing, or Level A (vapor protective suit) or Level B (liquid
    splash protection suit), plus positive pressure, full face piece, NIOSH-
    approved, SCBA.
Decontamination: STB slurry or DS2 solution, hot soapy water, M258A1,
    M280
Detection in the field: Draeger Phosphoric Acid Ester 0.05a tube, M18A2,
    M256, M256A1, M8 and M8A1 alarms, M8 and M9 paper
Carcinogen: No
Use: Quick-action casualty agent
CAS registry number: 50782-69-9
RTECS number: TB1090000
LCt$_{50}$: (respiratory) 100 mg-min/m^3 (resting); 30 mg-min/m^3 (mild activity)

PULMONARY (CHOKING) AGENTS

Pulmonary (choking) agents cause physical injury to the lungs through inhalation.
Membranes may swell and lungs become filled with liquid, and, in serious cases,
the lack of oxygen causes death. Phosgene (CG) was the big killer in World War I
(80% of all chemical fatalities) and is still around as a chemical weapon. Phosgene
is a corrosive and highly toxic gas that leads to what is called "dry land drowning"
through pulmonary edema (fluid buildup in the lungs) which can occur up to 48
hours after exposure. It is a serious skin irritant and may produce acute lesions
similar to those from frostbite or burns. Hydrochloric acid and carbon monoxide
can form through its decomposition, while thermal decomposition may release toxic
and/or hazardous gases. Diphosgene (DP) has the same physical effects as phosgene
(CG) since the body converts DP to CG. Potential pulmonary agents that might be
used as WMD are listed in a standard format below.

Pulmonary Agent Diphosgene (DP)

Vapor density (air = 1): 6.8
Vapor pressure (mm Hg): 4.2 at 20°C
Molecular weight: 197.85
Liquid density (g/cc): 1.65 at 20°C
Volatility (mg/m^3): 45,000 at 20°C
Persistence: Non-persistent

Median lethal dosage (mg-min/m^3): 3200
Physical state (at 20°C): Colorless, oily liquid
Odor: Newly mown hay, green corn
Freezing/melting point: −57°C
Boiling point: 127° to 128°C
Action rate: Immediate to 3 hours, depending on concentration
Physiological action: Damages and floods lungs
Required level of protection: Protective mask and clothing
Decontamination: None needed in field, aeration in closed spaces
Detection in the field: Odor
Carcinogen: Unknown
Use: Delayed- or immediate-action casualty agent
CAS registry number: 503-38-8
RTECS number: LQ7350000
LCt$_{50}$: 3000 mg-min/m^3 for resting persons. Since the effect of diphosgene are cumulative, the Ct does not significantly change with variations in time of exposure.

Pulmonary Agent Phosgene (CG)

Vapor density (air = 1): 3.4
Vapor pressure (mm Hg): 1.173 at 20°C
Molecular weight: 98.92
Liquid density (g/cc): 1.37 at 20°C
Volatility (mg/m^3): 4,300,000 at 7.6°C
Persistence: Non-persistent
Median lethal dosage (mg-min/m^3): 1600
Physical state (at 20°C): Colorless gas
Odor: Newly mown hay, green corn
Freezing/melting point: −128°C
Boiling point: 7.6°C
Action rate: Immediate to 3 hours, depending on concentration
Physiological action: Damages and floods lungs
Required level of protection: Protective mask
Decontamination: None needed in field, aeration in closed spaces
Detection in the field: Draeger Phosgene 0.25b tube, M18A2, odor
Carcinogen: Unknown
Use: Delayed-action casualty agent
CAS registry number: 75-44-5
RTECS number: SY5600000
LCt$_{50}$: 3200 mg-min/m^3

3 Biological Agents and Toxins

"Ring around the rosie,
a pocket full of posies,
ashes, ashes,
we all fall down!"

INTRODUCTION

The above nursery rhyme goes back to the fourteenth century when the Black Death (bubonic plague) killed over 25,000,000 people in Europe. Bubonic plague symptoms included painful, swollen lymph nodes, called buboes, in the armpit, groin, or neck, fever as high as 106°F, low blood pressure, exhaustion, confusion, and bleeding into the skin from surface blood vessels which produced a rose-colored ring.

Chapter 3 deals with biological agents that could be used by terrorists or others as weapons of mass destruction (WMD). There are all kinds of bacteria, viruses, rickettsiae, chlamydia, fungi, and toxins available in the world, but only a limited number may be used as weapons. Some bacteria are round (cocci), rod-shaped (bacilli), spiral (spirochetes), or comma-shaped (vibrios), and are capable of reproducing outside living cells. Some examples include anthrax, brucellosis, cholera, plague (pneumonic), shigella, tularemia, and typhoid. The nature, severity, and outcome of any infection caused by bacteria depend on the particular species, but diseases caused by bacteria often respond positively to the use of antibiotics. Viruses are tiny organisms that can only grow in the cells of another animal. More than 200 viruses are known to cause disease in humans. Antibiotics are not much of a help for virus-produced diseases, although viruses may be at least partially responsive to a few antiviral compounds that are available. Examples would include Crimean-Congo hemorrhagic fever, dengue fever, Ebola fever, eastern equine encephalitis, influenza, HIV (human immunodeficiency virus), and Rift Valley fever. Rickettsiae are small, round, or rod-shaped special bacteria that live inside the cells of fleas, ticks, lice, and mites and are transmitted to humans through bites from such pests. They are similar in one respect to viruses in that they grow only within living cells, but dissimilar in that treatment of disease caused by rickettsiae often includes the use of broad-spectrum antibiotics. Some of the world's worst epidemics such as scrub typhus, Q fever, and Rocky Mountain spotted fever have been rickettsial in nature. Chlamydia are microorganisms that live as parasites within living cells. Two species cause disease in humans: *Chlamydia trachomatis*, and *Chlamydia psittaci* (also known as parrot fever). Fungi are simple parasitic plants that lack chlorophyll and reproduce by making spores. Of the 100,000 known species of fungi, approximately ten cause disease in humans. Fungal infections tend to be mild but difficult to cure. Toxins are non-living poisons that come from living animals, plants, or microorganisms although some toxins can be produced or altered by chemical means. Examples include botulinum toxins, mycotoxins (e.g., trichothecene), ricin, and staphylococcal enterotoxins.

If you are a first responder, or hazardous materials response team (HMRT) firefighter, police officer, or emergency medical service person called to an incident, remember the following information about biological incidents: Except in unusual circumstances, you can't see biological agents; they are odorless, colorless, and tasteless. There is a delay in incubation; even the much feared Ebola fever, which has a moderate transmissibility from person to person, takes 7 to 9 days from exposure to the time when symptoms actually appear. How many persons could you expose during those 7 to 9 days? And how cheaply? No other weapon, including chemical agents, nuclear warfare, military ordnance, or bows and arrows can compare to the sticker price of biological agents: anthrax could cost you slightly under one U.S. dollar per casualty. Such weapons are easy to prepare. In some cases, you could even grow or produce your own if you have enough expertise to make home brew. Since many biological agents will be disseminated as aerosol, the enemy can be long gone before any reaction occurs. They can be almost untraceable.

A biological incident is very difficult to defend against because of ease of concealment, anonymity of the enemy, high lethal potency, ready accessibility, and relatively simple means of dissemination. A small quantity can do you in; a lethal aerosol anthrax dose could be as little as a millionth of a gram. As of today, bio response is beyond the capability of local government unless they have a fully trained and equipped Level A hazardous materials response team, and possibly beyond the ability of state government unless it has a top notch medical services program. A mass casualty biological incident may be even beyond the experience, training, and equipment levels of the national government. We won't know until we have one.

Currently, there is little or no field detection equipment for biological agents available to local first responders or local hazardous materials teams. Defense consultants are studying the possibilities, but delivery may take a long time. There is a very apparent lack of vaccines, antibiotics, assisted ventilation devices, experienced and trained medical personnel, detectors or meters for identification of biological agents, and funding for vastly important local programs. Federal programs will "assist" in biological response while local programs will be called upon to do the actual work of clearing, evacuation, control, triage, medical care, and urban search and rescue.

One of the central problems in the copycat anthrax threats, other than the very high cost involved in responding to such hoaxes, is that biological detection in the field is just not available to many local response personnel. The military may or may not have a couple of vehicle-mounted biodetectors that may or may not be currently reliable and past the developmental stage. "Human beings are a sensitive, and in some cases the only, biodetector," reports the Department of the Army Field Manual FM 89 (NATO Handbook on the Medical Aspects of NBC Defensive Operations, AMedP-6(B), Part II - Biological). Little or no field detection equipment exists for biological agents at the present time. The military does have BIDS (Biological Integrated Detection System), a vehicle-mounted system that can identify a limited number of biological agents through antibody-antigen combinations by exposing samples of air to antibodies. The process takes about 30 minutes and can currently detect the presence of botulism, anthrax, bubonic plague, and staphylococcus enterotoxin B.

Biological and chemical agent detection systems for field use are a key factor in the country's domestic preparedness programs for terrorist attacks, which may use WMD in light of the approximately 110 anthrax hoaxes that have occurred in the United States during the past six months. If local response forces do not have reliable field detection equipment, they could be forced to use unnecessary mitigation techniques ranging from decontamination to therapeutic drugs to vaccines. One of the problems with so-called field detectors for biological agents is a possible false positive reading on present detection equipment.

In at least one anthrax hoax, an incident that took place on February 18, 1999 at the Summit Women's Health Organization, 530 North Water Street, in downtown Milwaukee, WI, an abortion clinic received a letter stating the envelope contained anthrax. Fire department responders on-scene used a SMART® ticket detection system which reportedly indicated that anthrax may have actually been present. Therefore, responders on-scene went ahead with decontamination, hospital treatment, and provision of antibiotic medicine. Fire and medical officials said they would have taken such precautions even if the field test had actually been negative. It is not clear if the result was a false positive or an indication of the need for more adequate training in use of the SMART ticket system. According to the media, fire officials at the scene reportedly said the test was "negative." On March 1, 1999, local authorities confirmed that they did not know for certain until 12 hours later when they got results back from a laboratory in Maryland that anthrax was not present.

The situation was complicated by the quality of the instructional materials. The SMART ticket is supposed to turn red if anthrax is present. The color chart that firefighters had was reportedly in black-and-white, and there were words to represent various colors. The manufacturer faxed a real color chart. A second test, using a control solution, turned the ticket pink.

Environmental Technology Group, Inc. (ETG) located in Baltimore, MD distributes and markets SMART tickets as well as other domestic preparedness products. ETG is a prime Department of Defense contractor and an international supplier of military products. They deal in chemical and biological detection systems, as well as other products. The SMART biological warfare agent detection ticket employs patented immuno-chemistry tests for specific biological agents (including anthrax, plague, ricin, botulinum toxins, brucella and several others). They utilize antibody/antigen reactions. A reaction vial contains colloidal gold particles. If an agent is present in a sample, a complex forms between gold labeled antibodies and the agent. A dacron swab transfers this complex to a ticket where the complex is filtered and concentrated onto a membrane and becomes visible as a red spot. If an agent is not present in the sample, a complex does not form, gold particles diffuse through the membrane and are not visually detectable. To operate the system, wipe the suspected area with a swab, place six drops of buffer solution into the vial, tap the tube with your finger to mix pellet, place the swab into the vial, squeeze the swab against the vial wall to mix, and place the swab into the upper portion of the ticket and wait 5 to 15 minutes. For test results, observe the test spot; if any distinct red color appears which may be in the shape of a dot or crescent and is a stronger color than that in the negative control spot, the test is positive. The control spot must be free of any color other than a very faint pink. If a reaction occurs in the negative control spot, or the detection spot is difficult to read, place one drop of buffer on a clean swab and wipe the reaction area. Positive result will not wash away.

ETG also distributes the APD 2000 (advanced portable detector) which can simultaneously detect nerve and blister chemical agents, identify agents, recognize pepper spray and mace, and identify hazardous compounds. Sensitivity for V agents is four parts-per-billion (ppb) with a response time of 30 seconds. Sensitivity for G agents is 15 ppb/30 seconds, for H agents is 300 ppb/15 seconds, and for lewisite is 200 ppb/15 seconds. For high concentrations of these agents, detection time is 10 seconds. Selectable settings allow the APD 2000 to be used as a detector that automatically clears following an alarm or as a continuously-sampling monitor. A fixed site remote detector featuring the APD 2000 system can also be supplied for force protection, fixed installation monitoring, building installation monitoring, perimeter security, remote detector networks, or decon hot/warm zone monitoring.

Another detection instrument, the miniature chemical agent detector (Mini-CAM), can simultaneously detect nerve, blister, blood, and choking agents and warn responders or military troops through both audible and visible alarms. This detector is currently in use by U.S. and NATO forces. It weighs only eight ounces, and incorporates a replaceable sensor module which allows it to operate continuously for up to four months.

Contact: Environmental Technology Group, Inc., 1400 Taylor Avenue, P.O. Box 9840, Baltimore, MD 21284-9840; 419-321-5370; 410-321-5255 (Fax).

Biological agents may be alive; they can spread through infection; they may be able to duplicate themselves; some may be persistent, others may be transmissible from person to person. The United States closed down its offensive biological weapons program in 1969. There has since been a loss of knowledge, experience, and data on new developments in offensive biological weapons as knowledgeable persons sought other employment, retired, or died over the last 30 years.

Unlike chemical attacks which produce immediate casualties, biological attacks have delayed casualties, often delaying the realization that an attack has occurred at all. In a biological attack there is a delay due to the incubation period required by biological agents, occuring hours or days after the attack later when people arrive at hospitals with flu-like symptoms. We do not presently have the reliable disease surveillance programs necessary to identify the biological agent(s) and provide the correct treatment. Doctors, nurses, hospital personnel, and public health care workers would be the first line of defense against biological warfare. Many such people work for private firms rather than government agencies. Is the civilian medical community in all areas of the nation really ready to deal with a biological mass casualty incident? It will be the civilian health care system, plus local firefighters, police officers, and emergency medical technicians and paramedics that will manage and do the work required by a biological attack in the United States — at least in the first 24 hours when most of the life and death decisions will be made.

Biological weapons are old in the world but new in the United States. We have had many bombings and some chemical agent releases, but only a very few biological agent releases. There have been hoaxes and some actual attempts to use biological weapons here, but disregarding industrial incidents and releases of nuclear energy, only two incidents come readily to mind. At a Dallas, TX hospital in the fall of 1996, laboratory staff were sent e-mail messages inviting them to a free breakfast. A dozen of the 45 laboratory staff fell ill with severe intestinal symptoms. Inspection determined that muffins and doughnuts were treated with shigella which causes dysentery. In Oregon in 1984, two members of cult leader Rajneesh Bagwhan's religious group sprinkled salad bars in local restaurants with salmonella in an attempt to decide a local election. Over 700 persons were sickened, but there were no deaths.

The world has changed, and we are changing as well. The Army has begun training civilian responders in 120 U.S. metropolitan areas in chemical and biological agent response. The National Guard is funded to form "RAID" teams around the United States. The Marine Corps has a 400-person Chemical Biological Incident Response Force based in North Carolina for both domestic and foreign duty as necessary. The U.S. Army is forming a Chemical Biological Rapid Response Team (C/B-RRT). The F.B.I. has a hazardous materials response unit. The Federal Emergency Management Agency can supply USAR (Urban Search and Rescue) teams from 25 local fire departments in 18 states, and the U.S. Public Health Service is now developing Metropolitan Medical Strike Teams (MMST). Many other teams, such as the Army's Technical Escort Unit as well as Explosive Ordnance Teams have been around for years and are highly trained. What will these teams and local first responders find as threats facing them in biological attacks?

Biological agents are most likely to be disseminated by aerosol, (i.e., a fine aerial suspension of liquid; fog or mist) or by solid (i.e., dust, fume, or smoke with particles small enough in size to be stable). The perfect size for human exposure is between 0.5 to 5 microns (or micrometers) which are a unit of length equal to one millionth of a meter. Larger particles might be naturally filtered out by the inhalation process, while smaller sizes might be inhaled but not retained in an efficient manner. Aerosol exposure can also contaminate food, water, and skin. Although healthy, intact skin can resist the entry of many but not all biological agents, skin with wounds, cuts, or abrasions provides an opening for infection. Sometimes the threat is unknown and may be a single biological agent, a chemical agent with a biological agent (decon for the first before you decon for the second), or two biological agents with different incubation times. When the threat is unknown, protective clothing must be worn along with respiratory tract protection such as a mask with biological filters, or a self-contained breathing apparatus (SCBA) with positive pressure.

The best time for spraying aerosol is late at night or just before the first rays of dawn. The attackers want both security and a chance to get away with a dastardly deed; but they also need weather and atmospheric conditions as their unpaid assistants. They need a time when conditions offer minimum interference from ultraviolet radiation, and maximum assistance from atmospheric inversion which can assist a cloud to move along the surface of the land. As an example, the early morning hours tend to be a time of slowest wind speeds. The slower the wind speed, the higher the dosage, the smaller the area of coverage, and the higher the toxic effects. Dosage is a very important factor in relation to biological agents. Chemical agents have an "effective dose," the amount of a substance that may be expected to have a specific effect. Biological agents have a comparable term, "infective dose," which refers to the number of microorganisms or spores necessary to cause an infection. (Spores are a form taken by some bacteria making them resistant to heat, drying, and chemicals. In some circumstances, the spore may change back into the active form of the bacterium. Anthrax and botulism present examples of diseases caused by spore-forming bacteria.) For means of comparison, the average lethal chemical agents in storage today are thousands of times less lethal, by weight, than equivalent amounts of biological warfare agents. Because of very high toxicity, the lethal biological agent dose can be far smaller than that required from chemical agents.

Additionally, biological agents can be used against plants, animals, or materials rather than just against humans. Local responders will probably have no early warning of a biological attack, having fewer detection devices for biological agents than for chemical agents. It is entirely possible that local first responders (firefighters, police officers, and emergency medical personnel) will not even be called to the scene. Sooner or later, due to the incubation time delay, everyone will be "coming down with the flu." Always use the highest level of personal protective equipment available to protect the respiratory tract by using a full-face mask with biological filters or SCBA with positive pressure, at least until you know the specific threats.

POTENTIAL BIOLOGICAL WARFARE AGENTS

Anthrax

Planned release of anthrax would probably be done by aerosol since the spore form of the bacillus is quite stable. Anthrax is viewed as the single greatest threat for use in biological warfare; it is quite contagious with a high mortality rate (but is not contagious from person-to-person). Anthrax can easily be produced in large quantities, is relatively easy to weaponize, can readily be spread over a wide area, may be stored safely, and remains lethal for a long period of time. About 95% of natural anthrax infections are cutaneous; that is, they affect the skin. Additional routes of entry may be by inhalation or ingestion. It can also occur naturally; zebras are very much affected by anthrax. Anthrax spores can settle in the soil. Some herbivores may become infected in this manner, but humans are unlikely to be affected. Bleach will kill anthrax spores. When a terrorist or warlike act uses aerosol dissemination, inhalation-type anthrax will be the result — a much more dangerous disease than the natural form. It must be treated with high dose antibiotic treatment before symptoms appear. With anthrax, treatment must come quickly, within 24-hours, or most victims will die. Untreated, the mortality rate of inhalation and intestinal cases is about 95%, while untreated cutaneous (skin) anthrax can be up to 25%. A unique feature of anthrax is a treatment "eclipse" when patients start feeling better just before they die. At the present time, 2 million military personnel in the United States have been or are being vaccinated.

Medical classification: Bacterial

Probable form of dissemination: Spores in aerosol

Detection in the field: None

Infective dose (aerosol): 8000 to 50,0000 spores

Incubation time: 1 to 6 days (in this case, the incubation time between exposure and onset of symptoms is 1 to 6 days for anthrax which is not transmissible from person to person. Compare this incubation time with that of the virus, smallpox, which is 10 to 17 days. Smallpox is highly transmissible from person to person. After exposure to smallpox, a person could travel by air around the world a number of times and contaminate many people before developing any symptoms. However, naturally occurring smallpox has been eradicated world-wide since 1977.)

Persistence: Spores are highly stable

Personal protection: Protective clothing must be used as well as protection for the respiratory tract. Use a mask with biological filters or SCBA with positive pressure, at least until you know the specific threat(s). Also, time/distance/shielding.

Routes of entry to the body: Inhalation, skin, and mouth

Person-to-person transmissible: No

Duration of illness: 3 to 5 days (often proves fatal)

Potential ability to kill: High

Defensive measures: Immunization, good personal hygiene, physical conditioning, use of arthropod repellents, wearing protective mask, and practicing good sanitation

Vaccines: Yes. Michigan Department of Public Health vaccine

Drugs available: Yes. Ciprofloxacin, doxycycline, and penicillin

Decontamination: Soap and water, or diluted sodium hypochlorite solution (0.5%). Drainage and secretion precautions are necessary. After invasive procedures or autopsy, decontaminate instruments and surfaces with 0.5% sodium hypochlorite.

Botulism

A group of seven related neurotoxins (types A–G), botulinum toxins are typically found in canned foods. Such toxins block acetylcholine release in a similar manner to chemical nerve agents. Botulism can cause paralysis which can lead to respiratory failure requiring assisted ventilation until the paralysis passes. This toxin is not volatile and not dermally active. Botulism appears to be the most dangerous toxin available, but many botulinum toxins would not work on a battlefield. However, they can be effective assassination or terrorist weapons in closed areas such as subways or meeting rooms.

Medical classification: Toxin

Probable form of dissemination: Sabotage of food/water supply, or aerosol

Detection in the field: None

Infective dose (aerosol): 0.001 µg/kg

Incubation time: Variable (hours to days)

Persistence: Stable

Personal protection: Protective clothing must be used as well as protection for the respiratory tract. Use a mask with biological filters or SCBA with positive pressure, at least until you the know the specific threat(s).

Routes of entry to the body: Inhalation, mouth, wound

Person-to-person transmissible: No

Duration of illness: 24 to 72 hours (months if lethal). Therapy consists mainly of supportive care, such as intubation and assisted ventilation for respiratory failure.

Potential ability to kill: High

Defensive measures: Immunization, good personal hygiene, physical conditioning, use of arthropod repellents, wearing protective mask, and practicing good sanitation. Spores can be killed by pressure-cooking food to be canned.

Vaccines: Yes. IND (investigational new drug) Pentavalent Toxoid A-E.

Drugs available: Yes. IND Heptavalent Anti-toxin A-F (equine despeciated); also, Trivalent Equine anti-toxin A, B, and E.

Decontamination: Soap and water, or diluted sodium hypochlorite solution (0.5%). If contamination of foodstuffs is suspected, boil for ten minutes to kill toxin. Botulism is not dermally active and secondary aerosols do not endanger medical personnel.

Brucellosis

Natural infection of humans occurs through ingestion of unpasteurized milk or cheese, through aerosol present in farms and slaughterhouses, or by inoculation of skin lesions in people in close contact with animals. Intentional exposure would be likely by aerosol, or possibly by contamination of food.

Medical classification: Bacterial

Probable form of dissemination: Aerosol; sabotage of the food supply. Brucellosis is a "hindrance" bacteria; symptoms can take months to appear, and deaths are few and far between, even without medical care.

Detection in the field: None

Infective dose (aerosol): 10 to 100 organisms

Incubation Time: 1 to 4 weeks

Persistence: Long persistence in wet soil and food

Personal protection: Protective clothing must be used as well as protection for the respiratory tract. Use a mask with biological filters or SCBA with positive pressure, at least until you know the specific threat(s).

Routes of entry to the body: Inhalation, mouth, skin, and eyes

Person-to-person transmissible: No (except where open skin lesions are evident)

Duration of illness: Varies greatly

Potential ability to kill: Very low

Defensive measures: Immunization, good personal hygiene, physical conditioning, use of arthropod repellents, wearing protective mask, and practicing good sanitation. Avoid unpasturized milk products.

Vaccines: Yes

Drugs available: Doxycycline and rifampin

Decontamination: Soap and water, or diluted sodium hypochlorite solution (0.5%). Drainage and secretion procedures are necessary.

Cholera

Cholera is normally caused by ingestion of food or water contaminated with feces or vomitus of infected persons or with feces of carriers. A terrorist act would likely be the result of an intentional contamination of water or food.

Medical classification: Bacterial

Probable form of dissemination: Sabotage in food and water; aerosol

Detection in the field: None

Infective dose (aerosol): Has low infectivity to humans

Incubation time: Hours to 5 days

Persistence: Unstable in aerosols and fresh water; stable in salt water

Personal protection: Protective clothing must be used as well as protection for the respiratory tract. Use a mask with biological filters or SCBA with positive pressure, at least until you know the specific threat(s).

Routes of entry to the body: Inhalation, mouth

Person-to-person transmissible: Infrequent

Duration of illness: Equal to, or greater than, 1 week. Because of a common symptom of watery diarrhea, i.v. fluid supplies can be insufficient (fluid loss for one patient can exceed 10 liters/day). Therapy consists mainly of fluid and electrolyte replacement.

Potential ability to kill: Low with treatment, high without treatment

Defensive measures: Immunization, good personal hygiene, physical conditioning, use of arthropod repellents, wearing protective mask, and practicing good sanitation

Vaccines: Yes. Wyeth-Ayerst vaccine available in United States but provides about 50% protection lasting no more than six months. Also, Swedish SBL oral vaccine effective, but not available in United States.

Drugs available: Oral rehydration therapy. Tetyracycline, doxycycline, ciprofloxacin, and norfloxacin

Decontamination: Diluted sodium hypochlorite solution (0.5%). Personal contact rarely causes infection. Avoid vomit and feces, and wash hands thoroughly.

Crimean-Congo Hemorrhagic Fever

Medical classification: Virus

Probable form of dissemination: Aerosol

Detection in the field: None

Infective dose (aerosol): High

Incubation time: 3 to 12 days

Persistence: Relatively stable

Personal protection: Protective clothing must be used as well as protection for the respiratory tract. Use a mask with biological filters or SCBA with positive pressure, at least until you know the specific threat(s).

Routes of entry to the body: Inhalation of aerosol, tick bites, crushing an infected tick, or at the slaughter of viremic livestock

Person-to-person transmissible: Moderate

Duration of illness: Days to weeks

Potential ability to kill: High

Defensive measures: Immunization, good personal hygiene, physical conditioning, use of arthropod repellents, wearing protective mask, and practicing good sanitation

Vaccines: The only licensed vaccine is yellow fever vaccine.

Drugs available: Prophylactic ribavirin may be effective for Crimean-Congo hemorrhagic fever.

Decontamination: Diluted sodium hypochlorite solution (0.5%). Isolation measures and barrier nursing procedures are necessary.

Plague

There are two variations of plague, pneumonic plague and bubonic plague. Bubonic plague is the most common form, and has a secondary formation of large regional lymph nodes called buboes. Blood may clot in the vessels, and may show up in blackened fingers and toes. "Natural" plague most often is caused by the bite of a flea that had dined on infected rodents; a secondary source would be by sputum droplets inhaled from coughing victims. There is a limited incidence of plague in the southwestern desert of the United States. Usually, rodents there die of plague, fleas feed on the rodents' bodies, plague multiplies in the flea, flea becomes unable to bite normally, flea gets apprehensive and bites everything, and everything the flea bites gets infected with plague. Some plague victims in the United States have been infected by household cats. "Un-natural" plague, the result of terrorist or enemy action, could possibly be an aerosol, or less likely, a release of plague-carrying fleas — forms of dissemination that may develop into pneumonic plague leading to quick death. Pneumonic plague is an extremely virulent form, can be transferred from person to person, and seems unaffected by vaccine. During World War II, the Japanese established Unit 731 in Mukden, Manchuria and carried out experiments in biological warfare on prisoners of war from the United States, Britain, Australia, and New Zealand. They tried aerosolizing plague but were unsuccessful.

Medical classification: Bacterial

Probable form of dissemination: Aerosol

Detection in the field: None

Infective dose (aerosol): 100 to 500 organisms

Incubation time: 1 to 3 days

Persistence: Up to 1 year in soil; 270 days in bodies

Personal protection: Protective clothing must be used as well as protection for the respiratory tract. Use a mask with biological filters or SCBA with positive pressure, at least until you know the specific threat(s).

Routes of entry to the body: Inhalation, ingestion, flea bite

Person-to-person transmissible: High

Duration of illness: 1 to 6 days (usually fatal)

Potential ability to kill: High unless treated within 12 to 24 hours

Defensive measures: Immunization, good personal hygiene, physical conditioning, use of arthropod repellents, wearing protective mask, and practicing good sanitation. Utilize an insecticide as necessary to kill fleas on victims and their contacts; if local flea and rodent population becomes infected, institute control measures.

Vaccines: Yes. Greer Laboratory vaccine

Drugs available: Streptomycin, doxycycline, and chloramphenicol

Decontamination: Soap and water, or diluted sodium hypochlorite solution (0.5%). Removal of potentially contaminated clothing should be done by people in full protective clothing in an area away from non-contaminated persons. For victims with bubonic plague, drainage and secretion procedures need to be employed. Careful treatment of buboes is required to avoid aerosolizing infectious material. For victims with pneumonic plague, *strict isolation is absolutely necessary.* Heat, disinfectants and sunlight render bacteria harmless.

Q Fever

Q fever is a sudden feverish illness of the respiratory system caused by *Rickettsia burnetii,* as a result of contact with infected animals (particularly placentas of sheep and goats) or drinking infected raw milk, and ticks. There is a high degree of Q fever in Australia but not in neighboring New Zealand. It makes a good biological weapon because it's a high infectivity agent. Pneumonia is extremely common with Q fever, but patients generally recover.

Medical classification: Rickettsial

Probable form of dissemination: Aerosol; sabotage of food supply

Detection in the field: None

Infective dose (aerosol): 1 to 10 organisms

Incubation time: 14 to 26 days

Persistence: Months on wood and sand

Personal protection: Protective clothing must be used as well as protection for the respiratory tract. Use a mask with biological filters or SCBA with positive pressure, at least until you knows the specific threat(s).

Routes of entry to the body: Inhalation

Person-to-person transmissible: Infrequent or never

Duration of illness: 2 days to 3 weeks. A high fever could persist for three weeks or more, but treatment with antibiotics is usually effective within 36 to 48 hours. With treatment or without treatment, Q fever is generally a self-limiting illness.

Potential ability to kill: Very low (estimated at 1–3%).

Defensive measures: Immunization, good personal hygiene, physical conditioning, use of arthropod repellents, wearing protective mask, and practicing good sanitation. Persons who are regularly exposed to domestic animals should be vaccinated against Q fever.

Vaccines: IND 610 (inactivated whole cell vaccine given as single injection) is available through USAMRIID Fort Detrick, MD 21702, and Q-Vax (CSL Ltd., Parkville, Victoria, Australia).

Drugs available: Tetracycline and doxycycline

Decontamination: Soap and water, or diluted sodium hypochlorite solution (0.5%). Victims do not represent a risk for secondary contamination of medical personnel.

Ricin

Ricin is a poisonous protein in the castor bean. It has so far been used only as a weapon of assassination, but when used as an aerosol it could possibly lead to widespread illness and death among victims. It has been said that one milligram of ricin can kill an adult. Abdominal pain, vomiting, and diarrhea symptoms appear in a few hours. Within a few days, there is severe dehydration and a decrease in urine and blood pressure.

Medical classification: Toxin

Probable form of dissemination: Aerosol (plus many other exposure routes)

Detection in the field: None

Infective dose (aerosol): 3 to 5 µg/kg

Incubation time: Hours

Persistence: Stable

Personal protection: Protective clothing must be used as well as protection for the respiratory tract. Use a mask with biological filters or SCBA with positive pressure, at least until you know the specific threat(s).

Routes of entry to the body: Inhalation, mouth, sabotage of water supplies and foodstuffs

Person-to-person transmissible: No

Duration of illness: Days (death within 10–12 days for ingestion). No specific treatment exists.

Potential ability to kill: High

Defensive measures: Immunization, good personal hygiene, physical conditioning, use of arthropod repellents, wearing protective mask, and practicing good sanitation.

Vaccines: Under development as of 1996

Drugs available: No specific anti-toxin

Decontamination: Soap and water, or diluted sodium hypochlorite solution
(0.5%). Since ricin is not volatile, secondary aerosols are usually not a
danger to medical personnel.

Rift Valley Fever

The natural path of Rift Valley Fever is through mosquitoes that feed on infected
animals and then bite humans, but aerosols or virus-laden droplets could also be
used. The disease tends to be similar whether acquired through aerosol or mosquito
bites.

Medical classification: Virus

Probable form of dissemination: Aerosol, infected vapors

Detection in the field: None

Infective dose (aerosol): 1 to 10 organisms

Incubation time: 2 to 5 days

Persistence: Relatively stable

Personal protection: Protective clothing must be used as well as protection for
the respiratory tract. Use a mask with biological filters or SCBA with
positive pressure, at least until you know the specific threat(s).

Routes of entry to the body: Inhalation, mosquito bite, or other exposure to
virus-laden aerosols or droplets.

Person-to-person transmissible: Low

Duration of illness: Days to weeks

Potential ability to kill: Low

Defensive measures: Immunization, good personal hygiene, physical conditioning, use of arthropod repellents, wearing protective mask, and practicing good sanitation.

Vaccines: Yes

Drugs available: Prophylactic ribavirin may be effective for Rift Valley fever.

Decontamination: Soap and water, or diluted sodium hypochlorite solution
(0.5%). Removal of potentially contaminated clothing should be done by
people in full protective clothing in an area away from non-contaminated
persons.

Smallpox

Smallpox is a highly contagious virus with fever and a blisterlike rash. Smallpox
can survive for centuries, and antibiological drugs are not able to handle viruses. It
is caused by two species of pox-virus, *Variola major* or *Variola minor*. The disease
is only carried by humans. Since vaccination throughout the world, not one case of

natural smallpox has occurred. Only two samples of this virus are known to still exist in the world, one at the Center for Disease Control and Prevention in Atlanta, GA, and one at the Vector Laboratory in Novosibirsk in Kazakhstan. Smallpox has a death rate of about 30% for unvaccinated persons and 3% for those who have been vaccinated. Since smallpox has been exterminated in the world, civilians and military personnel have not been vaccinated for smallpox since the 1980s. You are unprotected if your vaccination is three or more years old. If you are old enough, you probably remember the smallpox vaccine scarification method; paint the vaccine on your arm and then get jabbed 16 times with a sharp needle. Pregnant women, and carriers of HIV or other autoimmune diseases should not be vaccinated for smallpox. Because the skin becomes pock-marked and sloughs off, smallpox can easily be confused with chicken pox, although lesions are in a smooth, orderly progression in contrast to chicken pox.

Medical classification: Virus

Probable form of dissemination: Aerosol

Detection in the field: None

Infective dose (aerosol): 10 to 100 organisms (assumed low)

Incubation time: 10 to 17 days (average equals 12 days)

Persistence: Very stable

Personal protection: Protective clothing must be used as well as protection for the respiratory tract. Use a mask with biological filters or SCBA with positive pressure, at least until you know the specific threat(s).

Routes of entry to the body: Inhalation by direct, face-to-face contact with an infected case, by fomites, and by aerosols

Person-to-person transmissible: High

Duration of illness: 1 to 2 weeks. Smallpox therapy is mainly supportive care, and no specific antiviral therapy exists.

Potential ability to kill: High

Defensive measures: Immunization, good personal hygiene, physical conditioning, use of arthropod repellents, wearing protective mask, and practicing good sanitation

Vaccines: Yes. Wyeth vaccine, one dose by scarification. (Pre and post exposure vaccination recommended if more than three years has passed since last vaccine.)

Drugs available: Vaccinia immune glorulin (VIG)

Decontamination: Soap and water, or diluted sodium hypochlorite solution (0.5%). Smallpox has great potential for person-to-person exposure. Removal of potentially contaminated clothing should be done by people in full protective clothing in an area away from non-contaminated persons. All infectious cases should be quarantined for 17 days following exposure, and strict isolation would apply to any victims. *All material used to treat victims or coming in contact with victims should be autoclaved, boiled or burned.* Patients should be considered infectious until all scabs separate.

Staphylococcal Enterotoxin B (SEB)

SEB is a common contributor to staphylococcal food poisoning and could be used by terrorists as an aerosol. However, since inhalation disease is not a natural phenomenon for SEB, intentional aerosolization could be easily recognized. Its action is rapid but not too serious. Inhalation following aerolization could cause up to 80% of those exposed to become ill.

> Medical classification: Toxin
> Probable form of dissemination: Aerosol, sabotage of the food supply
> Detection in the field: None
> Infective dose (aerosol): 30 ng (incapacitating) 1.7 µg (lethal)
> Incubation time: 1 to 6 hours
> Persistence: Stable (resistant to freezing)
> Personal protection: Protective clothing must be used as well as protection for the respiratory tract. Use a mask with biological filters or SCBA with positive pressure, at least until you know the specific threat(s).
> Routes of entry to the body: Inhalation, mouth
> Person-to-person transmissible: No
> Duration of illness: Hours, or days to weeks. Treatment is mainly limited to supportive care, but assisted ventilation may be necessary in serious cases, and fluid management is necessary. No antitoxin is available and antibiotics provide no benefit.
> Potential ability to kill: Low (less than 1%)
> Defensive measures: Immunization, good personal hygiene, physical conditioning, use of arthropod repellents, wearing protective mask, and practicing good sanitation
> Vaccines: Under development as of 1996
> Drugs available: No specific anti-toxin exists
> Decontamination: Soap and water, or diluted sodium hypochlorite solution (0.5%). Destroy food that may have become contaminated.

Trichothecene Mycotoxins/T-2

Trichothecene mycotoxins are a large group of toxins produced by several species of fungi. T-2 is one of the most stable of these toxins and therefore the most likely to be used in terrorist actions. This toxin was allegedly used as an aerosol, popularly known as "yellow rain," in Laos (1975–1981), Kampuchea (1979–1981), and Afghanistan (1979–1981).

> Medical classification: Toxin
> Probable form of dissemination: Aerosol; sabotage
> Detection in the field: None
> Infective dose (aerosol): Moderate
> Incubation time: 2 to 4 hours
> Persistence: Stable (for years at room temperature)

Personal protection: Protective clothing must be used as well as protection for
the respiratory tract. Use a mask with biological filters or SCBA with
positive pressure, at least until you know the specific threats.

Routes of entry to the body: Inhalation, mouth, and skin

Person-to-person transmissible: No

Duration of illness: Days to months. Therapy is mainly supportive

Potential ability to kill: High

Defensive measures: Immunization, good personal hygiene, physical conditioning, use of arthropod repellents, wearing protective mask, and practicing good sanitation. Mycotoxin-induced disease is not contagious but the
stability of this toxin in the environment is quite persistent.

Vaccines: None

Drugs available: No specific anti-toxin exists

Decontamination: Soap and water, or diluted sodium hypochlorite solution
(0.5%). Clothing of T-2 victims should be removed and exposed to a 5%
(not 0.5%) solution of hypochlorite for 6 to 10 hours, or destroyed. Skin
may be cleaned with soap and water, and eye exposure should undergo
saline irrigation. Isolation is not required. Regular disinfectants useful
against most other biological agents are often inadequate against the very
stable mycotoxins. After decontamination, isolation is not required.

Tularemia

Tularemia (also called "rabbit fever" or "deer fly fever") places particular risk on
rabbit hunters who skin and clean their catch. Under normal conditions, people get
this disease from blood or tissue fluids of infected animals, or bites of infected ticks,
deerflies, or mosquitoes, or less frequently, by inhalation of contaminated dusts or
consumption of contaminated food or water. For normally acquired tularemia, the
death rate would be 5 to 10% without treatment. With treatment, survival could be
total. It takes only a very small dose to make people ill. Any attack with dissemination by aerosol would be likely to spread typhoidal tularemia which could be
expected to have a fatality rate higher than the 5 to 10% apparent when the disease
is normally transmitted. Strict isolation of patients is not required.

Medical classification: Bacteria

Probable form of dissemination: Aerosol

Detection in the field: None

Infective dose (aerosol): 10 to 50 organisms

Incubation time: 1 to 10 days

Persistence: Not very stable (but can last for months in moist soil or other
media)

Personal protection: Protective clothing must be used as well as protection for
the respiratory tract. Use a mask with biological filters or SCBA with
positive pressure, at least until you know the specific threat(s).

Routes of entry to the body: Inhalation, mouth, and skin

Person-to-person transmissible: No

Duration of illness: 2 or more weeks

Potential ability to kill: Moderate if untreated

Defensive measures: Immunization, good personal hygiene, physical conditioning, use of arthropod repellents, wearing protective mask, and practicing good sanitation

Vaccines: Yes. LVS, live attenuated vaccine (IND)

Drugs available: Streptomycin, gentamicin, and doxycycline

Decontamination: Soap and water, or diluted sodium hypochlorite solution (0.5%). Secretion and lesion precautions are necessary, but strict isolation of victims is not required.

Venezuelan Equine Encephalitis

Children and the elderly are more susceptible to this virus than the average adult; however, pregnant women might develop serious problems. Most victims live through this disease.

Medical classification: Virus

Probable form of dissemination: Aerosol; infected vectors (carrier organism)

Detection in the field: None

Infectivity dose (aerosol): 10 to 100 organisms

Incubation time: 1 to 5 days

Persistence: Relatively unstable

Personal protection: Protective clothing must be used as well as protection for the respiratory tract. Use a mask with biological filters or SCBA with positive pressure, at least until you know the specific threat(s).

Routes of entry to the body: Skin

Person-to-person transmissible: Low

Duration of illness: Days to weeks

Potential ability to kill: Low

Defensive measures: Immunization, good personal hygiene, physical conditioning, use of arthropod repellents, wearing protective mask, and practicing good sanitation

Vaccines: Yes. TC-83 live attenuated vaccine (IND) or C-84: Formalin inactivation of TC-83 (IND).

Drugs available: No specific anti-viral exists

Decontamination: Soap and water, or diluted sodium hypochlorite solution (0.5%). Blood and body fluid precautions are necessary. Human cases are infectious for mosquitoes for at least 72 hours.

VACCINES

Anthrax Vaccine as a countermeasure to *Bacillus anthracis* has been licensed since 1971. The primary immunization dose schedule calls for a 0.5 ml dose at 0, 2, 4, 6, 12, and 18 months. Primate studies indicate that protection begins two weeks after a minimum of two doses in an abbreviated series given approximately two weeks apart. An annual booster schedule is required after the primary series, and preliminary data indicate boosters given up to two years after the initial abbreviated series may be effective. Antibodies against protective antigens of the organism develop in 85 to 95% of humans after the initial three doses, and in 100% after a 12-month dose. The vaccine protects against dermal exposure in an occupational setting, and, based on animal studies and occupational experience, probably protects against inhalant exposure. The vaccine may be less effective with overwhelming challenge of inhaled spores. The manufacturer is the Michigan Department of Public Health.

Plague Vaccine is a formalin-killed, whole cell vaccine against *Yersinia pestis*. The expected route of exposure may be inhalation, infected vectors (fleas), or secondary transmission. Experience in Vietnam where plague is endemic suggests the vaccine protects against intradermal exposure. However, recent animal studies cast doubt on the efficacy of the current vaccine against inhalant exposure. Approximately 7 to 8% of recipients do not show any antibody response to the vaccine. The immunization dose/schedule is 1.0 ml IM, followed by 0.2 ml IM one to three months later, and 0.2 ml IM at three to six months. Preliminary data indicates that a rapid immunization schedule of 0.5 ml IM at 0, 1, and 2 weeks produces similar antibody strength of solution as that observed after the third dose of the standard schedule. Recipients should receive three additional 0.2 ml IM doses every six months after the primary series, and then 0.2 ml every one to two years thereafter. Greer Laboratories in Lenoir, NC is the manufacturer.

Smallpox Vaccine is a licensed vaccine made by Wyeth Laboratories and available through a repository at the Center for Disease Control and Prevention, Atlanta, GA. Reliable data are limited as to efficacy and durability of protection. Indirect evidence seems to indicate a highly effective vaccine. Immune response in humans is greater than 95% in that primary vaccines develop neutralizing actions greater than or equal to 1:10. One dose by the scarification technique will provide protection within a minimum of 14 days. A booster dose is required every 5 to 10 years for protection against *Variola major* virus.

Botulinum Toxoid Vaccine, Pentavalent provides a countermeasure to five of the seven neurotoxins (Types A through E) produced by the bacterium *Clostridium*. It is in investigational vaccine status with expected availability in the year 2000. Primarily inhalation, or ingestion if used to contaminate food or water supplies would be the bacteria's expected route of exposure. This vaccine probably protects against exposure by inhalation and ingestion, although studies are lacking. Preliminary results from primate studies for *Botulinum A* indicate the vaccine is effective against inhalant exposure, but human studies of efficacy cannot be done. The primary immunization dose/schedule is a 0.5 ml dose given subcutaneously in a three-dose primary series at 0, 2, and 12 weeks. A few primate studies indicate that protection against inhalant exposure may occur after a minimum of two doses. A booster dose is required at one year. The manufacturer is the Michigan Department of Public Health and Porton Products, Center for Applied Microbiology and Research.

4 Standard Operations of Hazardous Material Response Teams (HMRTs)

BACKGROUND AND HISTORY

In the late 1970s, the first hazardous materials response teams were added to fire departments in the United States after a series of railroad wrecks caught the public's attention. On January 25, 1969, a Southern Railway Company train derailed with a fire and explosion in Laurel, MS leaving 2 dead and 33 hospitalized. On February 18, 1969, two trains derailed and a tank car exploded in Crete, NE killing 9 and injuring 53. A Toledo, Peoria and Western Railroad train on June 21, 1970 derailed with resultant fire in Crescent City, IL. Nine tank cars loaded with liquefied petroleum gas (LPG) ultimately exploded injuring 66 persons and causing the destruction of a number of buildings downtown. A photographer got a picture of the immense fireball that extended upwards about 1000 feet. The photo appeared in all the newspapers, and a blown-up version graced the wall of the National Transportation Safety Board in Washington, D.C. for many years. On January 22, 1972, a tank car loaded with liquefied petroleum gas collided with a hopper car in the Alton & Southern Railroad Company's Gateway Yard in East St. Louis, MO. The LPG in the tank car leaked to the ground and vaporized. A large vapor cloud was formed which ignited and exploded injuring 230 persons and damaging more than $7.5 million worth of property.

A derailment and subsequent burning of a Delaware and Hudson freight train in Oneonta, NY injured 54 persons by fire and rocketed parts of four tank cars on February 12, 1974. About 4:23 a.m. on the 16th of May, 1976, the locomotive and 27 cars of a Chicago and Northwestern freight train derailed and were struck by cars from another train headed the other way a little bit west of Glen Ellyn, IL. Fourteen persons were injured by the derailment and the release of anhydrous ammonia. On November 9, 1977, a Louisville and Nashville Railroad Company freight derailed and punctured two tank cars of anhydrous ammonia, resulting in a toxic cloud over part of Pensacola, FL that killed 2 persons and injured 46. Liability is completely different now from what it was after World War II when Melvin Belli was the first lawyer to win $100,000 for a client in a personal injury suit. Compare that to the $53,000,000 won by an attorney in the Pensacola case for a young boy who was caught in the toxic cloud of anhydrous ammonia after the derailment and suffered respiratory damage.

The weekend of February 22 to 26, 1978 had a death rate from two different incidents so high that a shocked nation demanded some sort of corrective action from the railroads and from government. At 10:25 p.m. on February 22, 1978 in

Waverly, TN, 23 cars derailed from a Louisville and Nashville freight. Two days later, on February 24, 1978 a tank car containing LPG blew up near downtown Waverly while workers were re-railing cars, killing 16 persons and injuring 46. About 1:55 a.m. on February 26, 1978 in Youngstown, FL an Atlanta and Saint Andrews Bay freight train derailed and chlorine from a leaking tank car killed 8 persons and injured 138. Florida State Highway #231 runs parallel to the rail track about 100 yards away. More would have been killed and injured had not both the head brakeman and the rear brakeman exposed themselves to the Poison A gas, certifiably heroic actions, by warning motorists that it was a toxic cloud across the roadway rather than a nighttime fog. When people had earlier tried to drive through the "fog," their car engines stalled due to the heavy concentration of chlorine in the cloud, which even at 9 a.m. was three miles wide and four miles long with a maximum altitude of 1000 feet.

Many argued that changes in tank cars allowed prior to 1978 had resulted in a disastrous safety record with dangerous couplers, lack of insulation on compressed gas cars, a three-fold increase in size of payload, lack of a continuous center sill, "rocketing," and other factors. Changes allowed in building tank cars over the years were not based on safety but for lightness in weight of the car which would allow more cargo capacity or payload. The new cars were presented as engineering marvels rather than inherently dangerous containers. It was said that they were "envelopes" — lightweight, weak, with no integral strength when confronted with a collision. As a result, public safety had not received priority attention. Tank cars were not the only problem. Railroads hired less help to reduce expenses which in-turn led to deferred maintenance, little or no observation of moving trains by crossing guards or other employees, and deteriorating roadbeds and equipment. Railroad incidents such as Waverly, TN and Youngstown, FL brought "Hazardous Materials" to the attention of the media, the government, and the citizens. However, no transportation mode or industry using chemicals could be considered guiltless.

On December 3, 1984 a release of MIC (methyl isocyanate) at a Union Carbide plant in Bhopal, India killed more than 2500 persons and seriously injured an estimated 150,000 more. This incident generated intense media and government interest throughout the world due to the massive loss of life. However, there was almost no outside interest just 19 days before the Bhopal incident on November 15, 1984 when 30 to 50 gallons of the very same material were spilled on the ground at an FMC Corporation plant in Middleport, NY. In Middleport, the MIC vaporized and was carried by a light breeze to an elementary school located less than 500 yards away. Approximately 500 students were in class as the school's ventilation system sucked MIC into the building. As the attorney general of New York noted in a report about the Middleport incident, "MIC is an extraordinarily destructive and dangerous raw material. A colorless liquid and therefore difficult to see, it vaporizes very readily even on cold days. MIC reacts vigorously with water, thereby increasing the danger of spills and making subsequent cleanup more difficult. Contact with common materials, such as iron and zinc, triggers violent reactions including generating great heat and under certain circumstances, explosions. Teflon and stainless steel are among the few materials appropriate to contain MIC. As a gas, MIC is not only extremely noxious, but is also highly flammable, thereby creating the risk of fireball type conditions if a spark source exists nearby."

Similar problems must be dealt with by HMRTs, trained, teamed, and equipped specialists employed by government agencies such as fire and police, commercial response contractors, and industrial corporations. Although the federal government has preempted the regulation of hazardous materials and hazardous waste, response to and control of incidents involving such materials has been left to local government. In the United States, methods and procedures for professional response to hazardous materials emergencies have been developed at the local level.

PLANNING FOR RESPONSE

There was a growth spurt in fire department, law enforcement, commercial, county, state, military, and federal HMRTs that started in the mid-1970s and has continued on to the present day as we prepare for terrorist incidents, weapons of mass destruction, and chemical/biological warfare within the United States. Haz Mat teams tend to use similar, standard operational guidelines as to what they do, and don't do, at a hazardous materials incident so that everyone is "singing from the same hymn-book." Hazardous materials response teams subscribe to and follow standards adopted by certain organizations including the National Fire Protection Association (NFPA) based in Quincy, MA and the U.S. Department of Labor Occupational Safety and Health Administration. NFPA standards 471, 472, and 473 reflect the latest technical knowledge and hands-on experience of people who actually work with hazardous materials.

> **NFPA 471**: Recommended Practice for Responding to Hazardous Materials Incidents covers decontamination and medical monitoring, methods of mitigation, chemical protective clothing, response levels, and site safety.
> **NFPA 472**: Standard for Professional Competence of Responders to Hazardous Materials Incidents deals with competencies at the awareness, operational, technician, and incident commander levels. The potential threat of terrorist activity led to a Tentative Interim Amendment (TIA) to Standard 472 during 1997. According to the NFPA, "a TIA automatically becomes a proposal of the proponent for the edition of the standard; as such, it is then subject to all the procedures of the standards-making process."
> **NFPA 473**: Standard for Competencies for EMS Personnel Responding to Hazardous Materials Incidents identifies the level of competence required of emergency medical service responders to Haz Mat incidents.

Peter A. McMahon, EMT is the director of the Grand Island Fire Co. located in Grand Island, NY. He is also chairman of the NFPA's technical committee for Haz Mat standards. "In 1996 we had finished with the revision of NFPA 472 when we became aware that the Haz Mat community were going to be ones that were called to deal with terrorism incidents involving weapons of mass destruction, especially if they involved chemical or biological forms or weapons, because we have the chemical protective clothing and the decontamination processes.

"We had not addressed terrorism in the standard to date because, historically, people in the United States have not had to face terrorist acts. 'It can't happen here,' had been our previous frame of reference. While doing a needs assessment at the National Fire Academy, it was noted that the United States in general isn't mentally prepared for terrorism. When the World Trade Center explosion happened, people thought it was a fluke, a one-time deal that was not going to happen again. Then came the explosion at the federal building in Oklahoma City, the bomb at the Olympics, and a couple of additional explosions in Atlanta.

"It became pretty apparent that we were not well prepared for other kinds of things that could be happening. If you set a package down in Harrod's Department Store in London or you set a package down in Macy's Department Store in New York City, two entirely different chains of event happen. In Harrod's, the security people will assume it's a bomb and call the authorities. They will evacuate the store, send in trained people to deal with the package, and some explosive ordnance person is going to find your box of underwear. If you set that box down in Macy's in New York City, someone is going to pick it up and take it home. It's going to get stolen. That is the frame of reference of not only the citizens but most of the emergency response community.

"So to raise awareness, we felt we needed to change the standard. We got together with the federal government and the military, medical personnel, and the technical committee. We looked at the minimum things we needed to teach the responders; the awareness, operations, technician, and incident command levels. We wanted to provide a frame of reference: this could be a terrorist incident, a criminal act, or a possible weapon of mass destruction posing a chemical, biological, or nuclear threat."

NFPA produces a number of standards including the following which deal with chemical protective clothing:

NFPA 1991: Vapor-Protective Suits for Hazardous Chemical Emergencies.

NFPA 1992: Liquid Splash Protective Suits for Hazardous Chemical Emergencies.

NFPA 1993: Support Function Protective Clothing for Hazardous Chemical Operations.

Contact: National Fire Protection Association, 1 Batterymarch Park, P.O. Box 9101, Quincy, MA 02269-9101; 800-344-3555; 800-593-6372 (Fax).

The Occupational Safety and Health Administration on March 6, 1989 implemented Code of Federal Regulations 1910, "Hazardous Waste Operations and Emergency Response" which required that all responders to hazardous materials/waste/substances incidents be properly trained and equipped. This OSHA rule mandates that all employers including fire departments, emergency medical and first aid squads, and industrial fire brigades conduct monthly training sessions for their employees totaling 24 hours annually. Additional requirements call for use of the Incident Command System (ICS), an on-scene command post, and an incident commander. Beyond these minimum requirements now applied to any organization that may be called upon to respond to Haz Mat incidents, OSHA requires employers — including fire departments, law enforcement agencies, and commercial response contractors — who utilize specially trained teams involved in intimate contact with controlling or handling hazardous substances to provide special training in such areas as care and use of chemical protective clothing, techniques and procedures for stopping or controlling leaking containers, and decontamination of clothing and equipment.

Personnel who participate in emergency response shall be trained in the following categories: First responder *awareness* level applies to individuals who are likely to witness or discover a hazardous substance release or who have been trained to initiate an emergency response sequence by notifying the proper authorities. They would take no further action after the notification was completed. There are no specific number of hours required for training at the first responder awareness level. The first responder *operation* level applies to persons who respond to releases or potential releases of hazardous substances as part of the initial response to the incident site for the purpose of protecting nearby persons, property, or environment from the effect of the release. They are trained to respond in a defensive fashion without actually trying to stop the release. First responders at this level shall have received at least eight hours of training or have had sufficient experience to demonstrate competency in six different areas (basic hazard and risk assessment techniques, selection and use of personal protective equipment at this level of response, basic hazardous material terms, performance of basic control, containment and/or confinement operations, basic decontamination procedures, and relevant standard operating and termination procedures).

The hazardous materials *technician* level includes individuals who respond to releases or potential releases for the purpose of stopping the release. They assume a more aggressive role than the first responder at the operations level in that they will approach the point of release in order to plug, patch, or otherwise stop the release of a hazardous substance. Technicians shall have received at least 24 hours of training and show they have competency in eight specified areas or categories. The *incident commanders* who will assume control of the incident scene beyond the first responder awareness level shall receive at least 24 hours of training, demonstrate competency in six areas, and have knowledge of nine procedures that must be performed during an emergency response. Overall additional requirements deal with training, refresher training, demonstration of competency of training course instructors, medical requirements, and emergency response plans.

The American Society for Testing and Materials (ASTM) has 132 technical committees to develop standard test methods, specifications, practices, terminology, guides and classifications for materials, products, systems, and services. The society has three standards for chemical protective clothing that might be of interest to readers.

ASTM F739: Test Methods for Resistance of Protective Clothing Materials to Permeation by Liquids and Gases.

ASTM F1052: Standard Test Method for Pressure Testing Vapor Protective Ensembles.

ASTM F1383: Standard Test Method for Resistance of Protective Clothing Materials to Permeation by Liquids or Gases Under Conditions of Intermittent Contact.

Contact: American Society for Testing and Materials, 100 Bar Harbor Drive, West Conshohocken, PA 19428.; 610-832-9500.

Another standard operational tactic for Haz Mat response teams is use of the Incident Command System (ICS), an organized system of responsibilities, roles, and standard operating procedures used to manage and direct emergency operations. The ICS is designed for any level of government, offers the possibility to expand as an incident grows, ensures that jurisdictional authority is not compromised, and uses standardized terminology. The purpose of ICS is to establish the responsibility for command on a single and specific individual, allowing for an orderly transfer of command, so as to ensure a strong, direct, and visible command from the very start of an incident. The ICS defines activities and responsibilities for all those who respond to an emergency incident. Tactical priorities for the incident commander are to protect life safety, stabilize the incident, determine objectives, protect and preserve property, provide for accountability, and utilize feedback. The incident commander's basic tasks are to assume and announce command, select an effective operating position, perform size-up to evaluate the situation, assign staff to ICS positions (such as, public information officer, liaison officer, finance officer, planning officer, logistics officer, operations officer, etc.), control the communications process, develop an "incident action plan" to identify overall strategy and select overall objectives, manage tactical objectives, and review/evaluate/revise.

The Occupational Health and Safety Administration on March 6, 1989 implemented Code of Federal Regulations 1910, Hazardous Waste Operations and Emergency Response which required that all responders to hazardous materials/waste/substances incidents be properly trained and equipped. CFR 1910 spells out that the incident commander has certain specific procedures he or she must follow in an emergency response:

- All emergency response personnel and their communications must be coordinated through the individual in charge or the incident commander.
- The incident commander shall identify, to the extent possible, all hazardous substances or conditions present.
- Based on the hazardous substance and/or conditions present, the incident commander shall implement appropriate emergency operations and assure that the personal protective equipment is appropriate for the hazards to be encountered. However, personal protective equipment shall meet, at a

minimum, the criteria contained in 29 CFR 1910. 156(e) when worn while performing firefighting operations beyond the incipient stage.

- Personnel engaged in emergency response and exposed to hazardous substances presenting an inhalation hazard shall wear positive pressure self-contained breathing apparatus until such time as the incident commander, through air monitoring, decides that the decreased levels of respiratory protection will not result in hazardous exposure to employees.
- The incident commander shall limit the number of emergency response personnel at the emergency site to those who are actively performing emergency operations. Operations in hazardous areas shall be performed using the buddy system in groups of two or more.
- Back up personnel and medical personnel (advanced first aid minimum) shall stand by with equipment ready to provide assistance or rescue.
- The incident commander shall designate a safety official, who is knowledgeable in the operations being implemented who has the specific responsibility to identify and evaluate hazards and to provide direction with respect to the safety of operations for the emergency.
- When a safety official judges that imminent dangerous conditions exist that may be dangerous to life and health of the emergency personnel, the safety official may alter, suspend, or terminate those activities. The safety official shall immediately inform the incident commander of any actions needed to correct these hazards at the scene.
- After emergency operations have been terminated, the incident commander shall implement appropriate decontamination procedures.

There are other ways in which experienced first responders to hazardous materials incidents tend to use similar, basic practices, procedures, and strategies. They tend not to rush in, but stop as soon as there is a visual sighting and, using binoculars if necessary, perform a cautious evaluation and size-up. It is generally agreed that hazardous materials incidents require a more cautious approach than do structural fire situations.

Hazardous materials incidents also require a command post — a location where persons having the authority to command and persons necessary to support the process, are brought together and provided with the necessary facilities. Operation of a command post tends to be viewed as an on-scene manifestation of extensive prior planning. If operational plans have not been written, problem areas identified and responded to, resources and expertise cited, and personality conflicts resolved ahead of time, operation of a command post will not solve your problems. A command post is an operational control device designed to maintain relationships previously agreed upon. Successful command post operation presupposes a written plan. The purpose of a command post should be to provide leadership a means to coordinate field operations in order to resolve a crisis situation in an orderly and expeditious manner. The most often asked question with regard to a hazardous materials incident is ... "Who's in charge here?" Operation of one central point where all information is processed, decisions made, and tasks assigned provides a strong indication to the public, emergency service agencies, and the media that someone is, in fact, in overall charge of the scene. Without a command post, control, coordination, and communication will tend to be hit-or-miss. It is not unusual to have to relocate a command post. Shifting winds, new information, desire for better facilities, problems with communications, need for additional space, or inability to provide security for the command post initially selected could all be reasons for relocation.

A staging area is often near to but separate from the command post. It is a marked area where responding personnel report with their equipment or apparatus to await direction. That is, not everyone reports to the command post. If fire, police, emergency management federal, state, industry, commercial contractors, military, medical and other personnel arrive on scene, they are directed to the staging area. The person in charge of a particular organization will report to the command post to make his or her equipment and expertise known, provide information, or stand-by for instructions. If a staging area is not used, a command post can easily be overrun with persons peripheral to the overall emergency effort if steps are not taken to control access.

All members of organized HMRTs must participate in medical surveillance in accordance with the Occupational Safety and Health Administration (OSHA) as stated in Code of Federal Regulation 1910. They must receive a baseline physical examination, plus a medical surveillance examination every year unless the attending physician believes a longer interval (no greater than bi-annually) is appropriate.

Emergency response plans are required for fire, police, and emergency medical service agencies prior to the start of an emergency operation according to CFR 1910. Such plans must be in writing and available for inspection and copying by employees, their representatives and Office of Safety and Health Administration of the U.S. Department of Labor. The emergency response plan shall include the following minimum elements:

- Pre-emergency planning and coordination with outside parties
- Personal roles, lines of authority, training, and communication
- Emergency recognition and prevention
- Safe distance and place of refuge
- Site security and control
- Evacuation routes and procedures
- Decontamination
- Emergency medical treatment and first aid
- Emergency alerting and response procedures
- Critique of response and follow-up
- Personal protective equipment and emergency equipment

HMRTs immediately and positively identify the chemical in order to learn its reactive characteristics and identify proper containment and control methods. Teams try not to commit personnel until they know what they are dealing with. This identification is based on at least two resource materials, and results may restrain their activity to initiation of evacuation and protection of exposures until positive identification of the hazardous material has been achieved.

HMRT members follow certain basic tactics. In making an approach they avoid visible concentrations of smoke, fumes, vapors, and liquid. They approach from upwind and upgrade using natural barriers for protection. They don't rush in as they may have been trained to do if they are firefighters, but rather take it slow and easy in evaluating the scene. Unmanned equipment such as monitor nozzles and "kelly coils" are used when and where possible. They recognize "no fight" situations where no life and minimum property are involved. Such responders use the buddy system, never send in one person alone, and initiate and maintain a suited back-up team as a safety measure for the primary entry team. They erect a windsock when vapors, fire, smoke, dust, or wind directional changes are present and use recognized non-verbal emergency signals such as hands and arms held directly over the head to indicate serious distress as would be the case when an acid/gas entry suit is ripped or penetrated. Simple hand signals, or chalkboards such as those used to communicate with race drivers, are often used to provide critical information under conditions of extreme noise or limited visibility.

They know that chemicals running amok in the street are completely different from chemicals in laboratory samples. They research chemicals at the scene, contact knowledgeable industrial contacts and chemists by telephone from the scene, identify the chemical(s), and establish a hot zone. HMRT members recognize that with flammable liquids, the vapor burns rather than the liquid. Vapor spreads unseen, and responders are careful to identify vapor concentrations and perimeters through careful, continuous monitoring. They do not extinguish flammable gas fires until the flow of vapor can be stopped lest vapors again spread unseen throughout the area to find a source of re-ignition. In working with closed containers, they are aware of both Boyle's Law (volume is inversely proportional to pressure) and Charles' Law (volume is directly proportional to temperature).

HMRT members wear different levels of protective clothing depending on the degree of risk present at a Haz Mat incident at a specific time; that is, they can "dress down" in levels of protective clothing as the risk level decreases. The four levels of protective clothing follow:

Level A: Vapor protective suit for hazardous chemical emergencies is the highest level of protection provided for the skin, eyes, or respiratory contact. Level A protection requires a totally encapsulated gas-tight suit, positive pressure self-contained breathing apparatus (SCBA) (OSHA/NIOSH approved), chemical protective clothing, chemical-resistant inner gloves, chemical-resistant outer gloves, chemical-resistant boots with steel toe and shank, two way radio communications which are intrinsically safe, with optional hard hat under suit, long underwear and coveralls under suit.

Level B: Liquid splash protective suit for hazardous chemical emergencies is used by responders when the highest level of respiratory protection is needed but when a lower level of skin protection may be required than that for Level A. Level B chemical protective clothing requires positive pressure, SCBA (OSHA/NIOSH approved), chemical splash protection clothing, chemical-resistant inner gloves, chemical-resistant outer gloves, chemical-resistant outer boots with steel toe and shank, two way radio communications which are intrinsically safe, with optional coveralls under suit and a hard hat.

Level C: Limited use protective suit for hazardous chemical emergencies provides minimum protection for respiratory and skin hazards and is used only where there are no skin absorption hazards. Level C protection includes use of a full face, air purifying respirator (OSHA/NIOSH approved), (optional) escape mask, chemical splash protective clothing such as a one piece coverall or hooded two piece suit, chemical-resistant outer gloves, (optional) chemical-resistant inner gloves, chemical-resistant boots with steel toe and shank, (optional) cloth coveralls inside chemical protective clothing, (optional) two way communications, (optional) hard hat.

Level D: The lowest level of protection provides no respiratory or skin protection and should only be used when there is no chemical hazard in the atmosphere. It is basically work clothes that provide little or no protection from chemical hazards.

Kappler Protective Apparel & Fabrics in Alabama makes Responder CSM® (chemical surety materials) garments that provide protection against a wide range of chemical agents including lewisite (L), mustard (HD), sarin (GB), soman (GD), tabun (GA), and the nerve gas VX. Responder CSM limited-use garments meet 29 CFR 1910.120 and are available in the following styles:

Level A: Totally encapsulating vapor protective suit, front entry, expanded back for SCBA.

Level B: Front entry coverall with optional overhood to protect SCBA.

Level C: Coverall for wear with filtered air respirators in low threat environments.

Chemical test data indicate that breakthrough time (the time it takes any one of the six chemical agents mentioned above to pass through the protective material of the clothing until first detected by an analytical device) equals or exceeds 480 minutes. Testing was conducted by GEOMET Technologies at ambient temperatures in accordance with MIL-STD-282 methods 208/209, with the test cell filled with agent to provide a contamination density consistent with the American Society for Testing and Materials standard F739. The breakthrough criteria for each of the six chemical agents were as follows: mustard (HD) = 4.0 μg/cm^2, lewisite (L) = 4.0 μg/cm^2, tabun (GA) = 1.25 μg/cm^2, sarin (GB) = 1.25 μg/cm^2, soman (GD) = 1.25 μg/cm^2, and nerve (VX) = 1.25 μg/cm^2. (Breakthrough criteria are based on cumulative permeation over the total test period.)

Contact: Kappler Protective Apparel and Fabrics, P.O. Box 218, Guntersville, AL 35976; 800-633-2410; 205-582-2706 (Fax).

The Trellchem High Performance Suit (HPS) is a fully encapsulating positive pressure self-extinguishing suit manufactured by Trelleborg Industri located in Trelleborg, Sweden that will be worn when Level A protection is required. It is made of a multilayer material consisting of a viton and butyl layer supported by a nylon layer, then a chloroprene layer. The final inside layer is a proprietary multilayer laminate film. The faceshield is 2 mm-thick, high impact PVC.

The HPS chemical suit material was tested by GEOMET Technologies of Germantown, MD. In the first test (done under test methods MIL-STD-282 method 204.1.1 and method 206.1.2 and MIL-C-12189H), blister agent distilled mustard (HD) yielded on endpoint time = 420–560 minutes. For nerve agent sarin (GB) samples, the endpoint time = 580 to greater than 960 minutes. In the second test (done under CRDC-SP-94010 method 2.2 and ASTM method F739, currently under consideration within an ASTM Technical Committee to become an ASTM standard), lewisite samples had no detectable agent penetration from 0 to 24 hours (not detectable (ND) = <0.0075 μg/cm^2 of lewisite). In the third test done to test decontaminability after exposure to blister agent, HD, material reached an airborne exposure limit (AEL) = 0.003 mg/m^3 between 37 and 45 hours. In the fourth test (done under test method CRDC-SP-84010 method 2.2) on seamed samples, blister agent HD samples with 10 1 mg drops of HD applied had no detectable agent penetration from 0 to 24 hours (ND = <0.1 μg/cm^2 HD).

Contact: Trelleborg Viking, Inc., Protective Products Division, 170 West Road, Suite 1, Portsmouth, NH 03801; 603-436-1236 or 800-344-4458; 603-436-1392 (Fax).

Experienced responders maintain incident vigilance and discipline in order to control the response effort. To control the scene, nonessential people are kept out of the area, approach and size-up are done with a minimum of personnel, additional forces are committed only as necessary, and reserve personnel are kept on remote standby. Vehicles, equipment, and personnel are kept out of sprays, run-off, residue, and vapor clouds and are positioned to permit retreat. Responders recognize that certain vapors entering a carburetor, such as oxidizers, can make an internal combustion engine take off like the space shuttle, while others, such as chlorine, will exclude oxygen and stall the engine. A fire engine that can neither pump water or be driven to safety suddenly becomes a substantial liability. Fearlessness at a hazardous materials incident may mean the daring person just does not understand the situation. Experienced Haz Mat personnel freely admit that some chemicals can be downright scary. For each team member inside the hot zone, an additional member is suited-up on the outside ready to go in. As responders learn the exact circumstances of an incident, they may back off from the initial level of personal protection. Inexperienced responders may call such stringent safety measures "overkill." Experienced responders use the term "vigilance and discipline" and warn each other to beware of the person who thinks he or she is immune.

Team members observe certain basic disciplines. They engage in detailed, prior planning to develop and implement a systematic approach to maximize effective use of available resources. Such prior planning attempts to identify unknowns, establish productive relationships, minimize variables, and identify responsibilities. Planning identifies what needs to be done and establishes a framework for getting the job completed. Hazardous materials pre-incident planning often considers the following components: development of a hazard analysis, creation of a resource inventory, accumulation of product information, implementation of a response capability, provision of training, and development of protocols or standard operating guidelines to guide conduct on the scene of an incident. Haz Mat responders sometimes use acronyms such as "P.P.P.P.P." (Prior Planning Prevents Poor Performance) and "S.O.F.S." (Shut Off Flow, Stupid) as reminders. They learn the mechanical environment and containerization used at identified hazardous materials facilities in order to know mechanical/electrical shut-off locations and operations ahead of time. Before attempting to patch or plug a leak, they first check for shut-off valves, electrical circuits, or other means of control built into the system that someone might have failed to activate.

Haz Mat team members utilize the expertise of industrial facility personnel, chemists, supervisors and plant technicians in order to better understand the systems and materials involved in an incident. They maintain call lists of persons and organizations with particular capabilities, obtain prior commitments for their assistance in an emergency, and identify alternative sources of supply. They call for help immediately if a need is even suspected. They obtain and train with specialized tools/equipment/materials, and have access to a variety of patching and plugging devices. The initial size-up or evaluation of a hazardous materials incident is a crucial component of incident response. Size-up includes detection and identification of hazardous materials; evaluation of fire, explosion, reaction, and health and environmental hazards; as well as rescue, request for outside assistance, estimation of danger areas, definition of potential secondary emergencies, identification of critical exposures, initiation of evacuation if called for, and determination of proper suppression and control activities.

Because hazardous materials normally become extremely dangerous only when released from some type of containerized environment, the ability to patch or plug leaks in a wide variety of containers has assumed great importance. A number of HMRTs around the country have assembled necessary equipment and materials that allow team members to control numerous types of leaks in literally hundred of different types of containers. They often use common, ordinary tools and materials in tandem with personal mechanical aptitude in controlling leaks. They point out that they do not try for the "ultimate patch," but rather work to temporarily stop the flow. Haz Mat responders often work under frightening conditions, but their basic stock-in-trade is the ability to temporarily patch or plug any leak in any vessel, tank, cylinder or drum. Mechanical aptitude is crucially important in selecting trainees for duty on hazardous materials response teams.

Members of major fire department HMRTs, by pure experience, should become more knowledgeable than members of industrial teams who have a limited number of incidents on a limited variety of chemicals. Such teams need continued daily training to maintain the skills of members. A portion of training has to be done by bringing all members together from all shifts rather than have all sessions be by single platoon shift only. Some teams train backup, reserve personnel along with regular members to maintain a full complement; such persons could serve as a pool for replacing team members. Training is often "hands-on" and practical; training scenarios are used to test members' comprehension. Haz Mat response teams must keep current on required training no matter how much time it takes. Years ago, no one sued the fire department. They were the "good guys." Nowadays, fire departments can and do get sued like everyone else.

Often a "team concept" is evident in HMRT operations. The entry team may not go into a hot zone without minimum staff present and the decontamination area established. The command structure of the HMRT may be somewhat different from that applied to a fire department. Sometimes, rank structure within the department plays no part within the team. Whether the member is a company grade officer or a plugman in the fire department, when some teams answer Haz Mat alarms every member can work any position on the team without regard to his or her rank in the fire service. A member may be a safety officer one day and an entry team leader the next day. On some teams, "go/no go" decisions — such as the specific tactics of the entry team in mitigating a problem within the hot zone — must be unanimous. Some teams have gripe sessions to air problems and express desire for changes.

Haz Mat responders ensure correct reporting of basic information such as chemical names because they know one misspelled letter or an incorrect pronunciation can dictate an incorrect response. The following pairs of chemical names sound alike but the proper response can range from somewhat to vastly different.

Acetyl iodide/acetaldehyde	Phosphorus trichloride/phosphorus trifluoride
Hexane/hexene	Benzol/benzoyl
Isopropanol/isopropenyl	Phenol/phenyl
Ethanol/methanol	Propionyl chloride/propyl chloride
Ethyl alcohol/methyl alcohol	Thionyl chloride/vinyl chloride

To insure proper transmittal of chemical names, responders "spell it out" and use the UN 4-digit identification number, or the Standard Transportation Commodity Code (49-series) 7-digit number, to be sure a chemical name is understood correctly.

With liquefied, compressed gases, responders are aware of liquid-to-vapor expansion ratios (propane = 270:1, chlorine = 470:1, liquid oxygen = 860.6:1, liquid nitrogen = 696.5:1, liquid argon = 841.2:1). That is, they recognize that if a tank car of chlorine viewed as "one volume" of liquid ruptures, it will release 470 volumes of vapor; 470 tank cars would be required to contain the amount of chlorine vapor released from one ruptured tank car of liquefied, compressed chlorine. With gas or vapor, HMRT members immediately determine its vapor density in order to understand proper monitoring and control procedures. Air has a vapor density of 1.0; a product with a vapor density of less than 1.0 (anhydrous ammonia at 0.6 for example) is lighter than air and will tend to rise. A product with a vapor density of greater than 1.0 (propane at 1.6 for example) is heavier than air and will tend to settle, finding its way into low spots such as trenches, depressions, or cellar holes. Thus, monitoring procedures for propane are significantly different from monitoring procedures for anhydrous ammonia. Many combustible gases are heavier than air and will tend to seep into low spots or travel at low levels for a considerable distance to a source of ignition and then flash back. When combustible gases are potentially present, many responders assign one person to continuously move throughout the area taking repeated readings with a monitoring device, paying particular attention to low spots.

Experienced responders often use a safety officer on scene, a knowledgeable individual whose overriding responsibility is personal and public safety related to placement of apparatus, evacuation distances and areas, adequacy of protective equipment and clothing, elimination of potential ignition sources, reactivity and incompatibility, and maintenance of incident vigilance and discipline. This person is kept free of other duties on scene so he or she can devote complete attention to ensuring utilization of proper methods and procedures. In addition, Haz Mat responders tend to train, practice, and learn as a team and learn to depend on each other. It just doesn't work if a person has to be told to pick up the shovel, dig the hole, and throw the dirt over there.

Haz Mat personnel develop an acute awareness of potential secondary and tertiary hazards. They expect something will go wrong. They expect to run out of air. They expect incorrect advice or information because they know bystanders want to be helpful and industrial personnel may want to keep an incident low-key. They have tremendous respect for "tire fires" on transportation vehicles until they see the shipping papers, for wheel rim holders, for "empty" tanks, for shock-sensitive materials, for alternative fuel in vehicles, for saddle tanks on propane delivery trucks, and for fumigants used in boxcars of agricultural products. They know that many times safety devices on tank vehicles have been altered, and that innocuous rail cars can carry significant quantities of fuel to run a reefer unit. They look for a passing train to crush them, a snapping cable to decapitate them, or a tank truck to turn over and bounce in front of them. They assume any liquid pooled on the ground is much deeper and more dangerous than it appears. They are aware of the phenomenon known as the synergistic effect whereby the joint action of chemicals working together can increase the effectiveness of each and thus cause far greater danger than if each chemical could be dealt with separately. They consider reactivity; will the patching or plugging material or device used react with the hazardous material? They consider incompatibility; are certain chemicals dangerously reactive if mixed accidentally?

Experienced Haz Mat responders are acutely aware of potential contamination of personnel, clothing, and equipment — particularly when working with oxidizers or pesticides. They immediately go through emergency decontamination with water, decon solution, and scrub brushes if contamination is even suspected. They remove contaminated clothing and equipment, demand that clothing and equipment suspected of being contaminated not be used again until decontamination has been completed, and discard clothing and equipment if decontamination cannot be adequately performed. They know that spare clothing maintained on the response rig is cheap insurance. Responders take care to recognize symptoms of chemical exposure, such as from pesticides, even though such symptoms may be nearly identical to those of flu, heat prostration, or smoke inhalation.

When exposure to pesticides is even suspected, they get the label and send it along with potentially contaminated persons in order to alert the examining medical personnel. They never give mouth-to-mouth resuscitation to persons known to have suffered contamination by pesticides or certain other hazardous materials such as freon. They know that water can make pesticides ultra-dangerous, that many pesticides can readily pass through the skin without causing any sensation or providing any warning, that solvents in some pesticides solutions can be absorbed even through rubber boots or other protective clothing, that massive prolonged exposure to a low toxicity pesticide can be as damaging as minor short-term exposure to a highly toxic pesticide, and that a number of pesticides are cholinesterase inhibitors that affect the ability of nerves to transmit impulses so that a pest, or a responder, will not recognize they are being poisoned (chemical nerve agents such as sarin, tabun, soman, as well as GF and VX are also cholinesterase inhibitors).

Haz Mat teams recognize the media can help them or hurt them, and are aware of the effect their actions, or lack of action, can have on media coverage of an incident. Obvious organization and control provide visible assurance that the job is being done correctly. They use formal, written news releases so the basic facts can be provided to an anxious public as quickly, accurately, and completely as possible. Their approach to the media is "facts, not fantasy." They use only one spokesman. The person assigned the duties of media liaison must have the authority and knowledge to know what is going on at all times. Media representatives have a right to expect someone who appears to be tuned-in and is routinely kept abreast of what is happening. Assigning a "junior assistant trainee" to duties as a media liaison can cause problems over which a junior person cannot be expected to maintain control. He or she must be able to provide access to information, and the incident commander must ensure that this is possible. An effective media liaison takes control of the exchange and does not wait for questions but begins providing information, in writing if possible, as soon as it is available. Media representatives are generally not out to antagonize anyone — they merely want factual, reliable information. The media liaison details the scope of the emergency so it does not become exaggerated, and reports what actions are being taken to control or contain the threat. He or she "goes easy" on blame or causes; these can only be determined later if at all. The media liaison controls the information given out and makes sure it is correct; misinformation provided to the media will come back to haunt you.

Media liaisons monitor television and radio news broadcasts. If they detect an error, they call the station immediately, ask for a correction, and log the call. They recognize accredited media personnel from prior dealings. If on-going relationships have not previously been established with local media, rapport, credibility, and confidence may be lacking. The media may be an incident commander's primary source of contact with the general public and should be viewed as a resource rather than a hindrance. Media assistance can be immediate and tangible if it becomes necessary to evacuate an area, close highways, maintain perimeter control, assure access of needed personnel through designated checkpoints, alert medical personnel, establish reception centers, or have the general public engage in activities necessitated by the emergency situation. If your knowledge of the situation is not displayed, the media may assume you do not possess such knowledge. If there is obvious organization, clearly evident preplanning, and observable incident vigilance and discipline, the media will recognize and report this. Remember, they have to report something even if everyone is milling about, no one seems to be in charge, and "nobody knows nothing." Reporters work with what they have. Counter this fact of life by clearly explaining the situation and what is being done to achieve control or containment.

A HMRT is an emergency response team; once the emergency has been mitigated and the victims cared for, the job of the team is done. Many teams do not do cleanup and do not transport/store/dispose of hazardous waste. Commercial response contractors transport, store, and dispose of hazardous waste rather than public safety agency Haz Mat teams, except in the case of "midnight dumping" on town or city property where there is no identifiable spiller or responsible party.

Whoever said, "They can't take it with them," never rode on a hazardous materials response vehicle. Such vehicles are often warehouses-on-wheels crammed full of absolutely necessary gear and material. Used bread trucks and Coca-Cola delivery trucks were popular types of vehicles for new teams because members could fill the bare space as they wished with shelving and partitions for maximum possible storage of cargo such as milk crates, special purpose kits, five gallon pails, padded boxes, color-coded chains, and other gear.

Many HMRTs keep a confidential medical file for every person on the team, provide a baseline medical evaluation upon entry to the team, and conduct a post-employment medical exam when a member leaves the team. They try to document every chemical exposure, either possible or confirmed. Some teams do an exposure report on all members who made an entry into a hot zone; others do an exposure report only in the event of a confirmed exposure. A team member who is super-sensitive to a certain chemical, or a person who is unable to wear a totally encapsulating suit for any reason, may be removed from the team for his or her own protection.

HAZARDOUS MATERIALS RESPONSE WORKSHEET

Every hazardous materials response team should perform a size-up, or incident evaluation, soon after arriving at an incident scene. Such an evaluation must be an automatic response by experienced responders. The following worksheet provides items for consideration.

1. Identification of the commodity:
   A. Shipping papers
   B. Waybills
   C. Placards
   D. Labels
   E. Tank configuration, stenciling, identification numbers, etc.
   F. Smell, sight, consistency, solid, liquid, gas, reactivity, toxic cloud, etc.
   G. Truckdriver, train crew, knowledgeable persons, bystanders, industry representative, CHEMTREC, chemist, medical personnel, etc.

H. Emergency tests (pH, combustible gas indicator, detector tubes, Geiger counter, chemical test(s), HAZ CAT, etc.)

I. Response manuals and handbooks (material safety data sheets, D.O.T. Emergency Response Guide, Bureau of Explosives Emergency Handling of Hazardous Materials In Surface Transportation, NFPA Fire Protection Guide on Hazardous Materials, U.S. Coast Guard Chemical Hazards Response Information System, EPA Oil & Hazardous Materials Technical Assistance Data System, etc.)

2. Identification of the specific hazards involved (circle the appropriate types):

Explosive	Flammable	Corrosive
Poisonous	Radioactive	Oxidizer
Reactive	Nonflammable	Cryogen
Irritating	Toxic vapor	Organic peroxide
Etiologic agent	Spontaneously combustible	
Environmentally hazardous substance		Chemical agent
Biological substance	Weapon(s) of mass destruction	

3. Chemistry of hazardous materials:

   A. Specific gravity
   B. Flashpoint
   C. Boiling point
   D. Ignition temperature
   E. Vapor pressure
   F. Vapor density
   G. Reactivity
   H. Flammable limits
   I. Pyrophoric materials
   J. Water solubility
   K. pH factor
   L. Lighter/heavier than air

4. Assessment of weather conditions on site:

   A. Wind direction & velocity
   B. Temperature
   C. Barometric pressure
   D. Rain, snow, or other moisture

5. Identification of primary and secondary exposures:

   A. High-residency dwellings (schools, nursing homes, hospitals, apartment complexes)
   B. Power lines, sewers, pipelines, etc.
   C. Industrial occupancies
   D. Storage areas for chemicals, gases, flammable liquids, etc.

6. Identification of rescue needs:

7. Evacuation requirements:

8. Run-off considerations:

9. Air-monitoring needs:

10. Placement of arriving emergency forces:

   A. Staging area
   B. Remote stand-by
   C. Commitment as needed
   D. Upwind and upgrade
   E. Use natural barriers for protection
   F. Protect equipment from run-off, vapors, sprays, residues, etc.

11. Control of access to the scene:
    A. Roadblocks, barriers, barrier tape, fire lines, traffic control, police guards, evacuation routes, shelter-in-place, street and neighborhood closings, etc.
12. Environmental considerations:
    A. Streams, lakes, ponds, rivers
    B. Sewers, water plant intakes, storm drains
    C. Groundwater
    D. Wells and other drinking water sources
    E. Fish and fowl
    F. Crops, vegetation, and cattle
13. Personal protection:
    A. Level A, vapor protective suit; Level B, liquid splash protective suit; Level C, limited use protective suit
    B. Level One Incident: Haz Mat Incidents that can be contained, extinguished, and/or abated using equipment, supplies, and resources immediately available to first responders having jurisdiction, and whose qualifications are limited to and do not exceed the scope of the training explained in 29 CFR 1910.
    C. Level Two Incident: Haz Mat Incidents that can only be identified, tested, sampled, contained, extinguished, and/or abated utilizing the services of a HMRT which requires the use of specialized chemical protective clothing, and whose qualifications are explained in 29 CFR 1910.
    D. Level Three Incident: A hazardous materials incident that is beyond the controlling capability of a HMRT (technician or specialist level) whose qualifications are explained in 29 CFR 1910; and/or which can be additionally assisted by qualified specialty teams or individuals.
14. Decontamination considerations:
    A. Wash-down
    B. Discard clothing
    C. Decontaminate equipment and apparatus
15. Extinguishing agents:
    A. Water (straight streams, fog, spray, mist)
    B. Dry chemical
    C. Foam
    D. CO_2
    E. Sand, tarpaulins, and other smothering agents
16. Command post operation:
17. Maintaining incident vigilance and discipline:
18. Call for help:
    A. CHEMTREC
    B. CHLOREP
    C. Manufacturer
    D. Shipper/carrier
19. Identification of additional resources needed:

A. Sorbent materials (sand, kitty litter, fly ash, cement powder, etc.)

B. Neutralizers or pH adjusting materials (agricultural lime, bicarbonate of soda, activated carbon, vinegar, etc.)

C. Earthmoving equipment, vacuum trucks, empty containers for off-loading, recovery drums, etc.

D. Pumps, skimmers, diking materials, patching and plugging materials, meters and "sniffers"

E. Manpower

F. Pump-off trailers, bulldozers, knowledgeable persons, etc.

20. Communications:

Cellular phones, pagers, satellite communications, FAX, telecommunications, computers, e-mail, GIS (Geographic Information Systems), websites, etc.

21. Control and containment ("Think Containment"):

A. Patch or plug the leak

B. Limit the area involved

C. Booms, dikes, sumps, dams, diversionary waterways

D. Shut-off the flow

E. Introduce vacuum by external means

F. Turn container over

G. Let it burn

H. Have knowledge of all mechanical and electrical systems in your area of jurisdiction (shut-off devices, operational controls and procedures, necessary tools, piping systems, tanks, transport vehicles, etc.)

22. Neutralization:

23. Secondary emergencies:

24. Briefing:

25. Plan the attack:

26. Carry out the attack:

27. Safety of personnel:

A. Safety officer position

B. Line of retreat

C. Exit route

D. Recognizable signal for retreat

E. Buddy system/constant backup

28. Clean-up:

29. Disposal of chemicals and hazardous waste:

30. Critique ("Lessons Learned"):

THE STATUS OF HAZARDOUS MATERIALS RESPONSE IN THE UNITED STATES

On November 25, 1969, President Richard Nixon visited Fort Detrick in Maryland and canceled the U.S. offensive biological weapons program. The entire program was ended within two years. In the interim 30 years, the knowledgeable people who specialized in the science and manufacture of offensive biological weapons in the

United States retired, died, or went into consultant work. Whether good or bad, there are not many U.S. experts in offensive biological weapons as we head into the 21st century. In addition, there is not a lot of defense available for a biological weapons attack within the United States. If Timothy James McVeigh could destroy the Alfred P. Murrah Federal Building on April 19, 1995 killing 168 persons and injuring hundreds more using fertilizer (ammonium nitrate), fuel oil, and his degree of formal education, think what the holder of a science or toxicology degree could do.

The U.S. government is lacking in both terrorist and chemical/biological weapons defense. This problem is particularly apparent for a biological attack which could be potentially devastating. Biological agents, ounce-for-ounce or gram-for-gram, can be a more deadly killer than chemical agents. At the time of this writing, bombs and explosions have been the favorite means of getting attention for terrorist groups in the United States. There is increasing fear within government that this situation may change, opening the way to chemical and biological incidents within our country. The big question is who will respond to such incidents if they occur.

The priorities in any such incidents will be to protect life, environment, and property; alleviate damage, loss, hardship and suffering caused by such incidents; restore essential government services; and provide relief to affected government agencies, services, businesses and individuals. Following long established procedure, the federal government stated within PDD-39 (Presidential Decision Directive-39) that response to chemical and biological agents would be handled by local government with the assistance of the federal government. Then, the federal government spent millions of dollars to purchase duplicative training, equipment, and materials to bring federal employees up to some kind of a standard of performance. Many of these federal employees had never responded to a hazardous materials incident before, let alone to an emergency situation. How are they going to assist state and local employees who have been doing hazardous materials response since the late 1970s and earlier? To bring the federal government up to some sort of a minimum standard of performance in response to terrorist incidents in the United States, the current cost during early 1999 is an expected $9.8 billion a year.

Federal response teams cannot be on-site in time to save lives or treat victims. Local response personnel need, but cannot afford, the same assets and training as federal employees are now receiving. Haz Mat mitigation is a local responsibility for the first 10 to 24 hours. Someone has to be on scene at once to isolate the area of a chemical or biological agent attack, evacuate and care for the injured and the dead, and deny further entry. Any and every chemical or biological attack will be a crime scene that has to be isolated and guarded. There is little or no field detection and identification technology or equipment available for a biological attack at the present time. There is a lot of funding for research and testing of biological agent detection and identification but not much delivery of working equipment. Now in service is the M34 Sampling Kit which can sample soils, surfaces, and water for chemical and biological agents. It contains two sampling kits, one vials container, two pairs of toxicological agents, protective gloves, and a set of instruction cards. There is also a "Biological Integrated Detection System" (BIDS) designed at the present time which consists of a vehicle and trailer which will be equipped with a high volume aerodynamic particle sizer (HVAPS), a biological sampler, a threshold workstation, a liquid sampler, and a flow cytometer. A biological detector as well as a Chemical Biological Mass Spectrometer (CBMS) biological detecting equipment will be integrated into the BIDS at some time in the future.

We must protect our local first responders (firefighters, police officers, EMTs, paramedics, ambulance and first aid crews, nurses, emergency medical physicians, emergency management agency personnel, military, commercial response contractors, and health department workers) from lack of proper training, equipment, and tactics in dealing with chemical and biological incidents. They are potential, and probable, victims. Defense against chemical or biological attack is unfamiliar and very difficult. For example, no one in the United States has any experience with a massive cleanup of anthrax spread by aerosol. The manufacture of biological agents can be easily concealed by civilian industry; anyone who has run an illicit drug laboratory can probably also make biological agents, not very much space is required, and the cost can be inconsequential. There is fear that genetically engineered organisms will be forthcoming that could defeat our present supply and assortment of vaccines and antibiotics.

The medical system will be overtaxed after any chemical or biological incident. The triage and treating of victims is the weakest link in the chain of preparations for response to terrorist actions or incidents. Hospitals and clinics will be visited by a lot of people who are not ill and have no symptoms. Control and triage must be provided. Triage can be different for various biological weapons depending on whether they were inhaled or cutaneous. People who have been contaminated by biological agent(s) might be better off being seen by medical students or newly licensed physicians since such medical personnel are still in the micro area of medicine and may better recognize specific symptoms. In some situations, there may be off-gassing of chemical and biological agents from patients that could contaminate medical personnel; many medical workers do not have respiratory protection as good as firefighters have available.

Military troops can be ordered to take vaccines and antibiotics; civilians need informed consent which they may hesitate to provide after hearing news items about military personnel reactions to vaccines and antibiotics known as the "Desert Storm illness." Anthrax vaccine was given to 150,000 troops during Desert Storm, while botulism vaccine was provided to 8000. Availability of vaccine is always a problem, and any vaccine could be overwhelmed by the dose of the biological weapon used. By the time biological attack victims begin showing symptoms, it is often too late to save their lives. However, many of the biological agents are not contagious person-to-person although there are exceptions: plague and smallpox are very contagious person-to-person. An intact skin is a very effective barrier against biological toxins except for mycotoxins (T-2 mycotoxins, such as in mold or fungus, can pass through intact skin).

5 Federal Agencies and Response Teams

"There is a need for an organization — manned, trained, and equipped — to counter the growing biological/chemical terrorist threat. The Marine Corps will have such an organization... manned with properly skilled and trained personnel... equipped with state-of-the-art detection, monitoring and decontamination equipment... suited for operations in a wide range of contingencies."

General Charles C. Krulak, 1995

INTRODUCTION

More than 200 state and local officials were invited to the federal capital in Washington, D.C. in August of 1998 by the Department of Justice, Office for State and Local Preparedness Support created in May, to assist cities and states to better prepare for terrorism. The key message from the local officials was that the federal government still has no coherent system for deterring or responding to terrorism. They asked the Clinton Administration to put a single government agency in charge of developing a new national plan within six months as the present system is often duplicative, frequently chaotic, confusing, and overly bureaucratic. An 8-page summary of comments was given to Attorney General Janet Reno. Only half the state and local officials on one panel said they knew that the President had revamped the terrorism system, and fewer than half had ever heard of Richard A. Clarke, the national federal coordinator for antiterrorism. The local officials seemed infuriated about the number of federal antiterrorism units including the Army's SBCCOM, the Marine's CBIRF, the F.B.I.'s DEST, the State Department's FEST, the Energy Department's NEST, and the Department of Health and Human Services' MMST. At least 40 different federal agencies or offices are to "assist" state and local response officials responding to weapons of mass destruction (WMD) and terrorist actions involving nuclear, biological, and chemical weapons. This chapter examines several of these units.

U.S. MARINE CORPS CHEMICAL BIOLOGICAL INCIDENT RESPONSE FORCE

The truism that superior thinking can overwhelm superior force has been the guidon for 350 young men and women with special talents who form the U.S. Marine Corps Chemical Biological Incident Response Force (CBIRF) based at Camp Lejeune, NC. In a terrorist incident in the United States or overseas that features chemical agents or biological substances, CBIRF will attempt to turn victims into patients through mass casualty decontamination. For chemical agents, CBIRF personnel have to be able to deal with blister agents (vesicants), blood agents such as hydrogen cyanide, nerve agents such as sarin, tabun, soman, GF and VX, and pulmonary agents like phosgene. Biological substances may include bacteria (anthrax, cholera, plague, typhoid fever, etc.), rickettsiae (epidemic typhus, Q-fever, Rocky Mountain spotted fever, etc.), chlamydia (psittacosis, etc.), viruses (dengue fever, eastern equine encephalitis, Ebola fever, lassa fever, smallpox, etc.), or toxins (botulinum, ricin, staphylococcal enterotoxin B, etc.).

The strategic mission of CBIRF is consequence management and force protection while the organization's operational mission is to turn victms into patients. Consequence management addresses the consequences of an incident, and involves measures to alleviate damage, loss, hardship or suffering. CBIRF has a command element with command and control of the organization during incident response, training and liaison. It is assisted by a group of consultants, called the "Electronic Reachback Advisory Group," who provide advice on organization, equipment, and required capabilities during the formation of the CBIRF and who act as a "virtual" staff of experts in support of CBIRF upon its activation. Eight nationally and internationally recognized civilian experts in science and medicine form the "Reachback" group.

Reporting to the command element are recon, decon, medical, security, and service support elements. The **recon** element handles agent detection and identification, chemical/biological sample collection, hazard area identification, and determination of a down wind hazard area. The **decon** element decontaminates personnel and equipment, including ambulatory and non-ambulatory victims, and can engage at multiple sites and depths of operations. The **medical** element consists of stabilization, collection and evacuation, and unit support sections and advises and assists local medical authorities, performs triage and emergency treatment, provides organic medical support, and undertakes initial epidemiological investigation. The **security** element provides incident area security and isolation, site evacuation, critical personnel and government property security, crowd control, and similar operations as required. The **service support** element provides advice and assistance to the on-scene commander, supports organic services, and mans limited emergency services.

The men and women who are CBIRF personnel come from a variety of military occupational specialties (MOS) selected from throughout the Marine Corps and the Navy, and are immunized against select weaponized biological agents. The MOS's represented include officer-in-charge, operations representative, NBC representative, clerk, intelligence representative, communicator, hazard coordinator, decontamination specialist, medical, supply, security, reconnaissance, medical officer (medical doctor or physician's assistant), corpsmen driver, engineer, contractor, and embarkation specialist. There is a 4-hour alert for a 120 person, limited-capability, "rapid response force," which is reinforced by the total force within 24 hours. The concept of CBIRF is task organized, formed, standing and ready, self-sustaining, and interactive with other organizations such as federal, local, foreign, academic, and industrial, resulting in a synergistic effect that features the joint action of organizations increasing each other's effectiveness.

Equipment costs for the CBIRF are expected to total $5 million, while annual operating expenses are projected to be $2 million. The Chemical Biological Incident

Response Force also is a testing unit for chemical and biological related equipment, techniques, procedures, and doctrine in the Marines Corps.

CBIR Equipment for NBC Detection and Identification

Radiological Detection and Identification Equipment

AN/PDR-75, DT-236: The DT-236 radiac detector is worn like a wrist watch and contains two detector elements. One measures total neutron radiation, and the second measures total gamma and X-ray radiation. The AN/PDR-75 computer indicator provides a readout of the total dosage of gamma and neutron radiation recorded on the radiac detector. Each radiac detector has a unique serial number to identify it to only one individual for the life of the detector.

AN/VDR-2: The AN/VDR-2's main feature is the detector probe, which contains beta and gamma sensing devices for radiological monitoring. The radiac meter can also integrate the dose rate count and display cumulative dose on command. It can automatically range through the system's entire range (0.01 uGy/hr to 9999 cGy/hr) and includes the following features: audio alarms, self-test circuitry, low-level beta monitoring, and high accuracy and reliability.

Chemical Detection and Identification Equipment

ABC-M8 VGH Chemical Agent Detector Paper: M8 paper detects and identifies liquid V- and G-series nerve agents and H-series blister agents. It comes in booklets of 25 sheets, which are impregnated with chemical compounds that turn dark green, yellow, or red upon contact with a liquid chemical agent. A color chart in the cover of the booklet helps determine the type of agent.

M256A1 Detector Kit, Chemical Agent: The M256A1 kit consists of a carrying case, 12 sampler detectors, instruction cards, and ABC-M8 VGH chemical agent detector paper. The sampler detector is used to test for nerve, blood, and blister agent vapors in air. The M8 paper is used for detecting nerve, blood, and blister agents in liquid form. The kit is normally used to determine when it is safe to unmask after a chemical agent attack.

M-18A2: This item is a chemical test kit using colorimetric tubes to detect and identify toxic chemical agents in the air and vapors from liquid chemical agent contamination on exposed surfaces. The kit is also used to collect and forward samples of unidentified toxic material agents to a technical intelligence team or laboratory for classification. Agents detected by the M18A2 are cyanogen chloride, mustards (H, HD, HN, and HT), phosgene oxime, hydrocyanic acid, phosgene, lewisite, ethyldichloroarsine, methlydichloroarsine, and V and G types of nerve agents.

pH Meter: The pH meter is a hand-held, man-operated device for identification of the acidity level, the alkalinity level, temperature reading, and conductivity of an unknown liquid source.

CAM: The CAM is a hand-held device for monitoring chemical agent contamination on personnel and equipment. The CAM can detect nerve (G series) and blister (H series) vapors by sensing molecular ions of specific mobilities (time of flight), uses timing and microprocessor techniques to reject inferences, and displays the relative concentration. The CAM is strictly a post-attack surveying/monitoring instrument. Due to the radiological source, Ni-63, the CAM requires a Nuclear Regulatory Commission license. The CAM's response time is less than 60 seconds for 0.1 mg/m of agent.

XM88 ACADA: The Alarm, Chemical Agent Detection, Automatic, detects and warns against the presence of chemical warfare agent vapor in the surrounding air in real time. It indicates G-series nerve and H-series blister agents and the hazardous level of agent vapor. It is capable of clearing itself and responding to another agent within five minutes and can be operated on the move. Due to the radiological source, the ACADA requires a Nuclear Regulatory Commission license.

M21 RSCAAL: The Remote Sensing Chemical Agent Automatic Alarm is a two-man portable, automatic scanning, passive infrared sensor which detects nerve and blister agent vapor clouds based on changes in the infrared energy emitted from remote objects, or from a cloud formed by the agent. The RSCAAL is a stand-alone, tripod-mounted, chemical agent overwatch system to be used in a defensive role. It consists of a detector, tripod, M42 remote alarm unit, transit case, power cable assembly, and standard military power source. It can be used for reconnaissance and surveillance missions. It will search areas between friendly and enemy forces for chemical agent vapors, and provide advanced detection and warning of chemical hazards. Where possible, the RSCAAL will be employed in pairs (two reconnaissance teams) so that one RSCAAL can be used in the overwatch position when the other reconnaissance team is moving. The RSCAAL's detective range is 1.86 to 3.1 miles.

PPID: The ToxiRAE Pocket Photo Ionization Detector continuously monitors hazardous and toxic gases or vapors in low part-per-million (ppm) concentrations. The PPID provides fast response and real time readings compared to operator programmable alarms. It is available through a direct link with a PC and provides datalogging for history/survey missions. The PPID can operate continuously for 12 hours with a rechargeable Ni-Cd battery.

HNu: The HNu is a powerful microprocessor-based photoionizer/data logger that has low-end sensitivity. Users may store calibration data, eliminating the need for individually calibrated probes. Advanced data storage provides data storage for up to 256 sites with data logging of date, time, concentration, and site information.

Draeger Tube System: The tube(s) measure air concentrations of toxic chemicals. The system draws air through the tube with a mechanical or hand pump. The tube will then change colors to show the concentration level of the agent detected. The system uses colorimetric tubes specific to each agent. There are currently 160 different tubes that detect and identify chemical agents.

Draeger Multi Pac: The Draeger Multi Pac gas personal monitor continuously and simultaneously measures ambient levels of oxygen, combustible gases, and two preselected toxic gases. The CBIRF currently has CO (carbon monoxide), NO (nitrous oxide), SO_2 (sulfur dioxide), NO_2 (nitrogen dioxide), Cl_2 (chlorine), H_2S (hydrogen sulfide), NH_3 (ammonia), HCN (hydrogen cyanide), and a broad band toxic sensor. The gases being monitored are displayed concurrently on the alpha-numeric backlit LCD. If any of the gases reaches the preset safety limit, the audible and visible alarms are activated immediately. The Multi Pac has an internal datalogger for storing all gas and instrument configuration parameters. All information can be downloaded to a PC for analysis. All data, including graphics, can be printed to create a permanent record.

HAZCAT: The Hazardous Categorization Chemical Identification System is based on a series of field tests which are used to identify liquid and solid unknowns. The HAZCAT system can identify over 1000 agents. Most unknowns can be identified or categorized by hazard class with as few as four to five tests. The reagents are stable in most environments and have a shelf life of at least one year, although many will last indefinitely.

Viking Gas Chromatograph/Mass Spectrometer (GC/MS): A two-man, portable chemical detection and identification platform, the Viking is capable of a one touch operation even in gloves and protective clothing. Data is logged both electronically and as a hard copy if printed. Software developed specifically for chemical warfare agent identification is included for identifying a chemical threat. An industrial chemical library containing approximately 75,000 compounds can be used to analyze compounds other than chemical warfare agents.

M93 FOX NBCRS: The FOX Nuclear Biological Chemical Reconnaissance System is a field detection and protection platform, equipped with a mobile mass spectrometer capable of detecting, identifying, and quantifying up to 60 chemical agents simultaneously. This system is also capable of collecting biological samples of solids and liquids. FOX vehicles are equipped with an overpressure system, six-wheel drive, seats for a crew of four, and are capable of both land and water operation. The FOX is equipped with an industrial chip that allows detection of 115 industrial chemical agents.

Biological Detection and Identification Equipment

Hand-Held Assay (HHA) ticket: The HHA tickets use immunochromatographic reactions to determine the type of biological agents present. There are currently several tickets available, with more in the research phase. Agents include anthrax and botox. A sample is collected, mixed in a buffer solution, then placed on the ticket. Operators will see either a positive or negative reaction within five minutes.

Biological Sampling Kit: The BSK is required to perform three types of biological sampling: surface, liquid, and solid. The kit contains the required equipment for monitor/survey teams in the field to collect and forward biological samples needed by medical facilities.

Forma Scientific Biological Dry Shipper: The shippers are designed for transporting small quantities of biological materials at cryogenic temperatures through the use of liquid nitrogen. Storage temperatures inside the shipping cavity can remain at −300°F for up to 30 days.

Smart Air Sampling System (SASS): The SASS is an air sampler designed to collect and concentrate biological aerosols into a liquid media for subsequent analysis. The collected liquid sample is then provided to a bio-detection field device to determine whether biological warfare agents are present. The SASS can take a 5 to 7 cc sample in 10 minutes, and is battery-operated for up to 8 hours.

Recently Procured Equipment

Mobile Modular Laboratory: The MML provides the capability for positive identification, within 20 to 25 minutes, for either an air or liquid sample. Once a positive
identification is made, the MML will provide information on medical treatment,
decontamination, containment, and the potential change in force protection requirements. The MML has an infiltration system that allows it to perform its mission in
a non-contaminated environment.

Inficon: This instrument is designed for easy, on-site analysis of industrial chemical
agents, weighs about 35 pounds with batteries, is completely automatic, is based on
Windows software, and has the capability to detect 78,000 volatile organic compounds (VOCs).

Cascade System: The cascade system is capable of filling multiple SCBA cylinders.

Litton Rebreathers: The system provides a duration of use for up to two hours,
weighs only 30 pounds, and is both NIOSH and OSHA approved. The Rebreather
system has positive pressure which will always keep gases away from the user's face.

LSCAD: The lightweight, standoff chemical agent detector is an interferometer
designed for on-the-move detection of chemical vapors. The LSCAD is capable of
detecting known nerve and blister agent vapors at a distance of 5 km.

Portal Shield: This device is a biological detection system that is placed strategically
around a working area and can work as an entire system remotely linked to one
computer. An integrated weather system allows the portal shield to determine the
difference between biological dissemination and an isolated alarm trigger. The unit
can identify very specific biological agents.

Captain Jeff Schwager is an intelligence officer with CBIRF who deals with
maps, imagery, the terrorist threat, and support to the operations sector. He works
with a number of different agencies such as counter intelligence and the Federal
Bureau of Investigation. "We have computers where we can download classified
information. If I needed information on a terrorist group, it would be of a classified
nature, so I would have to use a classified database to pass on that information. The
intelligence in CBIRF is a lot different from basic Marine Corps intelligence operations because we have a domestic mission as well as an overseas mission. A lot
of what we do is force protection. When we talk about terrorism and counter
terrorism, one of the things we worry about is the safety of our troops. You have
the initial incident, but there could be a secondary incident set up specifically to
injure our personnel. We try to keep track of hostile groups that could try to thwart
our attempts to affect their mission."

Chief Warrant Officer W3 Douglas Davis is the decon element commander. "We
have a thorough mass casualty decontamination process which means we take the
victims out of all their clothing, scrub them down, and prepare them for transport
to local hospitals in a condition where they would not contaminate the hospital or
the patients within it. First, they are triaged (the sorting of and allocation of treatment
to patients, particularly in warfare or disasters, according to a system of priorities
according to the urgency of their need for care designed to maximize the numbers
of survivors) in the hot zone. There is another triage in front of the decon area so
we always see the worst case scenarios first. We run three lines simultaneously for

ambulatory patients, non-ambulatory patients, and first responders or our own personnel.

"The decon site(s) take about 16 people to run and can be set up in seven minutes. For one decon site, we've estimated that we have the capacity to handle 25 to 35 non-ambulatory patients per-hour, between 70 and 100 ambulatory patients per-hour, and our force personnel line will go faster than that. That's just one site. We have two sites that are exactly the same that can be set up if needed. When we come to a large city, we have five of what we call 'hospital sites,' each of which can be staffed by five Marines who have been decon trained with a small site that is able to do ambulatory and non-ambulatory decon at a hospital site for those victims who make it out of the incident site and come straight to the hospital without being decontaminated. We have a roller system in a 21-foot movable shelter for non-ambulatory patients, and two shower systems. The local hospitals will close down their emergency rooms when word is received that there has been a chemical or biological attack and won't allow anyone into the emergency room until they have been decontaminated. Our whole purpose is to work with the local hospital staff to ensure that the emergency room is not contaminated.

"I have a platoon of one officer and 32 enlisted. These Marines have been trained at the Nuclear, Biological, Chemical Defense Specialist School operated by the Army at Fort McClellan, AL. They have a good background in weapons of mass destruction and decon when they come to Camp Lejeune. It's important to realize that when we go to an incident scene, we will be working for and with *local* emergency responders. We train with a lot of civilians including a SORT team (special operations response team) from Charlotte, NC. They are civilians who work for the federal Public Health Service under the direction of Dr. Lou Stringer. We also have SORT teams in Denver, CO and Los Angeles, CA.

"When we come into an incident scene, we work for the civilian incident commander. We don't come to take over the site. The local people tell us where they need us, and that's where we set up. We have two sites and five hospital sites, so we can be pretty flexible as to where we go and whatever the needs are. We practice decontamination training all the time. This is our full time job, so we are good at it because that's all we do. There is a lot of synergy within CBIRF. With 32 people, my group is going to get tired at a major incident so we cross-train with the security platoon. We cross-train with everybody in the 350-person CBIRF so everybody has a basic knowledge of decon and can come up here and replace the force as needed.

"We're chartered to go to all Department of Defense facilities anywhere in the world. Our 120-person Rapid Response Force is on call 24 hours a day and within 4 hours we can be ready to fly out, although we can often move faster than that. When we travel, we are self sufficient and have the ability to tap into civilian water sources. We bring all wrenches we may need and about 1000 feet of hose we can run from a hydrant or a standing water supply such as a swimming pool.

"We developed the decon sites with certain ideas in mind. I told the Marines that their families might have to go through decon. 'How would you like your mother treated; how would you like your family to be treated?' We separate the males and the females; the children go with their mothers. We have curtains to hide the people being decontaminated from view; the tents are heated. We provide gowns when they come out of the decon site, and blankets as necessary, before the patients go back to medical holding. All the decon processing is done inside a tent so nothing is seen by outsiders."

At one of the first anthrax hoaxes, the April 17, 1997 incident at B'nai B'rith, a national Jewish service organization in Washington, D.C., the firefighter responders put up tarps around the patients to shield them from the eyes of onlookers, but a television crew had climbed to the second or third story of a building and filmed the potential victims getting undressed for the world to see. "That is not going to happen here," according to Doug Davis, "not on our decon site."

David Shoemaker is the preventive medicine officer for the CBIRF. "One of my jobs is epidemiology," he says. "Another is looking at the different agents involved in symptomology and trying to determine what agent(s) are involved when we are at an incident site. I am also a liaison officer where I travel around dealing with all the federal agencies involved with consequence management. Together, we have become a major part of the federal response plan, and we want to continue that.

"Regarding epidemiology, if we did not know what was happening at an incident site we were called to, I could do an epidemiological investigation, study various factors, and try to identify the agent(s), and the time of onset," continues Shoemaker. "If we could tell early enough what was going on, we could possibly start immunization or prophylaxis. We can also do detection for biological agents; I can't say which ones or how many, but we can detect with a number of different active or passive means. However, we do air sampling and suck in several liters of air and check for aerosols. We also have equipment to detect biological agents in soil, water, on the skin, and on mucous membranes on direct reading instruments so we can get a 'yes' or 'no' on the spot. We do have to send samples in for confirmation just to make a double check, but we can get a pretty good idea on what's going on immediately. To keep our own personnel free of disease, we run an active immunization program. We also have a prophylaxis program (protective or preventive treatment); if we were going into a certain area, we could pull up the proper drugs and start the Marines on them while they are enroute to ensure that they are protected. We do have such programs, but (for security concerns) I can't say what we're immunized against or which prophylactics we carry."

Marine Staff Sergeant Jeff Toohei is a meteorologist with CBIRF, and was building plume models based on different agents when the writer talked with him on a training exercise at an old auxiliary airfield in Atlantic Beach near the Cape Lookout National Seashore on the outer banks of North Carolina. Sergeant Toohei entered the Marine Corps in 1987, and spent three years as an air traffic controller. He spent two years as a weather observer at Cherry Point Marine Corps Air Station, went to weather forecaster school at Chanute Air Force Base, and returned to Cherry Point as a weather forecaster for 3½ years. During this period, he served two six-month tours in Italy forecasting for the pilots flying over Bosnia, and has been with the CBIRF for over two years.

"I am employing a plume modeling program provided through the Defense Special Weapons Agency located in Alexandria, VA. On this model, you incorporate the local weather including service base weather, upper air winds, temperature, humidity, and boundary layers. I'm running the biological agent anthrax at the present time to learn if the anthrax agent plume would or would not reach nearby Morehead City.

"A plume will take a few minutes to appear depending on which agent you may be using. For the anthrax plume we are using now, the buoyancy works in conjunction with the local area air. The red or pink area observed on the computer screen is the percentage of the lethal content of the plume within that specific area. In that lethal area, as many as 90% of the people could die. This is just an estimate. We try to work with a worst case scenario; in this manner we have an idea of what *could* happen. We have both high resolution and low resolution maps of Morehead City which can be depicted on the computer screen. At the current time, this anthrax plume is showing winds coming from the northwest at 7 knots.

"This computer is part of a different weather system, called the Oceanographic Support System, that is not on this exercise site where we are today. It consists of a receiver that gives us chart data and some satellite pictures; we also have two satellites that go with the system. We receive direct satellite pictures pretty much anywhere in the world, so we have a large capability to get weather information. Just prior to your coming in, I was getting ready to export the file to a PowerPoint graphics program so I could make an overlay to put on the map over there so that we can show everybody what the present circumstances are."

Navy Chief Petty Officer Brad Grandy is a medical administrator with CBIRF. He keeps track of how many casualties are removed from an incident site, when they came through decon, if they were ambulatory or non-ambulatory, and other information when they arrive at medical stabilization. "We keep track of the casualties and consider what assets we have available from the local community; we call the local emergency medical system for transport of casualties to local hospitals or military base hospitals. Keeping track of such statistics allows us to know how many people go down because of exposure or injury, where the victims are at different stages in care or treatment, and how many casualties we have."

H.W.R. Dalton, M.D. with the Navy, is an emergency medical physician who explained the multiple triage performed on patients who have been contaminated by chemical or biological agents. "The majority of patients complain of burning eyes; often they can't see where they are going, they just lose control. Because of reduced eyesight, patients are instructed to grab the shoulder of the patient in front of them when they walk. We do three triages at the incident site. When a patient is first seen, somebody is making the decision about his or her condition. Next, when a patients gets to the decon line, staff are taking another look. If that patient is destabilized — he or she may be 200 to 300 yards from the incident site by now — we don't want to put that person on the decon line and have him or her crashing. There may be five or ten patients behind the destabilized patient who could be saved if they went immediately through decon. We want to prioritize who's going through and in what order. Triage works this way everywhere. Once the patients get to me, I look at them again. The people doing the triage down range may say that this patient is a '2' but by the time he gets to me, the agent has kicked in more and now he is in worse shape, and I may say, 'No, he's not a 2, he is now a 1.' In the case of a different patient, the antidote may be kicking in and this guy who was having a lot of respiratory problems at the start of his contamination now is doing fine, so I'm going to have to say he is a candidate for a delay in treatment. We certainly need to get him in, but there are other people who are more urgent that we will ship in ahead of him. That is, down range, the people doing triage are deciding who gets to go to decon first; at decon the people doing triage are deciding who goes through the decon line first; and here, we are deciding the order that we will ship them off to local hospitals, and to an extent, the hospital each patient will be going to. Hopefully, we're going to be interacting with the civilian side, saying things like 'These guys are really serious and they need to be transported to an acute care center; these other patients can go to a community hospital since they have some mild eye and/or air way irritation.'

"We have three physicians with CBIRF, and depending on how many are on scene we think the physicians are most useful back here where we can do the most. There's a limit to what we can do down range. To start with, at the triage line we have an emergency nurse who is very good at triage. If he has somebody who needs to be assisted by a physician, he is going to yell for us. That is part of the reason I am dressed in MOPP gear (mission oriented protective posture, the protective clothing used by members of the U.S. military who engage in nuclear, biological, and chemical warfare. MOPP gear provides a flexible system requiring personnel to wear only that protective clothing and equipment appropriate to the threat level or work rate imposed by the mission, temperature, and humidity) even though we are in a safe area. All I have to do is put on a mask, run down to the triage nurse's area, and help out. Generally, the triage line is the first place those who are contaminated are actually going to see a physician.

"Here in this safe area, a physician will do basic life saving techniques. My specialty is emergency medicine, so I start with treatment that can make a difference right now. We could provide niceties that don't make any difference right now, but we take care here of things that should be done immediately or the patient is going to get worse. The other thing we are doing here is further triage. Triage falls into two categories. Triage in the ER (emergency room) is based on the concept that we have adequate capability of taking care of everybody who is coming to our door, so people are then triaged on the basis that the most urgent patient is seen first. Everybody will be seen, but we take care of the worst ones first. However, in a mass casualty situation where you've got more people than you can reasonably take care of, triage means doing the most good for the most people. If you have somebody who is so acutely ill or traumatized that it's going to take all your facilities for the next half hour, while you have other patients who can be saved by a 10-minute procedure, the very ill patient is going to be set aside while we take care of these other people. At the end of a half hour, if the very ill patient is still with us, then that patient moves up in priority because then we have the resources to expend on him or her. It all depends on what's still coming down the line.

"One of the other decisions that we are making here, particularly if the ambulances are not here and ready to go, is who we are spending our energy on at that point," says Dr. Dalton. "A bad trauma case would occupy my full attention in a normal ER for a considerable amount of time. If I've got 20 patients lying here who are having respiratory symptoms and depending on me for help, I may not be able to spend that full time with that bad trauma case. We'll do minimum stuff. Certainly, I'm going to start an i.v. on most of these guys if warranted, but there is going to be a limit on what we can do for them until we get more time and more resources to expend on them.

"The results of triage are scanned by little tags on the patients' triage tickets, and that information is sent back to the command center via a wireless device. We provide nerve antidote kits containing atropine (sometimes used as an antidote for nerve agents; it inhibits the action of acetylcholine by binding to acetylcholine receptors) and 2-PAM-Cl (trade names protopam chloride, or pralidoxime chloride; 2-PAM-Cl can be used in the treatment of nerve agent poisoning), basically we stabilize the patients so they can be transported.

"What we are trying to do is be ACLS (advanced cardiac life support) and ATLS (advanced trauma life support) capable, to provide emergency care for people coming out of the incident site. In the incident site, we've got a bunch of contaminated people. Our goal is to take the contaminated people and turn them into decontaminated patients we can then ship them off to civilian facilities where they can be cared for. After they come through our decon area, we take a quick look at every patient; anyone who needs further emergency care will get it, whether it's for the chemical or biological agent they were exposed to, for any trauma involved, or, if in all the stress that was going on, they suffered a heart attack. Those are the types of things that we are set up to care for. We rely on the civilian population to provide the ambulance services to take the patients off our hands as fast as we can stabilize them since we are not set up to do long-term, ongoing care. It's kind of foolish for us to hang onto the patients when there is a general hospital 15 blocks away that's capable of providing a much higher level of care than we can.

"It's strict emergency care for nerve agents," continues Dr. Dalton. "We are going to be evaluating the extent of exposure. Most likely, they will have been given an antidote down range. If patients are still having marked symptoms, we will start antidote i.v.s as opposed to a shot which takes a little longer to get into the system. We want to get the antidote into the system, stop the symptoms that are occurring, and stabilize the patient. If there is trauma from open wounds or broken bones, we're going to take care of that at this point by cleaning the wound, stopping the bleeding, applying dressing or splints to the limbs, and getting i.v. fluids into the person.

"For a biological agent, it is a little different. Probably the most we would do is start a prophylaxis of either doxycycline or ciprofloxacin for everybody depending on the agent we think is involved, and assuming the person is not allergic to it. We would give patients their first dose right then, before we ship them off to a hospital. That is, we would get this started as early as possible, giving the patient the quickest protection we could."

Chief Warrant Officer W3 Mark Fishback is the reconnaissance element commander for the Marine Corps/Navy CBIRF. He was an enlisted man for 15 years, attaining the rank of E-7 before being promoted to warrant officer. He has been in the chemical field for 22 years, and has been to about every school the United States has to offer related to NBC (nuclear, biological, and chemical) concerns.

The recon element for CBIRF is made up of 20 Marines, 10 Navy corpsmen, and 1 medical officer. Its job is detecting, classifying, and identifying all known chemical and biological agents. If field detection equipment carried by the recon element fails to identify an agent, members are able to collect samples for other agencies to identify. The unit is also organized to provide emergency casualty evacuation teams of two Marines and one Navy corpsman capable of stabilizing and extracting casualties from the hot zone. The recon element are also the first members of the CBIRF to enter a hot zone at a chemical or biological incident.

How did CWO W3 Mark Fishback come to be assigned to the CBIRF? "Just through experience," he says, "attempting to do a credible job for the U.S. Marine Corps and building my credibility by being extremely well-read in the various jobs I was assigned. I was very fortunate early on to be sent to Fort McClellan in Alabama where the U.S. Army Chemical School is located. I taught nuclear and chemical target analysis, and also taught at the radiological safety officer's course at the radiological laboratory there. It was unique for a Marine to be able to do that. The Marine Corps actually has a detachment at Fort McClellan including students for the Marine Corps military police as well as all our NBC Marines who are trained there.

"I worked there for nearly 42 months in the early to mid-1980s when we had only 15 to 20 chemical officers in the Marine Corps. However, the first class I taught had 36 officers which almost doubled the NBC corps for officers. I taught about 80% of the officers, and 60% of the senior officers that we have today. As luck would have it, many of the senior officers knew me because of the time at Fort McClellan.

"Over the past 24 months with the CBIRF I have met a lot of firefighters, police officers, and emergency medical services people and it has been quite a learning process for me. I started my initiation with local response personnel at the National Fire Academy in Emmitsburg, MD by talking at a weapons of mass destruction conference. As Department of Defense forces, particularly the Technical Escort Unit of the Army as well as the Marine Corps CBIRF, grow together, we are also growing with the civilian infrastructure, and the learning curve have been tremendous for us.

"One of the things I found interesting in coming into this business is that when we were dealing with recognizable chemical agents such as VX, GB (sarin), or soman, we weren't really too concerned about them in the military. We've seen that before, and dealt with it before. We have devices that will detect it, and we know what our protection levels need to be. Tell civilian firefighters to go in and handle a 'nerve agent,' and they will be concerned. Tell civilian firefighters to go in and handle a Department of Transportation Class 6 hazardous material similar to an organophosphate and they will not think it is such a big deal. Conversely, tell a military team that they will have to deal with vinyl chloride and that's scary to us. The more we work together, the more comfortable we become in the Haz Mat arena. I say 'comfortable' even though we share a very healthy fear common to all. The people who are really good remain avid students of hazardous materials response.

"Two years ago when I arrived at CBIRF, we had a very limited military detection capability. To be specific, I could either tell you I was dealing with a blister agent or I could tell you I was dealing with some type of nerve gas. I really didn't have a biological collection or detection means, and I had no means to detect industrial chemical agents. We call them 'tics,' toxic industrial chemicals. Today, we have the ability to analytically identify about 135,000 chemical agents. We also have the ability to collect and identify a variety of biological agents. We can actually class type biological agents into spores, vegetative bacteria, or toxins. We still have problems currently with viruses; problems in that area will remain for quite some time. One of the greatest joys I've had was to work with wonderful technologies that are now present where I see rapid progress.

"Just the other day I went to California and worked with Jet Propulsion Laboratories on something known as an 'infrarometer,' a stand-off device that will allow detection of many industrial chemical agents up to a range of 10,000 meters in a straight line fashion. We've looked at this technology and how we could bring it to bear for the Department of Defense, even how we could detect certain things from space. The annual budget of the local Haz Mat response team, or the fire chief, is something the Department of Defense would probably spend in a month. So we took a very hard look at the situation, something very hard for me to come to grips with. What those firefighters are looking for today is one phone number they can call. Their next priority is education. They are telling us that they need to understand more about the military weapon agents. The more they understand these agents, the less fear they have. It's a healthy respect, but not a fear.

"I've visited a lot of fire chiefs around the country. They would like us to assist them in being able to get such things as Draeger tubes (The tubes measure air concentrations of toxic chemicals. The system draws air through a tube with a mechanical or hand pump. The tube will then change colors to show the concentration of the agent detected. The system uses colorimetric tubes specific to each agent. There are currently 160 tubes that detect and identify chemical agents.) and HAZCAT. (The Hazardous Categorization Chemical Identification System is based on a series of field tests which are used to identify liquid and solid unknowns. The HAZCAT system can identify over 1000 agents. Most unknowns can be identified or categorized by hazard class with as few as 4 to 5 tests.) Fire chiefs want something that is cheap, easy to maintain, and easy to use that going to help them detect and identify some of the weapons of mass destruction that could be a credible threat in the country today."

The strategic mission of the Marine Corps Chemical Biological Incident Response Force is consequence management and force protection. Consequence management addresses the consequences of an incident, and involves measures to alleviate damage, loss, hardship or suffering. "We will eventually be tying into what we call 'remediation,' looking to assist the Environmental Protection Agency, and perhaps the Centers for Disease Control, in this effort. Our overall mission is basically to mitigate the effects of terrorism; we do that best by deploying to a credible threat. More specifically, in my assignment as the reconnaissance element commander, my number one job is force protection. If we're dealing with an unknown, we go in with the highest level of protection that we have based on information of the situation. We will go in and determine if we can send follow-on forces with lesser protection into this area to continue work. That is, can follow-on forces operate in less protective equipment and still be safe? Second, we are looking for casualties. Third, we're looking for broad brush protection such as parts-per-million readings that can lead us to a more significant concentration where we can bring sampling equipment to bear."

How would a sample be obtained? "In the biological arena we use one air concentration sample," according to CWO W3 Mark Fishback. "We run air through a high volume sampler, say 250 to 1000 liters of air per minute. We introduce that air with a buffered pH solution and get a concentrated air sample in liquid form that we can apply, which gives some detection and identification means. We also look for electrostatic collection on computer screens. We do manual sampling and swabbing as well. There are certain area templates that we swab; it helps the microbiologists to plate these so we conduct sampling in this manner. We like to take a minimum of three samples. We will apply part of the sample against our field analysis detection equipment; are we dealing with a spore? A vegetative bacteria? Or something else? Two additional samples will be sent to confirmatory laboratories and they will get back to us in haste. They might say, 'Yes, you are dealing with a vegetative spore which is possibly anthrax; we've plated it and that's what we think it may be.' We are not a crisis management force; we do not take evidence. We take samples, which are followed by a chain of custody procedure.

"So that we can talk to local Haz Mat responders, and talk to local fire chiefs, my reconnaissance Marines are being trained in the National Fire Protection Association standard NFPA 472 (Standard for Professional Competence of Responders to Hazardous Materials Incidents)," concludes Fishback. "We have a resident train-the-trainer person within CBIRF, and the fire chief at Camp Lejeune is also train-the-trainer qualified, so we can get our credentials right here at Camp Lejeune which is very convenient for us. We think training in NFPA 472 is very important. We also think a local incident commander would like to know these Marines are qualified by standards that I recognize and that the incident commander recognizes."

Marine Captain John Pederson is the S-6 officer in charge within the CBIRF responsible for command control, communications and intelligence. He has a degree in economics from the University of Chicago and is originally from Scarsdale, NY. "Before this assignment, I was with the 2nd Battalion, 5th Marines at Camp Pendleton, CA and had just finished a deployment to Okinawa. I came to Camp Lejeune, NC to join the Chemical Biological Incident Response Force and was promoted to Captain shortly after I arrived at CBIRF.

"What we have here is essentially an organization which operates on about three different levels. We operate on regular voice communication level, regular low-level command, and control level, just to contain rudimentary situational awareness. Additionally, we have a wireless land network we set up during an incident in order to gather more detailed information about the incident as it progresses. This enables us to share large quantities of fairly routine information which can be difficult to handle just by virtue of its volume and enable us share that information between different key nodes within the organization. We then finally work to establish communication with headquarters through a satellite telephone system, or we can communicate with cellular phones or tactical radios. We are capable of going any place in the world and communicating with our headquarters, the U.S. Department of Defense, or similar American entities. If we had to support a department of state, a naval base, or an air field, we could do that. Essentially, that is what we do here."

Captain John Pederson walked through the large trailer that housed the COC (command operations center). "On the right side as we walk in, you will see a single-channel voice radio net and a lot of communications support functions for the COC, the actual radios, batteries, power cables, and things like that. These Marines are the radio operators who will maintain the net and gather the information that comes in off the circuit. When we go to the left hand side of the COC, you will see a site situation board, a map of the incident site as well as pertinent intelligence information posted on the board. On the right side board with maps, you can see three or four different computers that are hooked into the wireless land network that tracks information. We can put information into them as casualties are scanned down range; the casualties are bar coded, information about these casualties is collected and beamed automatically through the wireless land network back to the COC. Outside the COC we have generators that supply power; it takes about 11 kilowatts to power the COC."

Also outside the COC is a telephone unlike any the writer has ever seen. It looks like an umbrella which is actually the antenna to a satellite telephone. According to Captain Pedersen, "The signal comes in, bounces off the umbrella, and is concentrated on the focal point of the umbrella. Essentially, when we came here, there are a number of satellites which are basically like telephone entry points. We scan left and right with the signal meter on the umbrella. When the signal gets above a certain level, we know that we can establish a telephone call. Everything else is similar to a normal international telephone call including a special dialing sequence. Generally speaking, if you can do it with a telephone, you can also do it with this apparatus. If we know a person's phone number, we can call them from this location to anywhere in the world.

"Additionally, you'll see specialized directional antennas which are for the wireless land network. You'll see some additional boxes with cables coming out of them that are the hub signals that come into the antenna; the signals are converted in the hub so they can be transmitted through the wire and computer. A little farther away from the COC, you will see a group of three or four radios and some antennas. These are the primary tactical nets we use to control the operations. The voice command and control circuits are remoted away from the COC simply because of the amount of equipment associated with them. During the training exercise, the generators have just gone out of commission and the entire COC is operating on battery power which will last for about an hour. If these batteries die, it's going to affect our wireless land network. As part of the exercise, the engineer is over there now diagnosing the generator problem. Hopefully, he will be able to fix the problem. Even though we have completely lost power, we are still able to retain tactical control over the operation as it evolves. The automated data portion would be degraded if the batteries ultimately fail, but the raw command and control accounting for Marines, counting for casualties, and counting for operations will continue."

Contact: Lieutenant Steve A. Butler, Public Affairs Officer, Chemical Biological Incident Response Force, USMC, Marine Forces Atlantic, 1468 Ingram Street, Norfolk, VA 23551; 757-889-1581; 800-604-4734 (pager).

DEPARTMENT OF DEFENSE CHEMICAL/BIOLOGICAL RAPID RESPONSE TEAM

Public Law 104-201 signed in 1996 directs the Army to train local emergency responders in tactics and procedures in response to chemical and biological weapons, conduct training exercises to improve federal/state/local operations and cooperation, and provide expert assistance including a helpline, a hotline, a specialized equipment testing program, pre-positioned specialized equipment. It also proposes the establishment of a DOD Chemical Biological Rapid Response Team (C/B-RRT).

It is intended that the C/B-RRT, under certain circumstances, will assist state and local emergency responders in the detection, neutralization, containment, dismantlement, and disposal of weapons of mass destruction containing chemical, biological, or related hazardous materials. The C/B-RRT will profit from the long-standing capabilities of the Army's Technical Escort Unit's Chemical Biological Response Teams. In a graduated response, this unit would deploy within four hours and gather explosive ordnance disposal experts, analytical laboratory support, agent monitoring, and medical advisory assets as required.

The team membership of the C/B-RRT represents the following military agencies.

- U.S. Army Chemical Biological Defense Command, Aberdeen Proving Grounds, MD. Provides commander of C/B-RRT and reporting staff elements for operations, logistics, environment, communications, legal, medical, and public affairs.
- U.S. Army Technical Escort Unit. Provides chemical/biological initial response with technical expertise to render-safe, sample, monitor, detect, analyze, recover, decontaminate, transport, mitigate, and advise.
- U.S. Army 52nd Ordnance Group. Provides technical expertise in dealing with explosive components of a device.
- U.S. Army Medical Research Institute for Infectious Diseases, Fort Detrick, MD. Provides technical expertise in pathogenesis, diagnosis, identification, and decontamination of biological threat agents.
- U.S. Army Medical Research Institute for Chemical Defense, Aberdeen Proving Grounds, MD. Provides medical and scientific expertise in chemical matters.
- U.S. Army Materiel Command Treaty Lab, SBCCOM, Aberdeen Proving Grounds, MD. Provides on-site analytical laboratory capability with modular transportable laboratory equipment and scientists.
- U.S. Navy Medical Research Institute (NMRI), Bethesda, MD. Provides a transportable, biological field laboratory with rapid identification capability.

- U.S. Navy Environmental and Preventive Medicine Unit. Provides technical experts including medical doctors, industrial hygienists, environmental health officers, and scientists.
- U.S. Naval Research Laboratory (NRL). Provides microbiologists trained in chemical and biological agents.

NATIONAL GUARD RAPID ASSESSMENT AND INITIAL DETECTION TEAMS

In March of 1998, Secretary of Defense William Cohen announced the creation of the military's Rapid Assessment and Initial Detection (RAID) elements, a $50 million project consisting of ten special National Guard teams that will be dedicated to assisting local civilian authorities in the event of an attack within the United States by terrorists using WMD. Weapons of mass destruction are defined within the plan to include any "weapons or devices that are intended, or have the capability, to cause death or serious bodily injury to a significant number of people through the release of toxic or poisonous chemicals or their precursors, a disease organism, or radiation or radioactivity." (This definition of WMD does not really cover the fertilizer bombs of ammonium nitrate and diesel fuel used in the 1993 terrorist bombing of the World Trade Center in New York City that killed 6 or the attack on the Murrah Federal Building in Oklahoma City that killed 168 and injured hundreds. However, Timothy McVeigh and Terry Nichols were charged on one count of "Conspiracy to Use a Weapon of Mass Destruction to Kill Persons and to Destroy Federal Property," and one count of "Using a Weapon of Mass Destruction to Kill Persons," among other charges in the Oklahoma City incident, according to the F.B.I.'s report on "Terrorism in the United States 1995.")

The 10 RAID teams would each consist of 22 highly trained, full-time National Guard personnel and will be responsible for providing early assessment, initial detection, and technical advice to the local incident commander, and then initiating requests for additional state or federal response. The goal of a RAID team is to deploy rapidly and arrive quickly at the site of a domestic WMD incident. "As with any operation involving military assistance to civilian authorities," says a Department of Defense news release dated March 17, 1998, "the Department will play a support role. Local authorities will retain their overall jurisdiction and FEMA will retain its role as the lead federal agency for consequence management."

The release goes on to state that $49.2 million has been requested in the FY99 President's Budget, with the following recommendations:

- $19.9 million for 220 Active Guard/Reserve positions for Army National Guard RAID elements
- $15.9 million for patient decontamination and WMD reconnaissance element training (over two years)
- $6.9 million to establish and staff the Consequence Management Program Integration Office (apparently to oversee the integration of the Reserve

components into domestic preparations to respond to terrorist or other incidents involving WMD)

- $3.3 million to prepare medical personnel for operating in contaminated areas
- $1.8 million for additional Emergency Preparedness Liaison Officer training days
- $1.4 million to upgrade simulation systems with WMD-effects modeling

The report states that few military elements are currently focused, trained, or equipped to respond to WMD events. Hence the purpose of the RAID project is to increase the DOD response capabilities while developing the potential with Reserve component units. The plan actually develops capabilities of operational response to nuclear, biological, and chemical threats within the confines of the United States, its territories, and possessions. Such capabilities can be used outside the United States when required to support validated Commander-in-Chief requirements. The plan attempts to focus attention on filling a void in a state's initial assessment capability and the federal ability to quickly facilitate required help to a state. "Although immediate WMD response would be in a state status, under the control of the Governor, the unit's structure would also support homeland defense, military support to civilian authorities' missions, and provide secondary warfighting capability." The plan includes the following elements:

The National Guard RAID element forms the point of the military response spear in that it will, when formed, provide early assessment, initial detection, and technical advice to the incident commander during an incident involving WMD. The RAID elements assigned to each state/territory represent the first military responders and will have the capability to conduct recon, provide medical advice and assistance, perform detection/assessment/hazard prediction, and provide advice on WMD incidents and agents.

Information and planning element collects, processes, and disseminates information about WMD emergency to facilitate overall response activity.

NBC reconnaissance element provides NBC recon support to the local incident commander by search, surveys, surveillance, and sampling.

NBC patient decon element performs casualty decon near the incident site, prior to evacuation, or establishes decon/detection stations at community hospitals.

NBC medical response element provides medical advice to incident commander and local authorities on protection of first responders and health care personnel in an NBC environment. Provides advice on casualty decon procedures, first aid, and initial medical treatment. Provides medical threat information and characterizes the health risks to civilian and military population.

Triage medical response element provides triage support to the incident commander including the sorting and assignment of treatment priorities to various categories of wounded. Provides immediate emergency care.

Trauma medical response element provides expertise in triage, resuscitation, and damage control medicine near the incident site or at a definitive care location.

Preventative medicine element provides initial disease and environmental health threat assessment during early or continuing assistance stages of a disaster.

Stress management element provides initial stress management for military and civilian responders and incident survivors.

Security/law enforcement element (National Guard) provides support for the incident commander in accordance with state and local emergency response plans to assist in maintaining order, ensuring public safety, and providing assistance to law enforcement officials through access control, site security, civil disturbances, quarantine, and evacuation.

Mass care elements support the incident commander by providing shelter, feeding, emergency first aid, and bulk distribution of emergency relief supplies.

Mortuary affairs element provides mortuary support to include identification, processing, storage, and disposition of remains following a mass casualty WMD incident.

Communications element assures the provision of telecommunications support to the response forces following a WMD emergency.

Engineering element provides public works and engineering support, includes technical advice and evaluations, engineering services, construction management and inspection, emergency contracting, and emergency repair of wastewater and solid waste facilities.

Transportation element supports the incident commander in accordance with state and local emergency response plans and the Federal Response Plan to satisfy the requirements of federal agencies, state and local government entities, and voluntary organizations requiring transportation capacity (service, equipment, facilities, and systems) to perform their assigned response missions.

USCG National Strike Forces are regularly deployed throughout the United States on behalf of both the USCG and the EPA on-scene coordinators.

ARMY TECHNICAL ESCORT UNIT

The name, "Technical Escort Unit," does not really reveal exactly what this U.S. Army unit does. It is old-fashioned terminology from before the time when "hazardous materials" became a common media term. The U.S. Army Technical Escort Unit handles response to chemical and biological warfare material for the Federal Department of Defense, the Secret Service, the F.B.I., the State Department's Environmental Protection Agency, the Federal Emergency Management Agency, the Department of Energy, the Department of Health, and the United Nations. One segment of Public Law 104-201 calls for establishment of the U.S. C/B-RRT to assist state and local emergency responders in incidents of domestic terrorism where

WMD are threatened to be used or have been used. The Technical Escort Unit's Chemical Biological Response Teams are expected to be an important component of the C/B-RRT.

The unit's missions include worldwide response for escorting, rendering-safe, disposing, verifying samples, mitigating hazards, and identifying weaponized and non-weaponized chemical, biological, and hazardous materials. Soldiers and civilians who work for the Technical Escort Unit may be trained in explosive ordnance, identification and handling, radiography, military and commercial chemical handling, chemical and biological detection and monitoring equipment, hazardous materials, medical response, OSHA (Occupational Safety and Health Administration) requirements, and U.S. Department of Transportation and U.S. Environmental Protection Agency rules and regulations.

Although the Technical Escort Unit can and does operate in many areas of the globe such as destroying various chemical weapons in Iraq after Desert Storm, it has also responded to chemical and biological weapons within the United States. Major incidents include discovery of chemical agent ampoules at Fort Ord, CA, and Fort Polk, LA during 1997; bombs and chemical agent glass ampoules and grenades found in New Jersey in 1996; a cache of buried biological bomblets found at Wright Patterson Air Force Base in Ohio in 1995; glass ampoules discovered at Jackson, Mississippi State Fairgrounds; and World War I chemical munitions discovered at the Spring Valley housing district near Washington, D.C. in 1993. Potential chemical or biological incidents have required the Technical Escort Unit to be in attendance at national events in the United States including the Olympic Games in Atlanta, political party conventions, and the Presidential Inauguration.

William J. McKiernan is a Major in the Chemical Corps with the Chemical and Biological Defense Command at Aberdeen Proving Ground in Maryland. "The U.S. Army Technical Escort Unit was formed in 1944 and is the longest, continuously active, military chemical unit in existence," remembers Major McKiernan. "The unit was formed as a group of specialists to escort chemical weapons after the chemistry department at American University started the chemical warfare service, developing chemical warfare materials as well as defense materials. Early on, with the technology available at the time, 'disposal' meant the burial of a variety of chemical and biological weapons.

"The unit specializes in recovery of chemical and biological agents and munitions, and members are approximately 50% civilian and 50% military. The current authorized strength is 182 members with about 140 deployable at any time. Almost half the members are explosive ordnance technicians. The civilians are trained at a three week school while the military members are trained for eight months, so there's a significant difference in their levels of training. If U.S. Department of Defense ordnance is found in the country, DOD will handle the response. If the device is a liquid-fill type of round, a Technical Escort Unit local team will handle the response. Recently, the Technical Escort Unit has been involved in some other missions as well.

"In May of 1997, the New York City police raided a location where there were two cylinders labeled 'SARIN.' The Technical Escort Unit was requested by the F.B.I. and the City of New York. We deployed one of our 12-member 'Chemical

Biological Response Teams' up there, packaged the cylinders for transport, and shipped them down here to Edgewood Proving Ground in Maryland for analysis. They were both empty cylinders. It was a hoax. Unfortunately, for the person(s) involved in the hoax, the event was a federal offense and the F.B.I. got involved. The three 'Chemical Biological Response Teams" we have available are each authorized for ten soldiers and two civilians, and one such team is always on alert ready to depart within a 4-hour time frame. Members of these three teams are all trained to the 40-hour requirement under the Resource Conservation and Recovery Act and have gone through a three-week technical escort school.

"The soldiers and civilians work on Superfund (Comprehensive Environmental Response, Compensation and Liability Act) sites doing clean up and have been to Alaska, South Dakota, the Virgin Islands, Alabama, and a number of other places. At an Air Force base in another part of the country there were 92 biological bomblets that had been buried after President Nixon signed a declaration that the United States would not do any more biological work. We responded and did standard hazardous materials work which is very common to us; you respond, stop any leaks, mitigate any leaks, overpack the waste, and clean up the site."

The Technical Escort Unit is also the first response force for any chemical incidents on Aberdeen Proving Ground. "There is a standing team of 17 members which we call the 'Alert Team,'" continues William McKiernan. "If there is a spill at one of the surety laboratories on the site, the Alert Team responds. Chemical surety materials are toxic and super-toxic agents, basically warfare material. Defensive toxicology research is done here at Edgewood in the surety medical labs seen around the base that are enclosed with double strands of barbed wire. As an example, we are currently doing permeability studies for civilian hazardous materials protective suits. If there is a spill in a surety medical lab, a scientist slips or a beaker falls, the Alert Team will clean it up. There is a dual responsibility here at Aberdeen Proving Grounds. If the spill involves an *industrial* chemical which can kill you, the facility fire department hazardous materials response team will handle it. If it's a *military* chemical that can kill you, we will respond. A requirement that comes with that responsibility is that we have a standing decontamination team, an initial entry party, and a back up party or rescue team. We never send anyone down range without rescue workers there ready to respond if the need arises.

"It's very rare that there are any lab spills. Scientists know what they are doing and what they are dealing with," continues Major McKiernan. "However, every quarter we have a requirement to do what is called a 'Chemical Accident Response Activity' (CARA). Under Army regulation 50-6 that covers surety material, medical research chemical defense is an on-post, medical command asset. They do all the antidote work, have a surety laboratory, and are required to do an annual test of the different labs within the research development and engineering center. We then have the chemical agent storage yard which is one of the non-stockpile sites. Between these possibilities, we can do a quarterly response and get our training, whether it's a different lab, an ordnance item, or the chemical agent storage yard. In addition, we can use the 'Chemical Stockpile Emergency Program,' which is the Army's program to prepare installations in local communities for possible emergencies at one of the non-stockpile sites. We are required to do a major exercise annually to evaluate the local first responders as well as the installation's response capability, and this exercise is externally evaluated, which is good because we then have an independent agency validating the Technical Escort Unit's capability to respond.

"The independent agency's comments and questions are typically more concerned with the efficiency of the response and compliance with the regulatory requirements involved in the response. If you have an operating procedure that says 'you put your bucket here and your sponge there,' and you do it another way, they don't normally get into that. It's more a matter of: Did you have a timely response? Was it a safe response? Did you minimize the threat? Did you make the same mistake that you made the last time? It's a pretty common sense approach, although as a first responder, I never agree with everything the evaluators have to say. They are never happy with the speed of the response; we can always go faster in their opinion. I'm not going in until I am ready to go in, and I don't put anybody on air until I'm ready to put them on air. I'm more concerned about my workers than someone who is being evacuated five miles away.

"Considering a number of sites we have worked on, the backhoe happens to be the best chemical agent detector we have found. As an example, in Jackson, MS, workers were digging a sewer line two blocks from the state capital on the state fairgrounds and the bucket of the backhoe was dropping these little glass vials made just before World War II and into the 1960s. They were approximately 8 inches long and maybe an inch in diameter with a liquid inside. The fairgrounds had been a training area during and after the Second World War. The Army developed what were called chemical agent identification sets. The liquid is possibly pure phosgene, or it could be pure mustard gas or a percentage of agent and a chloroform carrier. They would have a group of soldiers stand downwind, break a vial, say 'take a good whiff and see what it smells like.' That how the soldiers were trained to identify chemical agents.

"In Jackson, MS when World War II ended, they buried the glass ampoules. Years later, a backhoe was working at the state fairgrounds, the vials were breaking, a haze began to form in the sewer hole being dug, and two workers went to the hospital with respiratory problems. The Health Department came in and said they would analyze the materials in the intact vials and suggested the contractor cease work in the meantime. Two more workers were sent to the hospital before work finally ceased. One vial analyzed was pure phosgene, another was pure mustard. A Technical Escort Team was sent to Mississippi to do recovery and found, I believe, 256 vials in the trash. We have found these chemical agent test kits in California, New Jersey, basically all over the country. There were several million manufactured and only a hundred thousand of so accounted for. These kits were developed before the war and contained phosgene, lewisite, or mustard. None of the nerve agents (sarin, VX, soman, GD, tabun, etc.) were ever put in chemical agent test kits.

"Aberdeen Proving Ground has a long history. During World War I, the base was a chemical warfare training area. Artillery soldiers, to be qualified in their specialty, had to fire three chemical rounds. However, during World War I, smoke rounds were considered chemical, but there is still a significant amount of phosgene out there, and some mustard rounds as well. I've heard estimates of up to 3 million rounds fired in one year with a 30% dud rate of World War I fuses. That's close to a million duds lying around. The base used to be called Edgewood Arsenal, and you can't dig around here without getting one of the teams to make sure you are not going to hit something. We found 28 smoke rounds buried where the new PX (post exchange) parking lot is located. They were all neatly stacked. The standard technology of the day was to bury it, or flush it down the drain. After the Second World War they had so much ordnance they scuttled ships full of it. Over the years at Edgewood, there were a significant number of rounds that either impacted or were buried, so even today you see signs all over the base that say 'Hazardous Area' for some ordnance still there at impact areas, firing points, or burial sites. Most of our responses on base will be for these types of locations."

Presidential Decision Directive 39 (PDD-39), "United States Policy on Counter-terrorism," was issued in 1995. Major William McKiernan was assigned along with others to provide a subcommittee report to the President that dealt with the criteria for Immediately Dangerous to Life or Health (IDLH) industrial chemicals with an IDLH level of 500 ppm or less. The IDLH level represents a maximum concentration from which one could escape within 30 minutes without escape-impairing symptoms or irreversible health effects. The subcommittee prepared a list of chemicals that could be a threat in a mass casualty chemical event. "It was the committee's scientific best guess, or whatever you want to call it, that the industrial chemicals were the higher threat," reports McKiernan. "That is, if you look at the big picture, industrial chemicals are more common, quite often just as lethal, and easier to get than military warfare chemicals." (The industrial chemical pentaborane has an IDLH level of 1 ppm, acrylonitrite has an IDLH level of 4 ppm, arsine has an IDLH level of 0.5 ppm, carbonyl fluoride has an IDLH level of 5 ppm, parathion has an IDLH level of 1.7 ppm. Methyl isocyanate, an industrial chemical that on December 3, 1984 killed more than 2500 persons and seriously injured about 150,000 more at a Union Carbide plant in Bhopal, India, has an IDLH level of 20 ppm. As a point of reference, ammonia has an IDLH level of 500 ppm. Certain military warfare chemicals, as a comparison, have IDLH levels as follows: hydrogen cyanide, 0.5 ppm; lewisite, 18 ppm; mustard/HD, 23 ppm; sarin/GB, 1.7–3.5 ppm; soman/GD, 0.67–0.9 ppm; tabun, 3–6 ppm; VX, 0.09–0.46 ppm.)

"Chlorine (IDLH of 25 ppm) and phosgene (IDLH of 2 ppm) are not new to Haz Mat responders who have responded to them for years. Fifty years ago they were chemical warfare agents, today they are industrial chemicals. Today, you can get a Haz Mat certificate on your CDL (commercial driver's license) and drive them up and down America's highways. The PDD-39 committee and the PDD-39 chem-ical agent subcommittee really thought the industrial agents we identified in the subcommittee report were as dangerous as chemical military warfare weapons. In the City of Baltimore last summer a sulfuric acid tanker leaked 50 gallons-a-minute. Had that been a phosgene tanker or a chlorine tanker leak, a good portion of the city would have to have been evacuated and there would have been significant casualties."

Major McKiernan was asked if the Technical Escort Unit responds to biological substances. "Not as often as chemical weapons, but yes, that is part of the mission," he responds. "I was trained in biological weapons. In fact, I've recently received a note to go back for more shots. Each of us is in a shot program at Fort Detrick,

MD and immunized against anthrax and botulism, as well as a standard shot program given to everyone else in the Army. If certain biological weapons are used, and we are exposed, we are less likely to be victims or casualties."

What on-scene tactics and mitigation techniques are used by the Technical Escort Unit? "There are two different types of response for us," says McKiernan, "On-post responses and off-post responses. Tech Escort has a detachment in Arkansas, and another detachment at Dugway Proving Grounds in Utah. If it's an on-post response, the ICO (incident control officer) will be the fire lieutenant or fire captain on site who is in command. He runs his response until the forward command post officer shows up and assumes control. We use an incident command system similar to what the fire service uses. If the incident is on-post and is a lab spill, typically the fire department will get there first, set up a hasty decontamination site, and process whatever casualties there are, letting the emergency medical services people take victims to an appropriate medical facility. The Technical Escort Unit would arrive and stage. Typically, that's me, a driver, and a safety person. I usually have Sheldon, an occupational safety health specialist, accompany us when we are called out. He's the best there is, and you always want to have your ace in the hole with you; it doesn't make sense not to.

"We take a briefing to get the big picture, assume control of the site, and then request forces as necessary to respond. Meanwhile, back at the technical unit proper, everyone in the unit has a pager. Our system can page up to 80 people at a time, and we start paging detachments whose members report to our alert room where the vehicles and the decon gear are kept. Members would be mustered into teams, sign for the necessary gear, and ready the vehicle(s). The First Sergeant of Charlie Company musters teams as necessary. I would call forward different teams: an initial entry party, a back-up/rescue team, and a personal decontamination station team. As the forward control point officer, I would be representing the installation commander of Aberdeen Proving Ground, would control the fire department people and security personnel, and generally would deal with the installation side of the house. There is a deputy, either a captain or a major in the escort unit, who runs the down range (hot zone) area.

"We wear OSHA Level B protective clothing (liquid splash protective suit) and OSHA Level A (vapor protective suit), both of which include SCBA. Prior to moving to OSHA level protection, we used what was called a 'modified Level A.' The Army has seven levels of protection that don't necessarily correspond with the civilian classifications. A modified Level A was an Army TAP (toxicological agent protective) which is basically a one piece butyl rubber coverall but not a pressure suit. With the mask, it was a Level C suit (limited use protective suit), but we modified the suit by adding a SCBA and called it a modified Level A which we wear when responding to mustard. The problem with mustards is that they are carcinogenic and you have to wear a SCBA regardless of the concentration."

The occupational safety health specialist mentioned previously was instrumental in getting the Technical Escort Unit to add commercially available protective clothing to their equipment list noted Major McKiernan (a Commercial Protective Clothing Joint Working Group at Edgewood is responsible for reviewing proposals submitted by chemical agent users desiring to use commercial chemical protective clothing in lieu of standard issue clothing. Approval of these proposals does not prevent the use of existing Army chemical protective clothing but rather, increases the options Army employees, contractors, and others involved in toxic chemical operations have to address the wide mix of chemical hazards that exist both on and off Army installations). "With the military, you must get approval to wear commercial suits because the suit has to be tested against the agent. We wear some commercially available suits, but we don't buy them 'off the shelf.' They are specially modified for our use. The commercial suit we use has reinforced seams and a extra layer of tape over the seams because when the Army did their tests, the seams did not pass military safety requirements."

When asked if there were any basic differences in the way the Technical Escort Unit decontaminates military chemical agents compared to the way firefighters decontaminate industrial chemicals McKiernan replies, "The principles are all the same, but each specific chemical may have a decon agent that works a little better with it. The most commonly used military decon agent is probably 5% chlorine bleach. It works extremely well on biologicals as evidenced by an exercise we did at Dugway Proving Grounds on SEB (staphyloccal enterotoxin B) which is a toxin. One percent bleach is an outstanding decon agent. Such caustics are extremely good, but then you have the problem that your decon agent is just as nasty as the contamination you may be dealing with.

"After every mission, the Technical Escort Unit goes on what we call an AAR, or an 'after action review,' either at the command level or the worker level. My criterion has always been that everyone gets out safely. We spent quite a bit of time and effort training these folks, and they are people I work with on a daily basis. The first question we ask is was it a safe operation and did we get everyone out safely? The number two question is, did we accomplish the mission? The number three question is, did we minimize the risk? Then we start working on whether we accomplished the mission as well as everyone wanted us to, making everyone happy, and minimizing the effect on the environment. But number one is always human health, and my first responders are the first priority. If you rush in, you can lose your first responders, the problem gets bigger because you lost first responders, and you're hauling someone out rather than dealing with the leaking valve to which you are supposed to be responding."

Major McKiernan was asked whether there is an easy way to dispose of a chemical agent or ordnance. "Not when it comes to chemical warfare material," he states emphatically. "If it's an ordnance item and it is fused, and the firing train is in a position where it's unsafe to store, there are protocols you must go through to get permission to destroy it. It can go all the way to the Pentagon, and the Department of Health and Human Services has to approve the destruction plan. If it's not a detonation threat, then it goes into semi-permanent storage until someone determines what they are going to do with the round they find. If it is not a phosgene round, or the industrial chemical phosgene, you can have the hazardous waste people take it away as long as there are no explosives involved. Often the volumes we are dealing with are so small that in the verification process analysis you can often get permission to destroy the chemical.

"We will always encounter chemicals we can't identify," continues McKiernan. "The chemicals out there are numerous, but we specialize in a small niche and we've got very good detectors for everything in which we specialize. It's very hard to identify an unknown, so to try to prove something is not there is extremely difficult. A dumping place can take a long time and a lot of analysis to identify what chemicals are actually there. If we are escorting a chemical shipment, we know what we have. If we are responding we typically know what we're looking for, or at least the population of the chemicals that are in our threat scenario. We typically have an easier time identifying what we are looking at than would a civilian Haz Mat team simply because we're focused on what we do.

"If we're called to an outpost response in a chemical agent storage area and we find only one chemical stored there, that's pretty cut and dried. If it's in a lab, their controls are stringent enough that you know what class of chemical they have available. We have one building that has several chemicals on hand; you may not know which one is causing a problem, but the inventory is limited to half a dozen chemicals.

"Our 'unknowns' come when it's an ordnance item that has been buried for the last 60 years. In that situation, it could be any number of things. We may default to phosgene as the worst case chemical simply because it's got the highest vapor pressure and the most down wind threat. If a chemical like phosgene got off post, that would cause a real threshhold of concern. If you stay in the installation, it's the installation's problem; if you cross the installation's boundary, then it becomes a community problem. You do everything you can not to have either situation, but if you are going to have an incident, keep it on the installation.

"With ordnance items, it is a problem if you don't know what is inside. Now we have the PINS Chemical Assay System which uses a californium-252 neutron source to irradiate an item causing the emission of gamma rays. A germanium crystal which has been super-cooled with liquid N_2 converts this energy to electrical energy which is then transferred to a laptop computer which will produce a graphic display or spectrum of the energy levels as well as conduct an analysis of this spectrum. Based upon the elements present, the operator can determine with a high level of confidence which chemical warfare material is inside the container. Basically, if the decision is that there is sulfur and chlorine inside, the container probably contains mustard. If it has an extremely high phosphorus peak, it is probably a white phosphorus round. Using the PINS, we have been able to decrease the number of rounds and responses. Put in temporary or permanent storage, it's a very good tool. It's expensive (about $60,000), not necessarily the easiest tool to use, you need to have liquid nitrogen to super-cool the crystal, and it has a radioactive source material so it comes with accompanying problems. However, it tells you what elements are inside a cylinder without drilling a hole and without exposing anybody so it saves of lot of installation problems.

"As I mentioned before, we do have a number of civilian employees who are toxic materials handlers with the Technical Escort Unit. I believe it was about 1988 when the government started to down-size the military and realized that quite a bit of experience exists in the civilian community. We have a population of civilian toxic materials handlers in the detachment down to the unit level. Civilians in the unit give us a tremendous capability because rather than being a soldier who rotates in and has a learning curve of several to many months and then has a stabilization for a few years, a civilian has got that learning curve and then they can reinforce the training through an entire civilian career. First responders are like any other first responders in the country, the only difference is that Uncle Sam pays them rather than the local community or a contractor. They have some outstanding experience and skills, and they bring quite a value to the Technical Escort Unit."

What special equipment does the unit have? Draeger tubes are commonly used by both the Technical Escort Unit and other first responder agencies such as fire departments.

TABLE 5.1

Draeger Tubes Used by the Technical Escort Unit

Chemical	Draeger Tube	Concentration
Chloroform	Chloroform 2a	2.0–10 ppm
Chloropicrin	Carbon tetrachloride 1a	1.0–15 ppm
Cyanide	Cyanide 2a	2.1–15 mg/m³
Cyanogen chloride	Cyanogen chloride 0.25a	0.25 ppm
GB	Blue band tube	0.5 mg/m³
HD	M256	3.0 mg/m³
Hydrazine	Hydrazine 0.2a	0.5–10 ppm
Hydrazine	Hydrazine 0.25a	0.25–3 ppm
Hydrocyanic acid	Hydrocyanic acid 2a	2.0–30 ppm
Lewisite	Yellow band tube	10 mg/m³
Nerve precursers	Phosphoric acid esters 0.05a	0.05 ppm (qualitative)
Organic arsenic compounds	Organic arsenic and arsine	3.0 mg/m³
Arsine	Organic arsenic and arsine	0.1 ppm
Phosgene	Carbon tetrachloride 1a	1.0–15 ppm
Phosgene	Carbon tetrachloride 5c	5.0–50 ppm
Phosgene	Phosgene 0.02a	0.02–0.6 ppm
Phosgene	Phosgene 0.05a	0.04–1.5 ppm
Phosgene	Phosgene 0.25b	0.25–15 ppm
Sulfuric acid	Sulfuric acid 1a	1–5 mg/m³

The unit also uses the PINS Chemical Assay System to nonintrusively identify the elements within a liquid-filled container; a Microtip IS-3000 which is a hand-held microprocessor-controlled air monitor/photoionization detector (PID) that measures the concentration of airborne photoionizable gases and vapors and automatically displays and records these concentrations; an AIM 3300 Multigas Detector, a hand-held instrument with sensors for electro-chemical and a combustible gas monitoring designed to warn the user of hazardous atmospheres; an AirPro 6000DL Personal Air Sampler which is a battery operated vacuum pump used to draw a potentially contaminated air sample through a sample collection device; and a Miniature Continuous Air Monitoring System (MiniCAM) which is designed primarily for the rapid determination of the 8-hour, time-weighted-average (TWA) of chemical warfare agents and simulants, but can also be used to determine IDLH concentrations of these and other chemicals.

"The MiniCAM is a very specific piece of equipment that can detect 173 different chemicals," says McKiernan. "However, with that device you need a standard to put in so you can quantify and qualify what the sample may be. So, depending on what optical filter you are using, you will get different peaks telling you 'something' is there, but without a standard you don't know what it is. It is similar to a Microtip PID; without calibrating it to a certain chemical, you don't know what the sample may be. Within the U.S. Army Soldier and Biological Chemical Command, we have the capability for making standards for all nerve and blister agents. So, if we are doing a certain response, and we know we are responding to a nerve agent spill, our operators can quantify and qualify down to the parts-per-billion level. They can tell exactly what it is and how much there is. A fire department can't do that. It's impractical because firefighters deal with so many chemicals that it would be next to impossible to have a standard for everything.

"When not doing response, we have significant training requirements that come with the chemical surety program. We also have weapons training since we escort chemical surety materials around the country. On remediation sites or base closures where people are digging up rounds or doing service clearances, the teams will be standing by 'just in case.' Teams rotate in doing chemical escorts for contractors and installations around the country that require small volumes of surety materials for their own standards or for tests for one purpose or another. Teams also do outpost response or undergo mandatory Army training which can include anything from equal opportunity and sexual harassment to subversion and espionage against the Army. The soldiers work in the Technical Escort Unit get chemical hazardous duty pay, an extra $150 a month.

"I can't get into the details of escort duty for security purposes, but unit members will do a route reconnaissance, make sure the road is safe, identify possible ambush points, check the cargo to make sure it is not leaking, staff a response team if a leak does occur, and basically be the cargo custodian, ensuring the cargo is what it's said to be, and signing it over to the signer at the other end. We don't escort like we used to. Back in the 1960s, it was in trainloads. I've seen records of 100,000 rounds of this or that. Now, it's typically a liter or less, but since it's a dangerous chemical the Army protects it very well. That's why our assessment that I mentioned before (with regard to terrorist attacks) was that you're much more likely to have an industrial chemical problem than you are to have a chemical warfare material problem. With the Internet and all the different information resources, it's not hard to make this stuff, but it's pretty hard to stay alive once you have made it."

Major William J. McKiernan has a master's degree in microbiology. He worked within the Army with a program called "Training with Industry" and spent 12 months with the U.S. Environmental Protection Agency, Hazardous Management Division, at Region Eight in Denver, CO.

Contact: Suzanne Fournier, SBCCOM Public Affairs Officer, U.S. Army Soldier and Biological Chemical Command, Technical Escort Unit, Edgewood Area, Building 5101, Aberdeen Proving Grounds, MD 21010; 410-671-4345; 410-671-5297 (Fax).

FEDERAL BUREAU OF INVESTIGATION HAZARDOUS MATERIALS RESPONSE UNIT

Started in 1996, the F.B.I.'s Hazardous Materials Response Unit (HMRU) is meant to be a response to potential terrorist action involving NBC weapons. It is an expansion of the F.B.I.'s readiness and proficiency for crime scene and evidence-related operations where NBC materials, and chemical and biological wastes may be involved. The HMRU will train, equip, and certify F.B.I. field office employees for hazardous materials operations. The overall mission of the F.B.I. Hazardous Materials Response Unit follows:

- Provide/coordinate laboratory and field forensic, scientific, and technical response to F.B.I. investigations of criminal acts involving the use of nuclear, biological, and chemical materials, and environmental crimes.
- Provide research and development leadership within the scientific, medical, technical, law enforcement, and intelligence communities.
- Serve as an advisory group to the F.B.I. and other federal, state, and local agencies on the scientific and technical aspects of NBC incidents.
- Provide emergency response and NBC training for the F.B.I. as well as other domestic and international organizations.

Within the total organization of the HMRU, are the following divisions: research and information systems, nuclear program, biological program, chemical environmental crimes program, operations and logistics program, emergency response and training program, and the health and safety program. The main concentration in this chapter will be on the emergency response and training program which falls under the F.B.I.'s Scientific Analysis Section.

Steven Patrick is program manager for the emergency response and training program. "My background is primarily in the fire service," says Patrick. "I started in 1984 with the Newport News Fire Department. As a member of the Haz Mat team, I was involved not only with hazardous materials response but also some emergency medical services issues such as toxicology, pharmaceutical caches, and operating procedures for the EMS paramedics assigned to the Haz Mat team. After seven years, I accepted a position as a hazardous materials officer with the Virginia Department of Emergency Services for the northern section of the state.

"At the same time I was involved in writing training programs and teaching hazardous materials courses. This progressed into doing work for the National Fire Academy while working on terrorism issues for the State of Virginia. I was co-author of a document called 'Emergency Response to Incidents Involving Chemical and Biological Warfare Agents,' did some work on 'Emergency Response to Terrorism, Basic Concepts,' and was also involved in teaching nationally on the subject of response to terrorism. I became involved in teaching because, while working in the regional response program in Virginia, I worked specifically on environmental crimes. The issues I learned in environmental crimes combined with my experience in hazardous materials response were the start of what we really needed to know about terrorism because it covered all the legal aspects of the subject. The only different aspect was dealing with chemical and biological warfare agents.

"Obviously, with PDD-39, the Bureau had a leadership role in dealing with chemical and biological warfare agents in terrorism, but they were not prepared to do hazardous materials types of work. The F.B.I. began an initiative to respond to chemical and biological warfare agents and that's when I became involved in helping the F.B.I. At the time, I was involved with preparing first responders to deal with terrorism, and David Wilson was tasked with putting a team together for the F.B.I. that could do a couple of different things. They would have to be able to enter a contaminated environment safely while integrating into existing command structures on the scene. They would also have to acquire a sample for forensic purposes as well as assist the first responders in immediately identifying the problem, then decon the sample and get it appropriately analyzed.

"The F.B.I. understands the fact that there will be first responders in such environments," continues Steve Patrick. "If they were to acquire a sample and were to identify it, that's wonderful. The problem would be if no first responders were around. A good example would be a search warrant type of situation where the F.B.I. was going in. Say we had credible evidence or probable cause to believe that someone was actually manufacturing chemical or biological warfare agents. Say we were actually going to do a search similar to a drug enforcement administration drug lab search. You get involved with evidentiary issues, legal sampling, and testifying in federal court. You could also have to deal with secondary devices or booby traps meant to kill responders; it does take a very specialized team to do that type of work.

"The intent of the unit and the intent of the F.B.I. is not to replace what the first responders are doing. When we arrive on the scene, whatever the first responders have already accomplished will be taken into account. First responders have a very specific mission — public safety within their community. The F.B.I. mission is to help them fulfill their public safety duties and to take care of the evidentiary side. We also run into situations where, under PDD-39, the F.B.I. is the federal command agency on the scene, which is not much different from when the Environmental Protection Agency has the authority to be the federal on-scene commander.

"The levels of expertise vary throughout the country as do opinions about what is appropriate in different parts of the country. One Haz Mat team team may take an aggressive response to a spilled chemical, while in other parts of the country you could have an identical situation and the first responders would take a completely hands-off approach. However, from the F.B.I. standpoint, we need to have our own internal advisor so we can listen to what the first responders have to say, take that into account, and look at our own policies and procedures to see how we are going to proceed. For all these reasons, the F.B.I. needed a hazardous materials response component. Keep in mind that we do not intend to duplicate what first responders do. The F.B.I. hazardous materials response unit is very different from a true Haz Mat team that's out on an incident scene. We are a very small unit intended to do a very specific mission.

"From my background as a firefighter for a number of years, and my position in the F.B.I. since late September of 1997, I have noted there will always be problems," adds Steve Patrick. "There are certain responsibilities that public safety first responders have, and there are certain responsibilities that forensic folks have. It's very difficult to have them meet in the middle. In dealing with the incident itself, some evidence will always be destroyed or contaminated, and that always needs to be taken into account. The first responders need to understand they can do their job without destroying evidence and remove themselves from the scene. Among the problems you encounter are situations where law enforcement individuals want nobody in there whatsoever. On the other side, there are first responders who want to go beyond their normal job in public safety. If law enforcement takes an extreme approach, public safety cannot be appropriately addressed from the first responders' standpoint. If first responders completely destroy every bit of evidence, people who perpetrate these acts are not going to be prosecuted. In light of the catastrophic results of terrorist acts, the American public will not tolerate this lack of accountability.

"The HMRU was established just prior to the Atlanta Summer Olympics when I first got involved with the F.B.I. The agency approached the Commonwealth of Virginia and requested assistance in training and guidance for the HMRU. The unit looked at a lot of different training programs and a lot of methods through which they could receive such training. They basically came back to Virginia for a couple of reasons. They were based here northern Virginia, and they were aware that Virginia had a full regional response program as well as a full statewide training program. They wanted to get training from actual first responders and did not want to go to a contractor for training."

How and why did Virginia have such a fully developed regional Haz Mat response program? "Back in the mid-1980s the General Assembly in Virginia passed some regulations to have regional hazardous materials response teams because of SARA Title III (Superfund Amendments and Reauthorization Act of October 17, 1986), the Winchester tire fire, and an incident in the Richmond area. They saw a need to establish a regional response program. They funded a program and identified where they wanted regional teams. Their goal was to have the state emergency operations center dispatch a trained and experienced response team that could get to you within two hours in the event of a Haz Mat problem any place in Virginia.

"The teams themselves operated under signed agreements with Virginia Emergency Services and established an employer/employee relationship. People from among the regional response teams were brought in and wrote the training programs. The state was fortunate that such leadership existed within Virginia Emergency Services, which produced demanding training requirements. If we needed 80 hours to teach a technician course, the administrators would fight the political battles to make that happen. It takes 'x' number of hours to be a Haz Mat technician or a Haz Mat specialist, but that is the state's requirement. If you don't want to commit that level of time, you can't be on a regional team. They ended up with a fairly comprehensive program: 8 hours for the awareness level, 32 hours for the operations course, 80 hours for the technician level, and an additional 200 hours for the Haz Mat specialist level.

"At the same time, they saw a need for local government to be able to respond in an appropriate fashion, so they signed an agreement with every single locality in the entire Commonwealth of Virginia. They gave each locality a little money, and said, 'We will fund a defensive response capability if you promise to train a significant number of your firefighters to the operations level.' At the present time, Virginia has 13 regional response teams and a number of local firefighters around the state trained to the operations level with 32 hours of Haz Mat training."

Steve Patrick was sent from Virginia Emergency Services to assist the F.B.I. with their training and coordinated the first program with the F.B.I.'s David Wilson. "A couple of months prior to the Atlanta summer Olympics, the FBI adopted a requirement where they had to have Haz Mat training operators on the ground in Atlanta," remembers Patrick. "We only had a week to train about 60 individuals, so we trained them at the operations level. That is, under the emergency response phase, they could only operate in a defensive fashion, and when the emergency was over, they could go ahead and make entries to do evidence recovery. Under the OSHA standard they were trained at the operations level, a lower level than the technician level. Eventually, we will require technician level training for our current operations level-trained staff.

"In Atlanta, there was a pre-staged response. They staged a response team at an Air Force Base that was a combination of the F.B.I., the Technical Escort Unit from Aberdeen, MD, and the Marine Corps/Navy Chemical and Biological Incident Response Force from Camp LeJeune, NC. There were two of us from Virginia who were actively involved in this training, me and Steve Ray, the Haz Mat coordinator in Fairfax County and co-author of the fire academy book. Both of us were involved with urban search and rescue (USAR), so we had to go to Atlanta for one week with Virginia USAR Task Force 1 at the county fairgrounds. Then we were assigned for a week to the F.B.I., primarily to give them some guidance on taking the training that we had provided previously and incorporating field operation guides and standard operating procedures. The Atlanta Olympics was the first pre-staged event, after which the F.B.I. started building their unit up to its present structure. In January 1997, Steve Ray and I were contacted again by Dave Wilson for assistance with the other team during the inauguration. In the spring of 1997, a position with the F.B.I. became available, and the agency went outside and hired me."

The Federal Bureau of Investigation HMRU has built an infrastructure while responding to any calls for assistance from the F.B.I.'s field offices. They use a four-tiered system of response. Tier 1 provides on-call Haz Mat emergency response expertise under which any F.B.I. agent in the field can call the F.B.I. headquarters in Washington, D.C. and get a member of the HMRU to provide guidance whether it's a scientific- or response-related matter. Tier 2 is the deployment of an assessment, advisory, or escort team. Such teams are necessary when a field office thinks it may have a Haz Mat problem but is not sure, or when F.B.I. agents with a search warrant may encounter hazardous materials, chemical or biological agents, or WMD. Another situation might be if somebody has a sample of something that may be a chemical or biological agent, we have the ability to go out and get the sample rather than shipping it commercially. Tier 3 actually deploys a rapid response team, seven to eight Haz Mat technicians who can be assembled within four hours and can travel by response vehicles, including trailer trucks, helicopters, or airplanes, anywhere in the country or anywhere in the world. A Tier 4 response would employ the same type of response team as Tier 3 but includes a mobile analytical laboratory that can be flown any place in the world. One of the biggest efforts underway right now is development of this small lab which will conduct the most advanced scientific chemical and biological analysis available anywhere. The F.B.I. and some contractors at the Edgewood area of Aberdeen Proving Grounds are coming up with the required equipment and are building a complete fly-away lab package that, when flown to an incident scene, will be as sophisticated as any chemical and biological agent lab anywhere. For nuclear or radioactive materials, the F.B.I. has agreements with the U.S. Department of Energy to use their existing laboratories, acquire the material, and have it analyzed.

"At pre-staged events like the Atlanta Summer Olympics, the F.B.I. brought the lab to Atlanta and had it set up and running," says Steve Patrick. "If something happened, not only could we analyze it for forensic purposes, but more important for first responders, we would have the laboratory on-scene, be able to respond, grab a sample, and analyze any agent, although we would be primarily set up to look for chemical warfare agents. There are just two ways to identify biological materials: either you culture them or you do a biological assay. There are a couple of different types of biological assays, but they basically work like a pregnancy test kit. You usually analyze the biological materials using all the different methods, then positively confirm the result. There are seven biological warfare agents that we can positively identify with biological assay tickets. A good example was in Atlanta after the Centennial Park bombing when this unit processed samples from that scene, looking for chemicals as well as biologicals using the biological assay tickets. We took a sample, looked at it on the tickets to see if they identified any of the seven biological materials, and had the samples cultured. The process is fairly rapid. The biological assay tickets take about 15 minutes; culturing takes anywhere from 4 to 24 hours, and we would do both at the same time."

Does the F.B.I. hazardous materials response unit have any different entry equipment that the average fire department Haz Mat team would not have? "Most of what we use is consistent with what the fire service would have. However, because of our specific mission there are a couple of unique differences. One example would be that we carry oxygen generating, SCBA which would give us as much as four hours of breathing time, quite a bit more time than the average fire department SCBA could provide. We also use a considerable amount of Level C protective equipment. Once we make an entry and determine that the environment meets the criteria for Level C equipment, we will actually downgrade to Level C and use respirators. Recovering evidence could take from 6 to 24 hours depending on how large the crime scene is, and it is very difficult to keep people in SCBA for that long. That is, we use a lot of air purifying respirators when it's safe to do so, and we also use rebreathers. It's very similar to the equipment used in the movie, 'The Hot Zone.' We also have some very sophisticated detection and monitoring equipment that a standard Haz Mat team would not have.

Once an F.B.I. entry team enters a hot zone, they check the atmosphere to learn whether they can downgrade their personal protective equipment from a SCBA with Level A or Level B protective clothing to a respirator with Level C protective clothing. Their concept is that since most bacteria are large particulates, they can use a HEPA filter that will filter out the first level (HEPA filters remove up to 99.97 percent of dust, pollen, pet dander, smoke and bacteria down to 0.3 microns size). There is no one single filter cartridge acceptable for any and all atmospheres, so they must then select a respirator cartridge appropriate for the atmosphere and time in which they wish to work. Within chemically contaminated areas, they would probably use an organic vapor cartridge (in addition to the HEPA filter). One of the respirator types they use is the 3M Breathe-Easy® powered air protection respirator, including the Breathe-Easy 5 (clear vinyl hood) and the Breathe-Easy 7 (full-face).

"There are two sides to the HMRU, the laboratory side and operations side. The laboratory side coordinates with the U.S. Department of Energy and is building a response capability here within the unit. We are dealing with nuclear materials, as the fire service does, but will be using more advanced equipment, the type of equipment the Department of Energy would use. In the biology program they are establishing the equipment we need to have in the unit and establishing the liaisons we need with all the appropriate laboratories in the country. For example, if we happen to be working in one part of the country and don't have our own laboratory with us, we know which is the closest lab where we can get help. In the chemical environmental crimes program they are building the chemistry lab that we will be able to take with us. Coming from the fire service, you don't have to have a chemist on your Haz Mat response team; some do, but many don't. In the F.B.I. HMRU we have a Ph.D. chemist who runs our chemical program. She also deals with the detection and monitoring equipment. I don't have to worry about making sure the instruments are correctly calibrated.

"I work on the operations side as program manager for the emergency response and training program. It's my responsibility to serve as the unit's hazardous materials officer. Right now, I'm the only person in the unit who has a Haz Mat response background. The unit is brand new and we are still building it. The actual response team did not go on-line until the spring of 1998.

"There are two major initiatives in the emergency response and training program. The first is that we will eventually have hazardous materials awareness, operations, technician, and specialist level training courses for our operators as well as for operators in the field. We would like to have three separate shifts where F.B.I. folks would receive hazardous materials training and then they ride along with selected Haz Mat response teams. We're already starting this conversation with some of the larger metropolitan response teams in the country. Students would finish the training we offer, but their certification would be contingent upon riding for a certain period of time with a Haz Mat response team that we designate. This way they will gain a bit of experience as to what hazardous materials response is all about. More important to me, is that individuals who have different responsibilities understand each other. If a law enforcement individual goes to work with a fire service individual, or vice versa, and they spend some time together, they can understand each other's priorities a bit more. If something occurs in their city, county, or state, they will have at least started some relationships that can help the situation.

"We will be writing our own courses in-house, but we will be using a lot of fire service instructors throughout the country to teach part-time courses. Our courses will be similar to other Haz Mat courses throughout the country except that they will have more of a forensic twist and will cover items important to F.B.I. employees in the unit. Under the second initiative, we are developing field operating guides for all the operators on-scene, instructions on how to do what they have to do.

"On the operations side there is a health and safety program which is run by a special agent who is also a nurse. This program is similar to Haz Mat emergency medical services which would be responsible for the entire health and safety of the total unit, making sure we get our annual physical exams, checking any possible exposures we may have, establishing entry protocols and things of that nature. There is also an operations and logistics program within the operation group. Persons in this program are responsible for moving unit people around the country, making sure all the logistical requirements are met, and ensuring we have the appropriate equipment to do the job. We have a responsibility to respond any place in the world. We have a domestic responsibility, but also an international responsibility because under the laws of the United States, the F.B.I. investigates all terrorist-type crimes against American citizens, so even outside the continental United States we would have to respond as well.

"The operators for the entire unit are drawn from these programs; most are special agents for the F.B.I., most of them are law enforcement personnel. We do have some individuals who are not special agents who have specific other jobs in the F.B.I. The ones in my program are hazardous materials officers. The operations group will get us to the scene, and the Haz Mat officers will size-up the situation, coordinate with the first responders on the scene, conduct a hazard risk assessment to see if entry is appropriate, and select the proper protective clothing. The folks who will actually make the entry will be Haz Mat officers, but quite a few will be special agents trained as hazardous materials technicians and specialists. However, on the operations side, no one goes anywhere, no one moves, without an experienced hazardous materials officer there who can make judgment calls as to the safety of the entries."

How many people volunteer for this type of duty? "I'm finding that here in the F.B.I. it's really no different from the fire service. We have people who don't want to go near hazardous materials; the same was true in the fire service. But, you also have people who are intrigued by emergency response to hazardous materials. Most of the people who end up in this unit are special agents who have some sort of a science background. You have other people who enjoy working in challenging environments."

Officials have discussed plans to send hazardous materials response teams around the country. "WE need law enforcement type Haz Mat teams around the country, but not so much for terrorism. Like the Tokyo incident, it would be useful there, but that's not going to happen everyday. A thing that does happen frequently is when someone has possession of chemical or biological material and is using it for criminal purposes, not necessarily as a warfare agent. Perhaps the HMRU, or possibly some regional law enforcement Haz Mat team could be there to address such issues, usually from a search warrant standpoint or something of that nature. We're looking right now at developing five or six programs around the country where there will be small units similar to this one that exist in some of the major metropolitan areas.

"The F.B.I. HMRU is expected to have 35 employees when it is completely staffed," according to Patrick. "Currently, we are focusing on responding to chemical and biological warfare agents because of the vacuum that exists there, but in the long term, we will handle environmental crimes in addition to chemical and biological warfare agents. An environmental crime is a violation of federal environmental crime statutes such as the dumping of hazardous waste which would be a Conservation Recovery Act crime, or a Clean Water Act violation, or perhaps an air pollution violation. Back in the 1980s there was a memorandum of understanding that was signed between the Environmental Protection Agency and the F.B.I. that gave the F.B.I. joint jurisdiction with environmental crimes."

When an F.B.I. special agent who is in the hazardous materials response unit has HMRU training testifies at a trial, the combination of Haz Mat training with law enforcement experience can lend substantial credibility. It does not necessarily put them in charge however. Steven Patrick cautions that the commission is very specific: "We are not here to replace the EPA, the Coast Guard strike teams, or local or state Haz Mat teams. They have a job to do. The F.B.I. just has a responsibility for the forensic science part of the hazardous material, as well as helping other folks to do their jobs. The forensic science component of hazardous materials has really not been addressed at all in this country with the exception of a few areas such as New Jersey."

"One of the problems that you run into in dealing with terrorism is that there is a certain amount of the subject matter that is classified, and I think that protection is necessary for the U.S. government," according to Steve Patrick. "We have access to that classified information where a local Haz Mat team may not. Not that there is much out there that is super-secret, but there are certain things that the government wants to protect in the national interest. If we're to directly interface with a local team at an incident, we could provide such information to allow them to complete their mission. There is also an effort underway so that our operators on the HMRU will be cross-trained in both hazardous materials response and explosives."

Contact: Steven G. Patrick, Haz Mat Officer, Program Manager, Emergency Response and Training, HMRU, ERF, F.B.I. Academy, Quantico, VA 22135; 703-630-6556; 703-630-6784 (Fax).

DEPARTMENT OF ENERGY NUCLEAR EMERGENCY SEARCH TEAM

The U.S. Department of Energy Nuclear Emergency Search Team (NEST), like the Army Technical Escort Unit, locates and examines unknown WMD devices, renders safe an armed WMD device, and identifies or evaluates WMD agents. The unit is headquartered in a $75 million compound in a portion of Nellis Air Force Base outside Las Vegas, NV. It was formed in 1974, and stocks such equipment as satellites, radiation detecting aircraft, power generators, radiation monitors, secure telephone systems, vans that can be easily camouflaged to represent almost any type of business vehicle, and wireless ear pieces that can communicate with monitoring devices carried by an incognito agent to experts outside the immediate scene where a device may be hidden.

Contact: Chris Kielich, Public Information Officer, U.S. Department of Energy, 1000 Independence Avenue S.W., Washington, D.C. 20535; 202-586-0581.

DEPARTMENT OF HEALTH AND HUMAN SERVICES METROPOLITAN MEDICAL STRIKE TEAMS

A Metro Medical Strike Team, at the request of local and/or regional jurisdictions, will respond to and assist with the medical treatment, management, and public health consequences of chemical, biological, and nuclear incidents resulting from deliberate or accidental acts. Each of the 27 teams to be developed, most of which will be in major cities, will perform the following functions: medical treatment and management, hospital and public health coordination, incident command liaison, decontamination, phamacology, communications, public information officer/media relations, law enforcement intelligence and security, and training.

The U.S. Department of Health and Human Services (HHS) has implemented a systems approach for response to acts of domestic terrorism involving WMD which they define as any explosive, incendiary, or poison gas, bomb, grenade, or rocket having a propellant charge of more than 4 ounces, missile having an explosive or incendiary charge of more than 0.25 ounce, mine or device similar to the above; poison gas; any weapon involving a disease organism; or any weapon that is designed to release radiation or radioactivity at a level dangerous to human life. Two basic goals must be met by this effort. The first is to assist state governments, local governments, and private industries considered important assets and to gain the additional capability to appropriately and effectively respond to a local NBC terrorist incident. The second is to significantly improve federal capability to quickly supply, equip, and assist state and local jurisdictions in preparing for and responding to a major terrorist incident.

The federal government will bring to bear the following important services, supplies, and assistance: planning and exercises, development of three NBC response capable National Disaster Medical System (NDMS) teams, development of three NBC pharmaceutical caches, development of an infrastructure to give medical responders access to on-line diagnosis and patient management information, the study of key research and development patient issues, and enhancement in surveillance and laboratory support.

State and local jurisdictions will create Metropolitan Medical Strike Team(s) (MMST) in major cities of the United States similar to, but different because of local needs, to MMSTs that have been developed in Washington, D.C., and Atlanta, GA. MMST management and development will be guided by HHS Regional Health Administrators with the support and national guidance of the HHS Office of Emergency Preparedness (OEP). Training and equipping will be coordinated with the Federal Emergency Management Agency (FEMA), the Department of Defense (DOD), and the Department of Energy (DOE).

The U.S. Department of Health and Human Services report, "Building a Systems Approach for Health and Medical Response to acts of NBC Terrorism," provides a clear and concise route to the development of a number of MMST in major cities around the country. It first recognizes the problem: most local emergency systems need enhanced capability to manage the threat or use of WMD. Then it pursues two goals: improvement of local health and medical capability to respond effectively and improvement of federal health and medical capability to rapidly augment state/local response. The resultant strategic national counter-terrorism plan is meant to create local resources that could respond rapidly, develop partnerships to improve local health and medical system capability to respond effectively, and improve federal health and medical capability to rapidly augment state/local response.

State and local emergency responders in MMST must develop and maintain a number of competencies with regard to chemical/biological (C/B) weapons. They must know what hazardous C/B weapons are, know their behavior and effects on the body, understand potential outcomes of their use in terrorism, recognize the possible presence of C/B agents, and identify the C/B material, if possible. In addition, state and local responders must understand their role in the response plan, recognize and communicate the need for additional resources, perform hazard and risk assessment techniques, communicate the hazard, select and use proper protective equipment, understand C/B terms and C/B control and containment, know the risk of secondary contamination, take measures to protect life and safeguard property, and develop an awareness of crime scene and evidence preservation. They must understand C/B decon procedures, basic life support in a C/B incident, safe C/B antidote administration, patient assessment and emergency medical treatment as it pertains to a C/B incident, and proper patient transport.

They must be expert in providing triage and primary care, in C/B sampling/detection/monitoring, in handling the deceased, in understanding "all clear" procedures, and in relevant C/B response standard operating procedures. MMST responders must know how to implement an emergency response plan, know the classification/identification/verification of C/B materials using field survey instruments and equipment, understand the Incident Command System, and know how to implement the state emergency response plan. Finally, they must develop a site safety and control plan, understand the risks of operating in protective clothing, and know how to access the federal support infrastructure.

Robert J. Jevec is a medical emergency coordinator/planner at the National Disaster Medical System within the DHHS Office of Emergency Preparedness in Rockville, MD. The Office of Emergency Preparedness is designed to manage and coordinate federal health, medical and health-related social service response and recovery to federally declared disasters under the Federal Response Plan. Major functions involve coordination and delivery of emergency preparedness activities including continuity of government, continuity of operations, and emergency assistance during disasters or other emergencies; coordination of the health and medical response for federal government, in support of state and local governments, in the aftermath of terrorist acts involving chemical and biologic agents; and direction and maintenance of the medical response component of the National Disaster Medical System, including the capability of the Disaster Medical Assistance teams and other special teams that can be deployed as the primary medical response teams in case of disasters.

"We felt that most local emergency systems need to be enhanced," according to Robert Jevec. "They are not set up to manage releases of weapons of mass destruction. Some of the major areas where local systems need enhancement are agent identification, personal protective equipment, decontamination of both victims and the environment, and two types of treatment: the initial treatment given right at the incident scene, and definitive care treatment provided away from the scene. The Department of Health and Human Services' Support Plan for the Federal Response to Acts of Chemical/Biological Terrorism deals with improving the local health and medical capabilities to respond, and also to improve the federal health and medical capability to augment the state and local response. We recognize that all C/B incidents are going to be local. Therefore, we have to enhance the local capability. Incidents are local, and the federal government is not normally going to be able to get there quickly.

"When the federal government comes in, they always augment and support; they never take over the incident. The federal government will support a local incident commander, and bring in response teams, equipment, and pharmaceuticals. Metropolitan Medical Strike Teams are our method of enhancing local capabilities. As required, patients will be evacuated from an area into a network of pre-enrolled, non-federal National Disaster Medical System (NDMS) hospitals located in major metropolitan areas of the United States. To augment this capacity, medical support including federal medical treatment facilities such as the military services, Department of Veterans Affairs (VA), and HHS, will be utilized as needed. This federal partnership also includes the Federal Emergency Management Agency.

"Health and Human Services is basically the administrative headquarters for this partnership which has three components; medical response, patient evacuation, and a definitive care component," continues Jevec. "The medical response component consists mainly of our teams, Disaster Medical Assistance Teams (DMATs). We have general purpose DMATs and we have special purpose DMATs like pediatric, burn, mortuary, or veterinary teams. The patient evacuation component is when victims are taken out of an area that is overwhelmed and brought to hospitals outside the incident area. This transport is usually by air, and the Department of Defense is responsible for that component. The definitive care component consists of federal coordinating centers around the country, normally Veterans Administration medical centers or Department of Defense military medical centers, and they are the ones that sign up civilian hospitals to be part of the system and maintain and train their civilian counterparts.

"For example, patients would be evacuated to a federal coordinating center and from there to a local hospital that is under a letter of understanding which states they will accept disaster patients and the government will pay for their care. The NDMS can be activated for two reasons: a presidentially-declared domestic disaster response when federal help is needed, or in time of war when we have a major overseas conflict with many casualties coming back to the United States and there is not enough room for them at military or Veterans Administration hospitals. In such a case, we would need the overflow to go to a civilian hospital. For a domestic disaster, the system would be activated by the assistant secretary for health at HHS; during wartime, it would be activated by the Department of Defense assistant secretary of defense for health affairs."

Returning to the implementation of 27 MMST around the country, Robert Jevec dealt with health systems necessary for response to WMD incidents. "You need to be able to identify the agent, safely extract victims from the hot zone, administer the correct antidotes, and decontaminate the victims. Decontamination must be done before victims are transported in ambulances to a hospital. If that is not done, you will contaminate the ambulances and the hospital, and they will be out of commission. Some of the chemical agents are very persistent, and there could be off-gassing as there was in the Tokyo subway incident where about 10% of the responders were affected by off-gassing from the victims. The teams will also have to provide triage and emergency care, and may need definitive care through the NDMS. The MMST may have problems with the appropriate disposition of the deceased, particularly if they're contaminated. Then, they have to decon the incident site.

"The MMST concept was developed by state and local responders and tested in Washington, D.C. and Atlanta, GA to ensure the response could be immediate, appropriate, and sufficient. It was not developed by us at the federal government level. First responders, particularly at the Washington, D.C. metropolitan level, sat down around a table, and we presented them with a problem. In fact, they presented us with a separate problem. It was initially a letter from the Arlington County Fire Department, which happens to be the smallest county in the United States. However, there are many buildings and people within the county; for example, the Pentagon is in Arlington County. They were concerned about what happens if an incident occurs because they were not prepared. They asked, 'How can the federal government help us?'"

First responders from the District of Columbia area got together for a series of meetings over a year. They developed a system, a locally developed prototype system which was tested here in Washington, D.C. and in Atlanta, GA during the Summer Olympics. "These two areas were where an incident might occur, and the first responders from there knew best their own requirements," adds Jevec. "Also, a project has more credibility when it's locally developed than when it's federally developed and shoved down the throats of local responders. The local responders developed a viable system, and all the federal experts knew they were on the right track."

The Metropolitan Medical Strike Team system developed and tested in Washington, D.C. and Atlanta, GA was issued as a guide to 25 other large cities around the country. The guide is clear, complete, and detailed. "It's strictly a guide," continues Robert Jevec. "They don't have to do it the same way. We are not directing them how to do it, but we're giving them guidance, some funding, and there are contracts involved where we ask them to develop a plan customized for their city. The funding averages $350,000 per city, and there are deliverables that they have to give to us. Once the resultant plan is approved by the federal government, they can present an equipment list because the equipment has to match the plan developed for each city. After the equipment list is approved, the city provides its phamaceutical list to the project officers who sign off on it. There's funding for 27 MMSTs at the present time. Our goal is to have 120 to 150 MMSTs, but only the first 27 have been funded.

TABLE 5.2

The First 27 Metropolitan Medical Strike Teams to be Funded in the United States

Anchorage, AK	Atlanta, GA	Baltimore, MD	Boston, MA
Chicago, IL	Columbus, OH	Dallas, TX	Denver, CO
Detroit, MI	Honolulu, HI	Houston, TX	Indianapolis, IN
Jacksonville, FL	Kansas City, MO	Los Angeles, CA	Memphis, TN
Miami, FL	Milwaukee, WI	New York, NY	Philadelphia, PA
Phoenix, AZ	San Antonio, TX	San Diego, CA	San Francisco, CA
San Jose, CA	Seattle, WA	Washington, D.C.	

"Regarding the Metropolitan Medical Strike Teams, a system approach is used to develop federal/city/state partnerships and build upon the local health and medical emergency system, identify what the local critical needs are, and enhance that local system's capability with plans, equipment, supplies, training, and pharmaceuticals. The local first responders are enhanced because any response needs to be immediate, appropriate, sufficient, and obviously, timely, and only the locals can do that. The MMSTs must also develop a health medical concept for federal augmentation to an incident; in other words, the plan that each city develops has to show how the federal government can interlock and integrate with the city. This would be very important for a large incident which could be overwhelming.

"All 27 MMSTs, except the Washington, D.C. metropolitan area team, are considered local teams," says Bob Jevec. "We helped put them together; but it's their team, their equipment, their cash and they can use the team as they see fit. If they want to use a team for a strictly local incident, they certainly can. The main purpose of a team is to support the local incident commander who may be from the fire department, the emergency medical services, or some other agency. The local incident commander is still the person in charge. The only federal team is the Washington, D.C. MMST. They will not be deployed anywhere else because of the vulnerability of the nation's capital."

Contact: Robert J. Jevec, Medical Emergency Coordinator/Planner, Office of Emergency Preparedness/NDMS, Office of the Assistant Secretary of Health, 5600 Fishers Lane, Room 4-81, Rockville, MD 20857; 301-443-1167, Ext. 42; 301-443-5146 (Fax).

6 Local Haz Mat Response Teams

TAMPA FIRE DEPARTMENT HAZARDOUS INCIDENTS TEAM

Tampa, on the west coast of Florida, is distinctive in that all five modes of transportation run through the Tampa Fire Department's area of coverage: highway, water, air, rail, and pipeline. The hazardous incidents team is located in Station 6 near the harbor which is the ninth largest port in the country. Tampa has over 100 square miles with close to 300,000 people. During a work day, that number swells considerably.

"The formation of the hazardous incident team (HIT) occurred in 1985," begins Captain David G. Costello, the usual operations officer for the HIT, a trainer at the Tampa Fire Training Academy, and a K-9 dog handler and trainer (the Tampa Fire Department has 4 urban search and rescue dogs who, in the event of man-made or natural disasters, are used to locate trapped or lost persons or cadavers. 'Alex,' a female Weimaraner, lives at Station 6 and goes on all alarms in an air conditioned portion of the vehicle where temperature-sensitive items are stored).

"Some controversy exists as to whether the team's formation was a response to a specific incident. The incident that probably triggered a closer look at Haz Mats in general was what started out as a medical call on a ship. We had a couple of workers that were down in the hold of the tanker vessel and one of our own people was injured severely. That firefighter suffered major neurological damage because he fell and was exposed to a high sulfur content, Argentine crude oil. The exact nature of the material wasn't determined for quite some time after the incident. The incident occurred in 1983, the formation of the team began in 1985, and I believe it was in February of 1986 when the team actually went into service for the first time.

"The team is made up of Station 6 which is composed of Engine 6 and HIT 6. There is a rescue squad that is part of our team, Rescue 31, which runs out of the downtown station. We are backed-up by about 20 people per-shift who are stationed on various units throughout the city. Perhaps our biggest backup is Aerial 1, which also runs from the downtown station. Aerial 1 performs regular aerial company functions, but also acts as a technical rescue company. Most of the people assigned to this unit have undergone Haz Mat training and have their Haz Mat physical exams as well. All in all, we have somewhere in the neighborhood of 120 to 130 people who are trained Haz Mat technicians. We have at least 20 people on duty at all times who have been through Haz Mat training and had the necessary physical exams that allow them to respond.

"There are several reasons for having that many Haz Mat technicians available to us," says Costello. "We originally trained approximately 25 people in a course that was 240 hours long. There was a trip to the marine firefighter school in New Orleans, LA, and we did a lot of training in toxicology with a Dr. Vance in Arizona. That experience established a basis for future growth of the team. Some of the people in that original group became instructors, and worked closely with the U.S. Coast Guard strike teams. Because of promotions within the natural personnel cycle, we had to provide for the continuation of a Haz Mat team. It was our decision to train more people with a focus on the entry level firefighter. We now run a 160-hour Haz Mat technician course about once a year. We feel 160 hours is the absolute minimum required, just a foundation to start with that is followed up with more training as firefighters are assigned to the team.

"The Tampa Fire Department has trained Haz Mat teams for a number of agencies, for the military, for a number of large industrial organizations, as well as for other fire department Haz Mat teams. The Tampa Fire Training Academy is actually administered through the Hillsborough County School Board as an adult vocational education center. That gives us a mechanism to pay instructors for basically any topic related to emergency response. We train EMTs, paramedics, and firefighters. Haz Mat is a big field because the training is mandated by law."

"There are refresher training sessions in Haz Mat about every 60 to 90 days. These are formal sessions where we actually go to the fire academy classroom or take part in a drill that has been set up by the training division. In addition, we have daily training on fire suppression techniques and emergency medical procedures. We are responsible for doing special operations drills where we bring in other companies once a month. Once a quarter, we participate in what is called 'a multi-company exercise' that may involve 6 to 7 companies who conduct a large exercise.

"The HIT also visits facilities that are required to report Haz Mats to the fire department. We generally do 20 of these visits for each of the three shifts each year. We confirm that materials stored or handled there are documented on their report form, and we check the accuracy of the reports relative to the quantities claimed and the storage locations."

Tampa's port sees a steady stream of gasoline, fuel oil, and chemical tank trucks. "We have a major Haz Mat presence in the area, not because anybody is doing anything wrong, but because there are so many hazardous materials coming and going," continues Captain Costello. "For that reason, we feel that part of the solution is training people who deal with Haz Mat: the truck drivers, the people who work in the terminals, the Department of Transportation law enforcement personnel, everybody who's part of the Haz Mat response system. We feel the more training we can offer to these people, the safer response we are going to have when there is a problem. It's a community education effort focused on those people who are key players in the Haz Mat business."

Hazardous materials alarms come in a number of ways. "Sometimes, and this is pretty rare, we might actually get a call that says 'this is a chemical emergency,'" says Costello. "Usually, however, the dispatch center will receive a call through a 911 line, and the operator has key questions to ask. If the dispatch center can determine through the information received from the call that there is in fact a chemical emergency, then we automatically initiate a Haz Mat response which includes three engines, two aerials, two rescue units, a chief, a rescue supervisor, the ventilation vehicle, the Haz Mat engine, and the team's support vehicle. Based on what we find on-scene, we may downscale or upgrade the response.

"Our station is not a large station, but we have other ancillary vehicles that we call on to respond to Haz Mat incidents with us. We have a tractor trailer loaded with foam and a high expansion foam unit, both at Station 4. There are also two smaller units that each carry 500 pounds of dry chemical (potassium bicarbonate). In addition, we have a rather unique apparatus called the 'ventilation vehicle.' It's a huge, truck-mounted fan that moves about 80,000 cubic-feet-per-minute at idle. A lot of times when we have a gas air spill, our solution is to move a lot of air.

"One of our unique problems in the port area is that hazardous materials are located right next to our population centers. Our downtown business district, a heavily populated area, at least during the day time, is probably within a quarter mile to a half mile of most of our major hazardous materials concentrations. We have the second largest hospital in the State of Florida within this area. The Port of Tampa is the largest port in the State of Florida, larger than all the other ports in the state combined. We have everything from passenger ships to tanker vessels to roll-on/roll-off equipment that use the port. Being located in the port allows us to be alert to what is coming in and going out. The Tampa Bay area has about half of all hazardous materials incidents in the State of Florida.

"General cargo is probably our biggest growth industry here in the port, and general cargo is what gets Haz Mat teams into trouble. It is 'the great unknown.' If you break open a 55-gallon drum of an unknown material in a cargo container five or six layers down inside a general cargo vessel, then you may have a serious problem. Maybe you haven't dealt with that chemical before. We are really cautious with general cargo for we may not know what it is or may not recognize the signs and symptoms of exposure. We do extensive research, because even with vessels that have good documentation, it is sometimes difficult to locate documentation on a specific container. At that point, we back up, move very slowly, and do some very specific research prior to starting any mitigation efforts."

On a Haz Mat scene, the first unit to arrive is required by law to take command. "They will do a size-up and determine whether or not a full Haz Mat response is needed," says David Costello. "If required, the hazardous incident team will respond. By the time we get on the scene, there is usually a chief officer present. Haz Mat incidents require command by someone at the district chief rank or higher. Incidentally, all of our district chiefs are trained hazardous materials technicians, which is rather unique in a fire department of this size. The chief will take command of the incident, the HIT will come to the scene, and usually I am the operations officer for the incident and actually handle the mitigation phase. The first-in unit will have identified the product, made appropriate notifications, isolated the scene, and provided for the protection of themselves and the public.

"When you think of strategic goals like spill/leak/fire control, that is where the HIT operates. We are usually the operations sector at a major Haz Mat incident. Based on what we find, we may start calling in the other hazardous materials technicians who are located around the city. I am aware of only one incident in which we had to call back off-duty technician personnel. Usually, we have enough people on duty to handle anything that we run into. Also, we have a close working relationship with the Hillsborough County team. They, of course, were one of the pioneers in Haz Mat response and have a number of really knowledgeable people.

"One thing that works in our favor here in the City of Tampa is that most of the major hazards we deal with are products very familiar to us. The big three for us are chlorine, ammonia, and petroleum products. Some may be critical of our rapid approach to these three products, but when we get such a product, we do not wait until all the research is done before making a move. We go ahead. We know which suits we will need, what the symptoms of exposure are and how to treat them, and which mitigation techniques are acceptable for that type of incident. Through experience and study of these hazards, we know they are prominent within our jurisdiction. We are not going to sit and look up ammonia in our computer program and in three reference books before we do anything else. We already know what we are going to do about ammonia, we have done it enough times, and we already know the pertinent information about the material. Research in such an incident becomes a formality.

"We use the computer on the response vehicle to research chemicals as the need arises, and we also have a couple of computers here at the station. Another valuable resource is the poison control center. They have the 'Micromedics' computer program (TOMES Plus) which is very comprehensive. We have the 'CAMEO' program that is pretty much a standard within the industry (Computer Aided Management of Emergency Operations: a computer data base for storage and retrieval of pre-planning data for on scene use at hazardous materials incidents).

"We don't have all the bells and whistles you'll find on some other Haz Mat teams," reflects Captain Costello. "I've visited a number of teams, and a lot of them have a lot of really exotic toys. We don't. We are basically a petroleum, chlorine, and ammonia port. We know what we have, and we gear our equipment to these specific hazards. We've had a tight budget in this city for many years, and it looks like that is going to continue. So, we focus on what we need rather than what industry says we need. There are a lot of expensive toys out there in the Haz Mat world these days, and we do have access to them if we need them. We call a contractor and get them brought to the scene. We do not stock a lot of materials that we know we can get quickly from private vendors because it is just not cost effective to do so.

"The one thing I would say we have that a lot of departments don't is massive foam capability. Our foam truck carries somewhere in the neighborhood of 4,000 gallons of foam. We have at least that much available at our supply division. An Air Force base here has a large supply of foam, and there is also a large amount of foam available at Tampa International Airport. Our major use of foam is to extinguish petroleum fires, but it is also used for vapor suppression to prevent fires in the event of a spill or leak. We stock only the typical Class B fire suppression foams. We do not use Haz Mat foams."

With regard to medical monitoring and surveillance, all active Tampa Haz Mat technicians receive a physical examination at least every two years. If there is any documented exposure that occurs in the interim, they receive another physical immediately. If an individual exhibits symptoms related to an exposure, there is an additional physical. Before anyone enters a hot zone at an incident, Tampa Fire

Department paramedics do pre-entry and post-entry physicals. All the paramedics are cross-trained, dual-role paramedics; they are also firefighters and probably half of the Haz Mat technicians in the department are paramedics as well.

"We use Level B disposable chemical protective suits made from Tyvec and Saranex, and our Level A encapsulating gear is also disposable after a one time use," according to Captain Costello. "The reason we have restricted our inventory to disposable gear is cost plus a knowledge of the particular hazards in this city. We know that disposable suits provide adequate protection for our purposes. Also, because of our cost recovery ordinance, we are able to recover the cost of these suits after they are used in an incident. With reusable suits, it's very difficult to recover the cost of maintenance. In the City of Tampa, we bill the responsible party for the services of the HIT. Included are expenses for the incident commander, the safety officer, and EMS services used on the scene, as well as for any expendable goods (duct tape, suits, absorbents, plugging and patching materials, etc.) and damage to durable goods. If we have a meter that is damaged on scene, we'll bill for the cost of repairing the meter unless the damage is a result of our own negligence. We also bill for damaged hose lines or damaged turnout gear that belongs to firefighters. Basically, any cost impact that our department has suffered as a result of that Haz Mat incident, we recover. All these items are covered by the city ordinance. Whoever is responsible for the spill, or whoever owns the product that we worked with, is considered the responsible party.

"I don't know the exact figures on collection of our cost recovery monies, but I do know it is very high because of the way the ordinance is written. In fact, when a responsible party does not pay, we usually hear about it because it may hinder replacement of some of our supplies. Basically, I write a cost recovery report that details all the personnel who operated on scene and how long they were there, all expendable goods that were used, and any damage to durable goods that occurred. I submit the report to the fire marshal's office and they handle the billing. The invoice to the spiller is very detailed. In our process of documenting the incident report, we also detail the material that was used so there is a check and balance there.

"Prior to our team making entry, I always hold a briefing in which I cover certain things. Number one is what the situation is 'inside.' We try to make everyone aware of what the problem is before going in. Number two is the type of protective equipment the entry team is going to need. Also, we consider more than one emergency signal in case something goes wrong. We always try to keep our people in line-of-sight, but this is not always possible. Probably the most important part of the briefing is stating exactly what we want to accomplish by the entry, nothing extra. In other words, 'This is your job, the entry. If you see something else you think would be helpful, I am certainly open to suggestions, but for the most part the objective is the entry.' Prior to going in, the objective may be nothing more than to find out what you are dealing with. As far as debriefing goes, we like to draw a site map as soon as we come out. 'What did they find? If they did anything inside, what were the effects of their actions?'

"We do a post-incident analysis, or critique, on all multiple alarms in the city. The captain of the HIT is responsible for writing an operational summary which addresses the following eight strategic goals: notification, isolation, protection, spill control, leak control, fire control, recovery, and termination. Each goal is addressed. We ask ourselves, 'How did we address it? Did we run into problems with a specific goal? What will keep that problem from being repeated?' The report includes any other information that may be important to improving operations down the road. The special operations division has a chief officer in-charge that the operating summary is directed to. If there is anything pressing that needs to be changed from the policy standpoint, it is submitted to staff for review. Internally, we go ahead and make the changes that are necessary."

The unique feature of the Tampa Fire Department's HIT is the amount of training that is done. "We are a training-obsessed organization," stresses Captain David Costello. "Our fire academy is set up as a vocational educational center so the cost of instructors has no impact on our department budget. We maximize the state and federal monies that are available for Haz Mat training. All our training is generally done on shift. Our fire academy is located only about six blocks from here (Station 6). When we go for training there, we stay in service. We only run about two calls a day here including Haz Mat, fire suppression, and EMS first response. During a year we get about 280 hazardous materials alarms. We run only 20 to 30% medical calls; the rest of our alarms are fire suppression and hazardous materials."

Contact: Captain David G. Costello, Hazardous Incidents Team, Tampa Fire Training Academy, 116 South 34th Street, Tampa, FL 33605; 813-242-5410 (Office); 813-242-5384 (Station).

LOS ANGELES FIRE DEPARTMENT HAZARDOUS MATERIALS SQUADS

The City of Los Angeles is home to 3,485,557 persons spread over 465 square miles of area, an average of 7495 persons per square mile. The Los Angeles Fire Department has three hazardous materials task forces and hazardous materials squads which are suppression companies generally responsible for tactical operations at a hazardous materials incident. These tactical operations include: entry into a hazardous environment, identification of the substances, determination of the hazard, containment of the hazard, and decontamination procedures.

The first alarm response for a confirmed hazardous materials incident will include one Haz Mat task force with one Haz Mat squad, one battalion chief, one rescue ambulance, and one senior paramedic. Under the hazardous material incident command system, a hazardous materials group would include the following units: rescue unit, perimeter control and access, safe refuge unit, and a site control unit. The site control unit will include an entry team, and is responsible for decontamination and technical information.

Incident command system (ICS) training lays out specific basic duties for members of the hazardous materials group. As an example, the technical information leader supervises the hazardous chemical library and data base. This position maintains proper records and documentation, including accurate time for the entry team. He or she assists in the selection of specialized chemical entry suits, proper detection and control equipment, and decontamination procedures and solutions.

Complex decontamination procedures are a fact of life for hazardous materials response teams. Los Angeles City Fire Department training stresses that measures should be taken to prevent contamination of sampling and monitoring equipment to the greatest extent possible. Once contaminated, instruments are difficult to clean without damaging them. Any delicate instruments that cannot be decontaminated easily should be protected during use. They should be bagged with the bag taped and secured around the instrument. Openings are made in the bag for the sample intake. Wooden tools are difficult to decontaminate because they absorb chemicals. Once used, such tools should be discarded.

Breathing apparatus, respirators, reusable protective clothing, and other personal items must be decontaminated and sanitized before reuse. With breathing apparatus, certain parts, such as the harness assembly, are difficult to decontaminate and might need to be discarded. Rubber components can be soaked in soap and water and scrubbed with a brush. Regulators must be maintained according to manufacturer's recommendations.

Heavy equipment, such as bulldozers and fire trucks, are also difficult to decontaminate. They should be washed with water under high pressure and all accessible parts scrubbed. Particular care must be given to those components in direct contact with contaminants, such as tires and scoops. Swipe tests should be used to measure effectiveness.

In some instances, clothing and equipment will become contaminated with substances that cannot be removed by normal decontamination methods. A solvent may be used to remove such contamination from equipment if that solvent does not destroy or degrade the protective material. If persistent contamination is present, disposable garments should be used.

All materials and equipment used for decontamination must be disposed of properly. Clothing, tools, buckets, and all other contaminated equipment must be secured in drums or in plastic bags and labeled. Clothing not completely decontaminated on site should be secured in plastic bags before being removed from the site. Contaminated wash and rinse solutions should be contained by using step-in containers, such as plastic inflatable pools, to hold rinse water and decontamination solutions. The spent solutions are then removed from the site by a commercial clean-up company.

Mike Balzano is an apparatus operator who drives the hazardous materials squad and the aerial ladder at Fire Station 4 which is located in the heart of the city near Union Station. "I've been assigned to this station for 15 years, approximately 13 of which I have dealt with hazardous materials response. Before that, the Los Angeles Fire Department had a couple of inspectors who would respond to Haz Mat incidents when engine companies called them. Basically, all they had was a couple of Haz Mat entry suits and some litmus paper — that was about it. However, they had good chemical knowledge. One of the men is still there, and he will come out and assist us.

"Some years ago, the department realized that the amount of chemicals in commerce was increasing and that the expected increase in Haz Mat incidents would require more than just one van and a couple of inspectors. They decided to convert some stations to Haz Mat squads. Stations 4, 27, and 39 were converted to a dual function. The hazardous materials response consists of one captain, one apparatus operator, and two firefighters. This squad also responds to all major structural fires in its district, anything other than a single engine response. On our squads we have a dual function, both hazardous materials response and fire response.

"Our response vehicle has firefighting equipment which consists of breathers, chain saws, SCBA, axes, crowbars, etc. There are no pumping capabilities or ladders on the squad, so we are basically a manpower pool at a fire. We just do as directed — pull hose lines, ventilate a roof, or perhaps form a search and rescue team. A squad also handles Haz Mat response. For the first couple of years we were trained in an old basement that was an operations control dispatch center. We got some chemical protective suits and combustible gas indicators. We knew we would need to decontaminate people and used salvage covers and ladders as make-shift decon pools. We didn't have a whole lot of equipment. We even still carried some salvage equipment, so there was a transition and evolution in the way we responded to Haz Mat incidents. As time passed, it became just firefighting or hazardous materials response, and we did no more salvage work.

"Early on, the effort was called 'SCAT' for Strategic Chemical Attack Team, but now the name has been changed to hazardous materials squad. In 1992, Squad 27 was located in the Hollywood area while Squad 39 was in the valley and we were downtown. We had three Haz Mat squads for years, but due to budget cuts over the last couple of years Squad 27 was eliminated. Presently, Squad 4 covers everything in the Hollywood area and downtown, Squad 39 covers the valley, and Squad 48 covers the harbor area. These are the only fully Haz Mat-certified squads in the city." With hazardous materials in the City of Los Angeles, the fire department responds to abate an emergency incident. "We basically do not clean up. We are there strictly for emergency service," according to Mike Balzano. "We abate emergencies, rendering a threat static and nonhazardous, then we turn the situation over L.A. County Health, the Los Angeles Police Department, Fish and Game, etc. For example, the police department has a Haz Mat unit to enforce codes against illegal dumping. They have the power to cite people and know how to get to the people who can pay for the spill."

In addition to response by the Haz Mat squads, a hazardous materials unit downtown does inspections. They will enforce the municipal code, record all the chemicals a company has in its facility, and put the information into their computer. The squads can access that information at their own stations by calling the OCD (operations control dispatch) who can provide a building plan and get a list of chemicals for any business that handles chemicals.

Mike Balzano explains the different circumstances to which a Haz Mat squad might respond. "The first-in engine company will often decide if a squad should be called. They might say that they can handle it. Or, it might be a totally static situation with no reason to call a Haz Mat squad. If there is any doubt, the engine company will call dispatch. They may say they are not sure if they have a Haz Mat, but something is leaking a bit. Sometimes an engine captain will call the squad station and ask a question. Any time they call us on the phone, we automatically send the squad. The squad will call dispatch and report that they are going to investigate, and go to the engine company on scene. We will determine if we need further assistance. Ninety-five percent of the time, we can handle it with just the squad.

"We might call the county health department since they are the final authority in L.A. County. They can cite people, they can say, 'You are closed, you are open.' They're like gods. They can also get a commercial cleanup crew out to the scene and charge the owner of the facility for the cleanup cost. The city will first try to get the person responsible for the situation to pay for the cleanup, and the county health department is very good at this job. If the spiller can't pay, there may be county or federal funds available. The last resort is the city having to pay for cleaning up the spill."

Necessary training for Haz Mat qualification is done in-house and through a 32-hour, state-certified course given every two years. Fire department instructors who are certified by the state teach the course. On-the-job, the graduates of the course learn to use the tools necessary to get the job done.

"We cannot send paramedics to a Haz Mat scene unless they are also qualified firefighters," notes Balzano. "At one time in the city we had only civilian paramedics, but we are converting to only firefighter paramedics now. A lot of paramedics are still on rescue squads, but they don't want anything to do with firefighting. So far, there are three engine companies with paramedics who are also firefighters. They are there mainly to check our vital signs before we go into protective suits. If a firefighter goes down at a Haz Mat incident, we will pull him out ourselves, but we want paramedics there because the area might be a hot zone.

"If we get to a scene and find the situation a little too much to handle, we will call in another squad. If we foresee a lot of decontamination will be necessary, we might call in the whole taskforce. There is a decontamination vehicle located at Station 17, basically a big trailer with numerous shower stalls inside. If we have a number of civilians who need decon, or some first responder firefighters, we would use that unit. However, we don't normally use it to decon firefighters on our Haz Mat squads.

"The reason to have a hazardous materials taskforce respond with the squad is to get more manpower, most frequently for decontamination. We have only 'x' number of firefighters on the squad. Two will suit up as the first entry team. From the taskforce we will take two firefighters and suit them in the same and equal protection as the first entry team. They are to relieve the first team or go in to make a rescue if necessary. Other firefighters, perhaps three or four, will go to decon operations to set up the pools and get out the decon solutions. They will be suited up in Tyvec suits unless we know they are really into some nasty product."

When the first person of the entry team comes to the inside edge of the hot zone, he or she will be met by a decon person in the equal protection who will remove gross contamination. The entry person will then be allowed to step over the hot line and enter the contamination control area. The entry person goes through six phases within the decontamination zone: removal of outer gloves, air hook-up, decontamination wash with neutralization/decontamination solution, a wash and rinse cycle, a final wash and rinse cycle, and removal of the entry suit and a change into clean clothing. The entry person can then step across the contamination control line into the support area.

"In the squad vehicle there is a reference library and a computer with the CAMEO program in it," continues Mike Balzano. "In the reference library we find things we don't have on the computer: chemical dictionaries, material safety data sheets, railroad material, maps of all the diesel and gas pipelines, sewer pipelines, and storm drain maps. If a chemical gets into the storm drains or sewers, we usually have to notify the Fish and Game Department and the harbor authorities because it will eventually end up in the harbor.

"Many times enroute to a scene we will use the library rather than the computer to learn about a particular chemical because it is not good to use the computer while traveling. Normally, we won't run the computer until we get on scene. Our computer printer will produce a printout for the chief officer that covers eight basic areas of interest: general description, fire hazards, fire fighting, protective clothing, health hazards, non-fire response, first aid, and properties.

"All businesses that use chemicals pay a fee to the city and must list the chemicals and quantities they have at their facility. For a specific address, we can access the OCD and find what chemicals they show in their business plan. However, in many incidents there are unknown chemicals involved. We use the HAZCAT (Hazardous Materials Categorization Test) system to determine the hazardous characteristics of an unknown material. Not everyone is allowed to use the big HAZCAT kit. Several firefighters on each shift are qualified for that, but most of us can handle a shorter version of the system, a five-step procedure for determining if an unknown chemical is poisonous, flammable, etc."

Everyone in the department takes a physical examination every two years, but anyone who is assigned to a Haz Mat squad must take a Haz Mat physical every year. "We get a physical every year when everyone else gets one every two years," says Mike Balzano. "Our physicals involve blood work and extensive record-keeping. For any person who is exposed to toxic materials, the captain will file this information in the personnel records and send it to a medical liaison. All of our Haz Mat exams are done by physicians who work for the city. If these doctors determine that a disease or sickness was related to hazardous materials, the city would consider the response person as 'injured on duty' and pay an outside doctor to provide treatment.

"When the Haz Mat task force and squad respond to an incident scene, a battalion commander is present, and all individuals other than the captain will start setting up the decon pools. If we have an idea of the suits to be used, the entry team will begin to suit up. The captains and the chief will get together and form a plan of just exactly what we are going to do. They will come back to the task force and explain the plan. We don't rush when dealing with hazardous materials. We fight fire aggressively but have a less aggressive approach to Haz Mat for our own health and safety.

"The engine captain and the engineer are on the technical side and come to the squad to operate our computer. All the other firefighters are in suits or doing other functions. The squad commander, a captain, is the entry team leader but does not enter the hot zone. This commander finds the best vantage point from which to observe the team. Firefighters from the squad suit up as the first entry team. The apparatus operator on the squad is responsible for getting the entry team into the proper suits, providing all the test equipment the entry team will require, and standing by to change suits or fix breathers.

"The apparatus operator on the truck company is in charge of decon and will run that show. A firefighter will be taken from the truck company along with one or two from the engine company to suit up as a backup entry team. The rest of the firefighters are assigned to decontamination. The task force commander is basically in charge of all operations except those held by the chief officer who is the incident commander. The chiefs pretty much let the task force commander call the shots because they know that these commanders are more qualified regarding hazardous materials. There are three remaining individuals, two engineers and an engine captain. One of the engineers will set up his rig to provide water for decon operations, and the other engineer and captain, as mentioned before, will go straight to the squad and provide technical support. They will run the computer, utilize the on-board library, and keep track of the time the entry team has been on breathers and how long they have been in the hot zone. Before the entry team goes into any type of incident, the paramedics will get a baseline on them and do the same when they exit the hot zone."

The three task forces handle any Haz Mat incident that happens within the City of Los Angeles, including transportation incidents. "We handle incidents on the freeways," commented Balzano. "The California Highway Patrol is the official authority in charge of the freeways, but they always let us take on the problem. Our squads are equipped with drilling equipment to drill gasoline tankers that are turned over. In the city, we don't try to right a gasoline tanker that has a full load because of the structural integrity problem. The tank is made of aluminum, and the chances of its bursting open are greater than if you just left it alone and pumped off the cargo. If we can get to one of the tanker valves and pump it off through the valve, we will do it that way, but oftentimes it can't be done that way so we will drill the tank and put a stinger into it. We'll call a commercial response contractor to actually pump-off the cargo.

"We don't do cleanup. We have absorbent and other materials to dike spills and contain them, but we have no place to keep the contamination. The county health agency will have to call the company and get them to clean up the mess.

"Everybody on the squad has to be hazardous materials certified. There are people in the task force who are not certified who may be working a trade day or an overtime day, but they are never on the squad. Riding on the squad and being Haz Mat certified gets the firefighter bonus pay of about 75 cents an hour."

Contact: Michael Balzano, Los Angeles City Fire Department, Station 4, 800 North Main Street, Los Angeles, CA 90012; 213-485-6204.

CITY OF SACRAMENTO FIRE DEPARTMENT HAZARDOUS MATERIALS RESPONSE TEAMS

Division Chief Jan Dunbar is in charge of the hazardous materials division for the fire department in Sacramento, CA. "I joined the fire department in 1965, became an instructor about 1973, and obtained my California teaching credential," remembers Chief Dunbar. "I attended a college class in fire technology and hazardous materials at a community college, and was terribly disappointed in the quality of instruction. I got frustrated and inquired a little more and was given a challenge by Bob Schultz, an excellent deputy chief I worked for at the time. Basically, he said, 'What are you going to do about it?' I thought I could teach it.

"Then Haz Mat exploded. The whole spectrum from 1975 changed, and I began teaching at the local community college and writing lesson plans on Haz Mat for the development of our own hazardous materials response team that was formed in 1979. Ed Bent was the Director of Fire Training in the office of the state fire marshal. He asked me to do a lot of lesson plan writing under contract for the state fire training program and some of the first hazardous materials programs. About 1979, Ed Bent got an environmental protection agency grant to develop a class on pesticide fire safety. We developed a one-day, 8-hour, class. He demonstrated it to someone at the National Fire Academy, they liked it, and doctored it up a little. It became one of the early Haz Mat classes at the fire academy, and is still offered to this day."

Today, Chief Dunbar is heavily involved with the development of the Sacramento City Fire Department HMRT. "Our chief wanted a HMRT even though it would be terribly expensive. We were watching the development of Haz Mat teams on the east coast since that's where the concept came from. I started networking out there, but I couldn't get much information initially. It was like trying to invent a fire protection program from scratch. Where do you go for information? Over the years I met Jerry Grey of the San Francisco Fire Department, John Maleta of the Los Angeles County Fire Department, John Eversol of the Chicago Fire Department, and a few others. Here in Sacramento, we came on-line with our first hazardous materials response team in 1981. A Haz Mat team in Sacramento is really a truck company with an additional piece of equipment, very similar to other Haz Mat units. Our second HMRT came on-line in 1982, and our third in 1987. We have a fourth unit that does decon at major incidents.

"Currently, we have four HMRTs spaced in different strategic areas of the city. The reason we have so many is multifold. First of all, we knew we would need at least two for the City of Sacramento because of backup, the size of the Haz Mat problem, and all the staffing positions that are required in a full incident command system as well as the training required for all these positions. In Sacramento County there are also 18 separate fire districts, none of whom, nor all collectively, could duplicate a Haz Mat program. They came to Sacramento City and asked if we would contract with them for coverage. The first contract was signed in 1983, and since then we have provided hazardous materials coverage in its entirety to both the city and county."

Sacramento County has 998 square miles and a population of 1.2 million. It's largely an agricultural environment with pesticide manufacture, storage, and movement. "Every year for the last 15 years we have had at least one anhydrous ammonia tank truck rollover, and that tends to keep us on our toes," continues Chief Dunbar. "We don't have much in the way of flammable liquids but we do have other industries. We have a number of plants that handle carcinogens, and a couple of them are the largest in the nation. We have Aerojet, a basic part of the rocket industry, which has contracts for building the Titan rocket. Their appetite for rocket fuels is still very high. Nitrogen tetraoxide is both trucked and railed through Sacramento.

There are only 12 approved and certified nitrogen tetraoxide rail tank cars in the nation, and at any given time you will find a few at Aerojet going back and forth. The Aerojet facility is almost 20,000 acres, it's massive and spread out. They have ended the production of solid rocket-fueled missiles but that doesn't mean there aren't chemicals for that type of missile being there.

"There have been a number of fires. When rocket fuels catch fire, particularly solid fueled rockets, there is nothing you can do. You don't want to intervene. Just let it go and it will consume itself and obliterate everything in its path. The safety record of Aerojet is above the standard. They are very safety conscious. We train with them. That training is done entirely at the Aerojet facility where classroom settings and outdoor hands-on training equipment is readily available. We have our own propane tank car, decommissioned of course, but we have moved it about and have used it at conferences. The Aerojet engineers have re-engineered the dome cover to 'leak' compressed air, water, and/or CO_2 in a number of different ways. We have done this also with a number of other static, mothball-type devices that are perfectly safe to use for simulated exercises and provide a tremendous amount of realism. We manufactured a number of portable 'trees' on metal pallets that we can take all over our region, and even use them in the hotel where we have our annual conference that we sponsor in Sacramento.

"Our turnover rate every year runs about 15 to 20 people. You want your new firefighters to have aspirations and get promoted. That's good. But, it creates a constant turnover in the hazardous materials program. A number of the Haz Mat personnel are also members of the Sacramento Fire Department USAR (urban search and rescue) team. About five years ago, FEMA (Federal Emergency Management Agency) selected 25 locations for USAR teams around the country. Sacramento's USAR team was the second team activated to go to Oklahoma City after the federal building explosion, and arrived there about two hours behind the Phoenix USAR team.

"When our USAR team is alerted, they report to a prearranged military air base within a specified period of time. They then are air-lifted to the scene of the incident. Each member has a big duffel bag to carry personal items, a change of street clothing, and some special protective clothing such as Nomex jump suits and hard hats. There are about 40,000 pounds of more sophisticated protective gear all packed away and palletized — cutting saws, torches, hydraulic systems, air chisels, electronic listening devices, and an underground tubular electronic television camera the size of a fountain pen which is fed by fiber optics."

In the Sacramento City Fire Department HMRT, Division Chief Jan Dunbar has one captain working in his office. That captain is known as "Haz Mat One," and together they handle day-to-day administrative and management decisions such as general training, scheduling new training, replacing disposable goods, purchasing new goods, examining and writing specifications, testing equipment, and other typical duties of managing a division.

Since Sacramento is California's state capital, Chief Dunbar goes to a lot of meetings. "Because of our association with and our proximity to the state capitol, we are involved with a lot of state legislation," says Dunbar. "They solicit our opinions on legislation, and sometimes they want my opinion, 'yea or nay.' I spend a great deal of time at the state legislature testifying, bargaining, agreeing, or disagreeing. I also keep in touch with various state agencies such as the fire marshal's office, California EPA or California OSHA, all of which introduce bills to add this, modify that, or solicit an opinion. Based on that, we share information with the other fire associations such as CalChiefs and MetroChiefs. We keep everyone notified, and I am sometimes asked to read a bill and pass it on as a recommendation. I go to a lot of meetings."

Chief Dunbar is also a member of the LEPC (local emergency planning council) All the LEPCs send representatives to the CEPRC (California Emergency Planning and Response Commission). Through the CEPRC there are state subcommittees that have been formed, including the CEPRC equipment and training subcommittee which has been influential in promoting standardization of equipment and training.

"Another thing this subcommittee did was to take a look at hazardous materials response teams in California to ask if there was a way to categorize such teams with regard to their level of training, staffing, and equipment. We came up with a two tier system that took about two years to study and complete. The California Office of Emergency Services had been trying to answer the question of how to move Haz Mat teams up and down the state on a mutual aid basis, as when we fight large grass fires and bring pumpers in during our fire fighting season.

"To give an idea, four years ago there was the terrible Malibu Fire that I think everyone watched on television. From the time that fire started, about 8:00 or 9:00 at night, the State Office of Emergency Services activated the mutual aid response system and probably 250 pieces of fire apparatus and 600 to 700 firefighters from all over the state converged on Malibu. They came from virtually every community. It's a very efficient system, it's very fast.

"However, the moving of hazardous materials response teams was never considered until the Northridge earthquake. After the fact, we realized there was a need to bring in qualified Haz Mat teams, even though the Los Angeles basin already has an awful lot of Haz Mat teams. There had been no mechanism to specifically dispatch such units. It was necessary to 'plug' hazardous materials response into the existing system. Phase 2 of the effort was the rewriting of the State Office of Emergency Services mutual aid plan to include Haz Mat units. By the end of 1996, we had a system in place that would enable the movement of HMRTs in case of another major disaster requiring that type of expertise."

Jan Dunbar was asked what type of alarms the Sacramento team receives. "Perhaps more unique to the west coast than to the rest of the nation were the illegal drug labs of about 10 to 12 years ago when some 25% of our Haz Mat calls were drug-lab related. Nowadays, it is not that drug lab situations have gone down, they have actually increased, but our incidents have gone down because more and more the law enforcement agencies will intervene or have their own breakdown teams come in. This change took a lot of training and required a great deal of grant money. U.S. Department of Justice, California DEA, Federal DEA, and other law enforcement officers attended a two-week, intensive, drug lab training session held at the Sacramento Fire Department tower. We probably trained close to 2000 people both from the state and from the rest of the nation.

"They were able to equip themselves better, handle the intervention, and tear down major drug labs. Since then, we in the fire service have backed out of first level response to drug labs. That was the correct decision. But, when a first responder unit finds some unusual aspects to a drug lab, they will activate a system and call in the Haz Mat team. Illegal drug labs are still a very serious problem, and they are increasing, but the fire service does not respond to them as often."

In Sacramento there are three levels of hazardous materials response, a process that has been used since the team went on-line in 1981. A Level 1 is any incident that is completely within the capability and training of a first responder, taking into consideration what the first responder unit may have for equipment. Level 1 will include most vehicular accidents, leaks of innocuous chemicals, small low-pressure gas main ruptures, and fires including all of the above.

A Level 2 incident is one that a Level 1 response can't handle. If first responders arrive and determine that they have a lack of training, tools, or capability, the first responders can upgrade the incident to a Level 2. The Haz Mat team will respond with a team of four, and they will assume all on-scene responsibility for the handling of the hazardous materials. "They do not assume any incident command responsibility; that stays with the first responders," says Chief Dunbar. "That works a lot better. We tried in 1982 to allow the captain of the HMRT to take over the incident command, but it diluted the system. It took the captain away from the team. We keep the team integral and within the incident command system.

"A Level 3 alarm brings two more Haz Mat teams to the scene — a Haz Mat team, and a decon team. This is for a major incident that has 12 trained firefighters and a battalion chief present. That is pretty good depth and we still have one more HMRT in the firehouse capable of being called to that incident or handling other incidents simultaneously. We have had to handle three incidents at the same time, and things can get pretty thin.

"In mutual aid, we can go outside of Sacramento County which we have done a few times. The team can go up to 50 miles without asking for permission. If somebody needs us out in the 'boondocks,' or in some other sparsely populated area, we can respond, often automatically."

The Sacramento Fire Department bills the spiller of hazardous materials for their clean up expenses. "Early on we looked into cost recovery," relates Dunbar. "The person who caused the spill, or is most closely related to having caused the spill, is the person or agency that we concentrate on with our cost recovery efforts. If there is no known spiller, such as when somebody clandestinely dumps in a creek or in somebody's back yard, then you consider property responsibility. The creek can be the property of some canal district, irrigation district, or reclamation district. Unfortunately, that is who we're going to hold responsible. It might be on city, state, or private property.

"Probably 15 to 20% of our responsible parties are unknown, and that anonymity is a result of an awful lot of abandoned waste such as motor oil. There's a growing problem with the increase in 55-gallon drums abandoned at the end of a cul-de-sac during the night and nobody knowing how the drums got there. But this low percentage has allowed the Haz Mat division to recover a fairly high percentage of our costs. Starting from nothing five years ago, today we are up to around 78% cost recovery, not counting the unknowns. Why aren't we 100% effective? Some people just refuse to pay, some go to jail, some people we cannot find, and some are out-of-state land owners who are difficult to collect from. A case might be tied up in court for perhaps 2 to 5 years before we can get out money back, and this delay must be included in our figuring."

What does the Sacramento Fire Department HMRT bill for? "We bill a flat rate which is based upon the time of four personnel (wages and fringe benefits) plus a specified rate for each piece of equipment," continues Jan Dunbar. "If it's a ladder truck, it is so much. Different rates may be applied for an engine, a Haz Mat response unit, or a battalion chief's car. The program is the result of a city ordinance, and we get very good backing from the city collections department. The fee schedule should be rather reasonable and not be outlandish. I do take offense with other fire departments and Haz Mat teams who charge outlandish cost recovery bills. That should not be tolerated. For our Haz Mat team, with a crew of four plus vehicles, we would bill about $1800 for a 10-hour incident. We are not into response to make a profit, and the $1800 or so is very reasonable and a true reflection of our costs. We bill for actual costs in a tiered system. The total sum we have been able to recover in a five year program is about $200,000.

"We were able to develop an automatic tiered system of calls that go out during Level 2 and Level 3 Haz Mat alarms. We had found on a number of calls that the incident commander or the Haz Mat captain had to recall from memory every agency that needed notification. I think that people will agree that this 'shoot from the hip' type of dependency is going to mean you are going to forget somebody. By accident, we forgot to notify certain agencies or individuals. The computer now controls the calls so that anytime a Level 2 dispatch is activated, automatic notifications go out. The calls go to me, my Haz Mat captain, and the poison control center, as well as a medical doctor who works in conjunction with the poison control center and the Haz Mat team. The doctor is notified of the incident, is briefed on the chemicals that may be involved, and starts to network with the team. He can cell-phone the Haz Mat team directly if he feels it is prudent to do so.

"The call-down also includes Sacramento County Health which has responsibility for a health or environmental threat caused by the hazardous material(s). Only Sacramento County Health can release the scene of an incident. Another agency that may be notified is the California Highway Patrol because the incident may involve chemicals that were illegally transported in a vehicle even though they are now in the middle of a vacant lot. Neither the incident commander or the Haz Mat team leader has to remember to make these calls; it's done automatically.

"If the incident goes to a Level 3 Haz Mat call, the call-down list lengthens. It begins to include state agencies such as the California Office of Emergency Services, California OSHA, and the Sacramento County district attorney. The district attorney is called for several reasons. A limited number of cases may involve a crime requiring an investigation beyond the capabilities of the fire department and the health department on scene. We have learned through experience that the county D.A. can be very beneficial in assisting us with a continued investigation, and they respond on any Level 3 hazardous materials alarm. This is a bonus to the incident commander. The D.A's office will work with all investigative agencies on scene: highway patrol, city police, county sheriff, county health, state fire marshal's office, and any Federal agencies present such as the U.S. Department of Justice or the U.S. Drug Enforcement Administration.

About eight years ago in Sacramento, the Environmental Crimes Task Force came into being. "We meet once a month," explains Jan Dunbar. "We all agreed that something had to be done about the number of incidents that were crimes against the environment and to private property. The task force's success is dependent upon the degree to which each agency wants to back such an effort. You need to have the cooperation of all agencies involved. Otherwise, these environmental crimes will not be successfully investigated. We have an aggressive district attorney who sits as the chairman of the Environmental Crimes Task Force.

"When we meet we review all existing cases, cases that are pending, cases that still require more investigation, and cases where there is still outstanding paperwork. All of this information is shared and discussed among all participating agencies. This meeting puts us on the same wavelength, gives us a progress report, and reviews some of our past successes. The people at these meetings are pretty much the same players who end up responding to a Level 3 incident. At one extreme, there are cases where fines or penalties are levied quite quickly. At the other end of the list, some major cases are now $3^{1}/_{2}$ years old.

"To give a better idea of some of recent incidents that were investigated by the Environmental Crimes Task Force, I might review a couple that were quite convoluted. In September of 1995 there was an unusual incident involving a 4000-gallon, stainless steel tank typically used to transport everything from gasoline to fertilizers. When we got to the scene, the tank had ballooned. It did not disintegrate, but came close. It wreaked havoc on the building, and inside the tank we found the body of a 23-year-old male who had been asked to clean the inside of this confined space. He had received no training, no direction, and no sufficiently protective garments. He had been given tools that quite likely contributed to the ignition of flammable vapors inside the tank. This incident is being pursued quite vigorously by the task force, and the final outcome is still to be seen.

"About 18 months ago an incident occurred at a chemical plant that is a distribution center for large quantities of chlorine. The facility also handles a variety of other chemicals. Chlorine comes by train cars and is recontainerized in 1-ton cylinders and 150-pound upright cylinders. There were numerous 911 calls by neighbors complaining of irritating odors and a red gas that was drifting over the neighborhood. Not a single call came from the chemical plant. When firefighters arrived on scene, they saw a red vapor coming from the top of a warehouse. They were told by the plant people it was chlorine, but chlorine is not red.

"The responding companies called for Haz Mat teams for a Level 3 incident. The two teams at the scene suspected the company's employees lied because chlorine does not cause red gas. They learned an employee was asked to 'neutralize' old chemicals that accumulated over the years. He used various sulfate and nitrate compounds to get the chemicals to a neutral pH (7), then illegally poured the resulting chemicals into the city sewer system. Altering the pH of a chemical to 7 does not mean the chemical is inert. The employee was caught when he added something to a vat that triggered the release of red fumes of nitric acid.

"The teams donned Level A clothing and entered the building. A violent reaction occurred when a colorimetric sampling tube was inserted in a drum on a catwalk. The company never reported that it used concentrated hydrofluoric acid, an extremely dangerous and toxic chemical. Further investigation revealed at least 35 leaking drums of hydrofluoric acid that had been exposed to the elements on a loading dock for about five years. That meant the plant workers had been subject to injuries and illnesses for that period. The involvement of the Environmental Crimes Task Force was immediate because not divulging the truth about possible chemical hazards to a first responder is a criminal offense in California.

"An investigation the following morning found not less than 35 other drums of hydrofluoric acid left, exposed to the elements, on a back loading dock for the last five years. They were all leaking. Hydrofluoric acid is one chemical that you do not want to come in contact with. The working atmosphere at this plant meant workers had been subject to hydrofluoric acid injuries and illnesses for up to five years. This was unacceptable to say the least, and the case went to court almost immediately. A very hefty settlement resulted from punitive damages against the company.

"The guy who was neutralizing old chemicals probably did not know that firefighters would be able to dig as deeply and as quickly as they did. He probably also assumed that they did not know as much about chemistry as he did," explains Chief Dunbar. "We in the fire service are so often confronted with, 'What's a dumb firefighter know about chemicals?' Firefighters know a lot more about chemistry than they used to, and sometimes a whole lot more than the chemistry professor at a local college, particularly as related to safety. It is a criminal offense not to divulge the truth about possible chemical hazards to a first responder in California. You pay the consequences if you do not provide correct information."

Another incident from about three years ago also illustrates the value of the Environmental Crimes Task Force, according to Jan Dunbar. "On a routine fire inspection of a large warehouse, one of our first responder companies walked inside, and something didn't look quite right. Down the middle of the warehouse there were 1.4 million pounds of military 150 millimeter howitzer shells all palletized and on their way to Taiwan. We found that the sale, ownership, and transfer of this ammunition was indeed legal. That was not the problem. The problem was why this commodity was in the warehouse and how it got there.

"The company had asked the Sacramento Fire Department for a permit to store explosives, which is standard practice in concert with the uniform fire code we use in California. The permit was denied and was denied in writing one year later. The first responders very prudently called personnel from the fire prevention bureau. As soon as they arrived, they noted it was a Haz Mat incident. What ensued were the three longest weeks of an incident in our history.

"Not only did we find through records that the facility received the 1.4 million pounds of explosive, but they had actually done it a year earlier without our knowledge. The business operates as a large wholesaler of industrial grade chemicals that are used by custodians. These chemicals are quite concentrated and dangerous.

The facility ships them in and out in 55-gallon palletized units from a huge warehouse where we found leakers up and down the aisles. Obviously, this business was more profit-motivated than safety-motivated. Employees were never taught to stop and rectify a leak, or even bring it to the attention of the owner. We shut down the entire facility, and worked with fire department employees and the California Highway Patrol to remove the explosives, which took two weeks. The second problem was to clean up all those leaking containers.

"We made a contract for cleanup with a commercial response company, and we inspected the scene every day with the Sacramento Fire Department, the fire prevention bureau, the Haz Mat team, and other members of the Environmental Crimes Task Force. We suddenly discovered a third room we didn't know existed. The company had purposely blocked a roll-up door with pallets of materials. We gained entry and walked into a room of horrors: the company had been stockpiling leaking containers for the past five years. The owner had also been accepting hazardous waste from other people in the community, with a promise to them that he was licensed to do so. He was not.

"We probably counted 40,000 to 50,000 hazardous waste containers, and not less than 12,000 were leaking. We documented every single leaking container and videotaped the process. It took a week to do this job. There were Haz Mat teams working around the clock doing chemical determination, sorting, and reclassifying.

"Since there were serious violations of state law, the entire case was first turned over to the Sacramento County District Attorney who referred it to the state district attorney's office. They investigated further, and because the acts were also violations of Federal law, they turned the case over to the Federal District Attorney. The individuals have been arrested and arraigned, and the case is now in the courts. They face no less than 16 counts each of violations with a $250,000 fine possible for each count. It all started with the sharp eye of one company captain."

One of the three Haz Mat teams has an onboard computer; the two other teams will be upgraded with computers eventually. On the computer there are about 20 different hazardous materials software programs in DOS versions. They have the CAMEO program, but there are other stand alone programs that will quickly calculate wind plume considerations, display results, and print data independent of CAMEO. The program can do the same for water-born plumes. The teams also have several well-known manuals such as *Sax*, the *Chemical Dictionary*, and the *Farm Chemical Handbook* on CD-ROM.

Medical surveillance is mandatory for members of the Sacramento Fire Department's Haz Mat teams. "We started eight years ago with medical surveillance," says Jan Dunbar. "When you join the program, the poison control center located at the Sacramento Medical Center starts and maintains a folder on your condition. All personnel report there once each year for a medical evaluation which is treated as confidential information. If there is any exposure, like a breach of a chemical suit, we can notify the poison control center and send the firefighter there at once. Our records are not buried away in somebody's file cabinet. They are readily accessible to the doctors when their access is required. Other than that fact, the medical information is kept under very tight security. The city pays for the medical surveillance effort.

"There is also medical surveillance when people get into and out of chemical protective garments. Everyone has to wear a suit through an obstacle course. Before a firefighter even gets into the suit, we do full baseline testing on the individual. Again, this involves a doctor from the poison control center. We also work closely with coach Al Beta who is at American River College, as well as with a track and field coach for the U.S. Olympic Team. After 20 to 30 rigorous minutes on air in the suit, we monitor heart rate and blood pressure. The information is downloaded into a computer and from a print-out we can show the firefighter exactly what his or her body is doing under stress. We also work with the individual in a physical agility program. This is strictly voluntary with an 80% participation rate. Coach Beta works with them to maximize good physical condition."

Dunbar was asked about information that might be useful to other hazardous materials response teams. "I think there are too many agencies that have developed Haz Mat teams thinking it was the necessary thing to do, or the right thing to do, and have overindulged themselves. For as long as the Sacramento Fire Department has had a Haz Mat team, we have not had a single injury as a result of contact with a chemical. Overall safety should be everybody's goal. Unfortunately, this is not happening. The less and less professional training and coordination provided to Haz Mat response teams, the higher frequency of injuries. Some teams think they are as good as the most active teams in the nation, and that is not true.

"Some Haz Mat response teams have gone beyond the intervention of hazardous materials incidents, and perhaps have included a little of the cleanup function. I don't know any state in the union that makes that a requirement of, or a responsibility for, any fire department. We are providing a service of intervention, of reaching out and contacting other responsible parties and authorities to continue the response beyond the normal scope of what of what the fire department should be involved with. There is a problem in that some fire departments and some Haz Mat teams evidently have not made that distinction. They have crossed that line and are potentially allowing their personnel to get injured. That should never happen."

Contact: Jan Dunbar, Division Chief – Haz Mat, City of Sacramento, Department of Fire, 1231 I Street, Suite 401, Sacramento, CA 95814-2979; 916-264-7522; 916-264-7079 (Fax).

GAINESVILLE DEPARTMENT OF FIRE/RESCUE HAZARDOUS MATERIALS RESPONSE TEAM

William "Skip" Irby has been the Deputy Fire Chief, Operations Division, of the Gainesville Department of Fire/Rescue since August of 1996. He started with the Gainesville department in 1975 and was promoted through the ranks within the organization. Previously as a district chief he, along with the hazardous materials engineer, oversaw the hazardous materials program. During his career, he has been a firefighter, driver-operator, lieutenant, district chief, and deputy chief. He was also training chief with the department and the EMS program.

Gainesville is a city off of I-75 in north central Florida that is home to the University of Florida and 84,770 residents. In addition to responding to city emergencies, the Gainesville department provides certain services within Alachua County and the North Central Florida Local Emergency Planning District. A contract signed with the county in 1989 requires Gainesville to provide fire, emergency medical services, and emergency hazardous materials incident response in the unincorporated areas of the County. Gainesville and Alachua County are located within the planning district of the North Central Florida Local Emergency Response Planning Committee which includes the following counties: Bradford, Columbia, Dixie, Gilchrist, Hamilton, Lafayette, Madison, Suwannee, Taylor, and Union. In early 1989, the Gainesville department was designated as one of the regional hazardous materials response teams for the north central Florida area.

"Our total fire/rescue department's size is approximately 150 people," recounts Deputy Chief Irby. "For emergency medical services we provide ALS (Advanced Life Support) to the community in a nontransport capacity. All of our engine companies are ALS-capable, but we don't transport patients to hospitals. We do the first response, treat, and stabilize patients on the scene, and prepare them for transport. The county handles the actual transport. We do have paramedics on every one of our units, and all our personnel are trained at least to the EMT level (Basic Life Support).

"We started in the hazardous materials business in the late 1970s with a squad truck we called 'the flying squad.' The unit was primarily used for extrication and heavy rescue emergencies, and the personnel responded out of a bread truck. This flying squad eventually evolved into the hazardous material response team. They started serious training in the early 1980s to attain the operations level, and in the mid-1980s we hired a hazardous materials engineer who formerly worked with environmental engineering firms. We hired him to oversee our hazardous materials program and act as a consultant for us on scene. Our fire chief at the time wanted to get more involved in Haz Mat response, saw that we were called more often, and felt it was important to have somebody on board who had extensive background and knowledge. The hazardous materials engineer was actually a certified environmental engineer, had a chemistry background, and did a lot for us in the field of education. He left us approximately four years ago, and one of our hazardous materials lieutenants moved to fill the position as hazardous materials officer."

Gainesville Fire/Rescue currently has 28 certified Haz Mat technicians who received at least 80 hours of training taught by the International Association of Firefighters through a federal grant. Other technician-level training was obtained from Safety Systems, Inc. Other outside suppliers of training include the Florida Gas Transmission Company, Chevron, Gainesville Regional Utilities, and the University of Florida Center for Training, Research and Education for Environmental Occupations (TREEO). In-service training has covered chemical protective clothing, air monitoring, standard operating guidelines, and other topics.

Most of these personnel are assigned to the truck company (Tower 1) and engine company (Engine 1) located at Station 1. The truck company is the primary team that responds to incidents in Tower 1 and Haz Mat 6. All firefighters in the Operations Division have received the first responder awareness level and operations level training required by the EPA Worker Protection standard. A 40-hour course combining both levels was developed and presented based on the National Fire Protection

Association Standard 472, "Standard for Professional Competence of Responders to Hazardous Materials Incidents." Required annual refresher training is ongoing. Haz Mat emergencies currently account for slightly over 3% of the total call load for the department which is approximately 16,000 calls a year. There were 463 hazardous materials incidents in a recent 12-month period. Emergency medical calls account for 75% of total calls, typical for fire departments around the country that provide medical services.

"My guess would be that we are probably similar to a lot of other emergency services organizations throughout the country in that a lot of our responses are to LPG, natural gas, and gasoline," continues Deputy Chief Irby. "However, there are a couple of chemical companies in the area that have required significant responses. A recent large scale incident involved a release of trichlorosilane at a local chemical processing facility in June of 1994. This incident resulted in the evacuation of 600 residents near the plant. Hundreds more were advised to shelter in place; 148 persons were treated and released from local hospitals; and the local airport was closed for approximately four hours. More recently, there was an explosion and fire that totally destroyed the manufacturing plant of a company that makes pharmaceuticals. There were a lot of chemicals involved, and we did not attempt to extinguish the fire. We basically were there to protect against other exposures. The site has been declared an Environmental Protection Agency 'Superfund' site, and I don't know if they are going to rebuild at this location or not. We also have the University of Florida which has significant chemistry labs and storage facilities. We have not had a big incident there, but we have had some live spills and leaks.

"We try to maintain three hazardous materials technician-level personnel on the Haz Mat truck at all times, although we do have additional technician-level fire-fighters and operation training personnel that can back them up. Haz Mat technicians perform several services. Their primary response is as a truck company on the fire ground. They may be called for medical response as well, but, because of their specialized training, they are on first call for any hazardous materials alarm. We have a response protocol which will, depending on the incident, send other units in addition to the designated Haz Mat team. Of course, Dan Morgan, our hazardous materials officer, responds to significant incidents as well."

Gainesville is a union fire department. The city gives extra certification pay to EMT's, paramedics, and fire inspectors, but not currently to Haz Mat technicians. That would have to be a negotiated item in the union contract. Because no certification process for hazardous materials exists, Haz Mat technicians cannot receive certification pay. The total chain of command within the department is the fire chief and the deputy fire chief. On the scene would be a district chief, hazardous materials officer, lieutenant of the Haz Mat team, and firefighters. "We have just repositioned the hazardous materials officer's rank to the district chief level," adds Skip Irby. "Prior to that, we had the Haz Mat engineer who was not in the rank structure.

"We have certified state fire inspectors who do site safety inspections, while our Haz Mat team officer has the teams go out and identify target hazards and do preplanning. Enforcement actions and any violations noted would be turned over to our fire inspector in the fire safety management unit. We've got a good working relationship with the business community regarding compliance. We have done a lot of joint training with PCR Chemical, opened some of our technician-level classes to their personnel, and do joint drills at their facility which helps each side know their counterpart's capabilities. PCR has assisted us with their technology and their expertise in the chemical industry. We have just recently been dealing with the University of Florida on a chemical called pentaborane.* They have a 20-pound cylinder of the chemical that they will need to dispose of. This is a really unstable chemical, both poisonous and pyrophoric. U.S. EPA has a study on the history of this chemical, and the history is not very good from the disposal standpoint. EPA does have a pentaborane task force that will be trying a new method to neutralize this material."

When asked how the hazardous materials response field has changed over the last 20 years, Deputy Chief Irby responded, "The environmental awareness has undergone a great change. Twenty years ago, any fire department would have hosed a petroleum product off the street or into the ground. Today, we would not even consider that. Now, we contain it, stop the run-off, and pick it up with absorbents or by other means for proper disposal. Twenty years ago, a gas spill or diesel leak was not looked at as a hazardous materials incident. It was 'just a leak.' That viewpoint has changed considerably. Probably another area that has changed is the storage of chemicals at the University. Before, we never knew what they had, and they had no up-to-date inventory of all chemicals on the campus. Since that period they have developed a safety department that is tracking what they have in inventory, such as the 20-pound cylinder of pentaborane.

"A couple of years ago, a chemistry professor who retired from the University died. When his estate was settled, somebody found one of these mini-warehouses stacked to the ceiling with chemicals he had used in his work. That turned out to be a big hazardous materials incident. Simple awareness of what is out there has changed drastically over the last 20 years. The biggest change has been in how we handle the different incidents according to classification. It used to be just an emergency response; nowadays, many hazardous materials incidents become environmental issues as well."

Who is in charge of a Haz Mat incident in Gainesville? "It depends on the level of incident," says Irby. "If you are at an LP gas or natural gas leak, say a 20-pound gas grill cylinder or a line going into a house, the lieutenant on the Haz Mat team may be in charge. If an incident is a regular gas line, you are going to have a district chief on scene. If the product or chemical is unknown, you will have a district chief in conjunction with the Haz Mat officer. An incident could go all the way up to a deputy chief being in charge at some incidents. And, we've had incidents where the fire chief was in charge and setting up a larger command structure where evacuations were in progress."

A comprehensive medical surveillance program meeting the requirements of 29 CFR 1910.120(q)(9) was established in July of 1991, as were several other resources, primarily Chapter 5 (Medical Program) of the Occupational Safety and Health Guidance Manual for Hazardous Waste Site Activities prepared by NIOSH, OSHA,

---

* Refer to the "2000 North American Emergency Response Guidebook," or another manual, for the tricky handling characteristics of pentaborane.

USCG, and EPA. "Haz Mat team members have a baseline physical examination through our employee health program with city government, but we also send them to a private physician for blood work and X-rays," relates Skip Irby. "After that baseline exam is done, they have a more extensive health assessment annually than do regular members of the fire/rescue department. All of our department personnel have annual health assessments, but the Haz Mat team members have a lot more tests, and certain files are maintained on them for monitoring purposes."

Lieutenant John V. Mason is a Hazardous Materials Technician, Haz Mat Team Supervisor, Florida State Fire Safety Inspector, and an EMT for the last 12 years with Gainesville Fire/Rescue. Prior to that, he worked three years at the nuclear submarine base at King's Bay, GA.

"I've been stationed on this truck for the whole 12 years and came up through the ranks on the same unit," says Lieutenant Mason. "Originally, we just ran a little truck. Now we have evolved to an assortment of specialized Haz Mat response equipment. Many of our runs involve propane or natural gas leaks, and we're set-up for any type of semi tractor-trailer incidents because we have I-75 running through this area. A majority of our equipment is used for repairing regular and pressurized leaks. We have different types of clamps, rubber plugs, and patches. Our protective equipment includes Level A (vapor protective) suits, Level B (liquid splash protective) suits, and assorted sizes of Saranex and Tyvec suits.

"We function as both a tower company and a Haz Mat company so we use two vehicles and run them in tandem. When we run as a tower company, we leave the Haz Mat response vehicle at the station and four firefighters ride in the tower. On a Haz Mat run, two people ride in the tower and two in the Haz Mat truck. The most obvious function of the tower would be rescue in a multi-story building. Ground ladders do not reach over two or three stories. If a tower unit is not needed, the crew's job is to perform search and rescue and then ventilation. When an engine company pulls up to a fire, the members are dedicated to putting out the fire. When we pull up as a tower crew, we are dedicated to getting the people out. After that, we are basically manpower to do ventilation and to cut off the electricity and gas to the building. Pumper crews put out the fire, and we do everything else.

"On the Haz Mat truck, we have petroleum absorbent dikes since we have several lakes in our area of coverage plus a lot of other environmentally sensitive areas where run-off could be a problem. Our library for chemical research is in the truck, as is the decontamination gear. Getting involved with decon is a common occurrence for any Haz Mat team. Any type of chemical that can be hazardous to responders or citizens, no matter how minor the chemical may seem, requires a decon area and equipment. Whether or not to set up a decon area is one of the very first questions we have to answer at any incident. If you get into a situation where you don't have a decon area established, you can throw up a quick one with just a tarp and a booster hose without having the three wash-down pools. However, with the exception of products such as natural gas where decon would not be necessary, we try to set up decon on almost every incident we go to.

"While the decon area is established, we usually have a medical sector in operation. We usually use our Rescue 1 crew. They take the vital signs of both the entry team and the decon team," adds John Mason. "When the entry team comes out of the contaminated area and goes through decon, their vital signs are taken again. There are certain criteria applied here. If a person's heart rate is over 150, or if his/her blood pressure is at a certain point on the sheets the medical team uses, that person may not enter the hot zone.

"I've been scavenging for years," replies Lieutenant Mason when asked why the team has so many pipe fittings. "We have a fairly large chemical processing company, PCR, right here in town that we work with on training and leak stoppage. PCR probably has every size pipe in the world out there. They have literally miles and miles of piping, and there is no telling what kind of pipe we're going to come across. Some of this piping is not a basic format for what you need at some time. We talk to a lot of different organizations, and we pick up methods and equipment from them.

"We have some stock kits such as A and B chlorine kits for cylinders and ton containers respectively, soda ash for neutralizing acids, and sealers for the tops of the domes on tank caps. There is an MDT (mobile data terminal) aboard the truck, and usually one person is assigned to it. It provides a location of the incident and the past history of the premises, and allows us to keep in touch with other units such as main dispatch and other fire-rescue response vehicles. If I want Engine 1, I just type in my message and hit the transmit button. If we are 'available' or 'responding,' we hit the correct button. If we want dispatch to find out some information, or call somebody, they can respond to us with a printed copy. If we need data, dispatch can provide updates on wind direction or other factors. When radio traffic starts getting high, it is important that we can use the MDT rather than radio. The MDT also has a 'territory change.' In other words, if we go from one territory to another, I can hit the shift key and write what territory we are entering. If response preplans have been made for any area business, I can pull them up. I like MDT dispatch better than standard voice or radio dispatch. One obvious reason is that you often get a dispatch over the speaker that is not quite clear. When I use the MDT, the message comes out in printed format which is much better.

"I believe we have drastically increased our knowledge in the field of hazardous materials. We all learn from each other's mistakes. I've made a lot of mistakes. You come out of a situation and say to yourself, 'Man, that was close. I will never do that again.' Each and every incident is a learning experience. I learned a lot at King's Bay, and probably plugged more major chlorine leaks there than I have in 12 years here. I brought that experience with me. I'm pretty good at repairing leaks and tanks because of my experience. More important, significantly more safeguards exist now. You see a lot of chemical accidents because of equipment malfunction. You don't see as many caused by human error because the knowledge in quality control is getting better. Everything progresses. Ten years from now people will probably view our methods and tactics as barbarian."

Contact: William "Skip" Irby, Deputy Fire Chief, Operations Division, Gainesville Fire/Rescue, P.O. Box 490, Station 34, Gainesville, FL 32602-0490; 352-334-2590; 352-334-2529 (Fax). John V. Mason, Lieutenant, Gainesville Fire/Rescue, P.O. Box 490, Station 1, Gainesville, FL 32602-0490; 352-334-2594.

HICKSVILLE FIRE DEPARTMENT HAZARDOUS MATERIALS RESPONSE TEAM

The Hicksville Fire Department was the first volunteer fire department on Long Island to form a HMRT. The department started researching needs in 1979 and went on-line in January of 1981. The team's first incident was in February, 1981 involving an overturned tanker loaded with MEK (methyl ethyl ketone) in a parking lot between two industrial buildings containing bulk hazardous materials.

The Hicksville Fire Department is located in Nassau County, New York, and covers about seven square miles of territory with a population of around 50,000 people. The area is residential, commercial, and industrial with a large commuter railroad hub and a small freight yard as well.

The Hicksville Fire Department is volunteer. Eugene J. Pietzak's paid career is as a fire investigator with the Nassau County Office of the Fire Marshal dealing with both Haz Mat incidents and fire investigations. He also instructs at the Nassau County Fire Academy where 71 fire departments, almost all volunteer, are trained. Approximately 10,000 firefighters within Nassau county are volunteer.

"The Hicksville Haz Mat team celebrated its 15th anniversary this year," remembered Gene Pietzak, who has been a team leader on the Haz Mat team for more than ten years. "Back in 1981, we started seeing large volumes of hazardous materials transported through and into our community. Also, we had an incident where 15 to 20 firefighters went into a routine-looking fire in a building with no placarding, marking, or identification of what type of chemicals might be inside. Multiple types of chemicals stacked on shelves dropped down and became mixed. Firefighters were wearing SCBA and turn-out gear, but shortly after this operation, the firefighters became ill. Chief officers within the department started to formulate more in-depth preplanning with regard to hazardous materials.

"The HMRT comprises members from eight companies of the Hicksville Fire Department. It's a volunteer team but members have to be recommended to it by their co-captains. When considering a candidate for the team, we look at his or her professional background. We look for people who may be plumbers, electricians, or carpenters, or who have certain other skills that can assist us in the many tasks involved with hazardous materials incidents. Also, we consider their interest level, and whether they have attended Haz Mat training programs that have been offered by Hicksville or within the county.

"All HMRT members within the department also belong to an engine company, truck company, or the heavy rescue company, and they need a minimum of three years within the department. Currently, we have about 30 members on the team. Also included on the team are EMTs or paramedics, and a doctor who is with the department and is involved when we need him. We work closely with both the Nassau County fire marshal's unit which has a Haz Mat team, and the Nassau County police department which also has a Haz Mat team.

"For team staffing, we have three team leaders who are under the direct control of the chief's office. If a team leader is not present at an incident, then one of the other members of the team will take that position. The team leader position is similar to a captain/lieutenant with a regular fire company. They cover administrative aspects, assist with the decision making process, and act as liaison to chief officers at an incident.

"We follow an incident command system and have an accountability system. Other firefighters within the department can assist the HMRT in various ways. We have medical personnel from our rescue squad who are on the team handling medical duties. Other in the department are not team members per se but assist in setting up decon, suiting, etc. so that team members can get to other tasks where their knowledge is needed. The department has cross training in various tasks. All members within our department are trained to the hazardous materials operations level, and members on the HMRT are trained anywhere from the operations level up to the Haz Mat specialist level within applicable standards.

"We respond to anywhere from 30 to maybe 80 incidents a year. The most common incidents involve hydrocarbons, petroleum products, and propane. We also encounter various types of gases, different corrosives, and anhydrous ammonia. At one time a large electroplating company occupied several local buildings. It has since folded up and left town.

"Regarding specialized tools and equipment, we try whatever is out there. We have two members within our department, one a team member and the other a district mechanic, who are extremely talented. If there is anything we need made up, you name it, and they can do it. They have created a number of items for the team. When it comes to equipment, we are in pretty good shape.

"The team is dispatched through the department. The district has full-time dispatchers on duty 24 hours a day, 7 days a week. Members are alerted through paging units. We also have phone numbers and personalized, numerical and alpha pagers for contacting members who may be at work and may not hear a regular alert notice. Some members make arrangements with their employers or take time off to respond. Several of our members are civil service employees within the county who can get time off as well.

"As to non-response activities of the HMRT, we do preplanning, inspections, and public relations. The team does a great deal of public relations both inside and outside the district. For ten years now we have also been a part of the 'STOP' program (Stop Throwing Out Pollutants) where homeowners can bring unwanted chemicals, batteries, insecticides, etc. to a chemical recovery company that will dispose of them properly at no charge. We also have an open house yearly where we display our equipment, hand out information about hazardous materials incidents, and general information about our training. At a training session, we go through general business and cover various training aspects, critique an incident, or cover preplanning developments."

"Referring to contingency planning, standard operating guidelines, and protocols, we may have contingency planning through the LEPC (Local Emergency Planning Committee, a committee appointed by the state emergency response commission, as required by Title III of SARA, to formulate a comprehensive plan for its Office of Emergency Services mutual aid region) or do our own contingency planning with them. The plan we develop is then recorded in books and placed aboard the response vehicle so we can refer to it.

"For standard operating guidelines relative to scene management, evaluation, team concept, command post operation, and general operations at the scene, we work through the department training committee, the chief's office, and the Haz Mat team itself. We determine the type of incident we have and call in any additional resources we may need. As an example, the Nassau County Fire Marshal has enforcement powers for spill cleanups.

"At an incident, the team leader has discussions with the chief. What the best actions would be, how we go about it — these types of questions are discussed on a teamwide basis, so it is not really one person making the decisions. We work on a team concept. As information is gathered, the entry team and other team members will sit there and discuss what we have. What is the best way to deal with this situation? What are our options? How do we implement our plan? We have team leaders within the department who are captains or lieutenants. Once they respond and become part of the Haz Mat team, their rank of captain/lieutenant is gone. They are now team members, and every authority line goes through the team leader.

"When the team was first started, we had no rules and regulations. We worked on those at once. Over the past 15 years, the growth of the hazardous materials field has forced the volunteer fire service to meet special training requirements. Department members battle with us about receiving the necessary training. We have had departments within the county who have formulated Haz Mat teams that lasted only 5 to 6 years before folding. One of the reasons is the time that must be dedicated to meet the requirements of the training standards.

"Most of the fire departments within Nassau County train on a weekly basis. Hicksville Fire Department trains twice a week, on one night during the week and on Sunday morning. People are excused from required company training to attend specialized or team training. The workload for training can be very heavy at times. Civil service employees seem to fare the best because they are able to get time off, unlike members from private industry. We have learned it is tough to get everyone trained to required standards within a volunteer department. In our district, the chief's office and the district team have been open minded, have recognized the need for training, and have provided training opportunities. The majority of our members have been to the International Hazardous Materials Response Teams Conference in Maryland, trained at Safety Systems in Florida, and attended the New York State Fire Academy, and the Nassau County Fire Service Academy.

"There have been lots of changes and growing pains. Every day we continue to learn. We've seen changes in equipment, levels of protection, breathing apparatus, thermal protection, monitors, cameras, and types of suits. We are using all the technology available, limited only by the imagination."

Contact: Eugene J. Pietzak, Fire Investigator, Nassau County, Office of the Fire Marshal, 899 Jerusalem Avenue, P.O. Box 128, Uniondale, NY 11553; 516-572-1081.

TORONTO FIRE DEPARTMENT HAZARDOUS MATERIALS RESPONSE TEAM

To say that the City of Toronto grew by leaps and bounds is an understatement. In one day, January 1, 1998, it swelled from 29 square miles to 250 square miles enveloping what used to be six cities (Toronto, Scarborough, East York, North York, York and Etobicoke) to form a mega-city for the millennium.

Rem Gaade is chief of hazardous materials and special operations for the "old" Toronto Fire Department. "The mega-city structure disbanded six municipalities and one area government and replaced them with one city. Instead of having six fire departments as in the past, we now have one fire department. I can provide information on Toronto under the City of Toronto Act of 1837, but not on the new City of Toronto under the City of Toronto Act of 1997 (the new mega-city).

Gaade states that prior to the reorganization, "We probably were Canada's busiest full-time Haz Mat team with 600 calls a year. However, we don't exclusively do Haz Mat. We also run all multiple alarm fires in order to provide atmospheric monitoring and other safety-related functions. When our instrumentation tells us it is not safe to unmask, we do not want our people taking their masks off even though it *appears* safe to do so. We have always maintained a focus on firefighter health and safety within our Haz Mat program as well as responding to spills and other hazardous materials emergencies.

"We have been running Haz Mat alarms for a long time. I have a 1942 annual report that says the department responded to 44 ammonia leaks that year. Haz Mat is nothing new, but we didn't formalize our program until 1979, and we really didn't start equipping and training people to the best of our capabilities until the early 1980s when NFPA 472 (National Fire Protection Association "Standard for Professional Competence of Responders to Hazardous Materials Incidents") came along. We have a dedicated hazardous materials response vehicle built in 1985 and put into service in February of 1986. We originally staffed it with only a driver. It was like a mobile toolbox, and heavy rescue squads or foam pumper crews would do the Haz Mat work at the scene. Eventually, we started training some people to advanced levels. We had a department reorganization in 1994 and got permission from the administration to establish a permanent, full-time hazardous materials team. The Haz Mat truck in 1986, its first year, handled a total of 23 calls. By 1994 there were 134 calls. When we put on a full-time, trained crew in 1995 the total calls jumped to 407, and in 1996 it climbed to 552. In 1997, we had well over 600 calls.

"We believe the growth rate over the years can be traced to greater environmental awareness among the public and to dispatcher training protocols," explains Chief Gaade." We get a lot of Level 1 calls, but we don't count those in our Haz Mat statistics. On the gas washes and leaking propane calls, we just send the local engine company. Our Haz Mat truck currently responds only to Level 2 and Level 3 incidents. A Level 2 would typically be an incident where you would have a requirement for some diagnostic instrumentation, or maybe extra absorbents or some special knowledge. It does not require entry in Level A protective clothing or decontamination. What you get on a Level 2 response is the local pumper, the nearest heavy rescue, the Haz Mat team, and the local district chief. On a Level 3 incident you get your basic street full of fire engines. Its first alarm is two pumpers, an aerial, a foam truck, a heavy rescue squad, the Haz Mat team, and a district chief. Within the department, we have four aerials that have been equipped and the crews trained for decontamination. These aerials carry equipment such as containment basins and protective clothing. The crews have certain procedures whereby they lay out their equipment in a certain way and follow the guidelines that have been established by the Canadian Association of Fire Chiefs. We put decon crews into a level of protective clothing one lower than the entry teams wear.

"We ensure the health of our firefighters, foremost, through training. Our basic training for Haz Mat team members is 250 hours. Most of the people who take the training course, and that includes me, find at the end of 250 hours you want another 250 hours because you realize how much you don't know. It's one of those 'the smarter you get the dumber you become' situations. It really teaches you that you need a good network of contacts so that when an incident happens you know your own limitations. We can do a lot, but we can't do everything for everybody. Once we are qualified at a first responder operations level, every two years we have a one-week refresher course. We provide opportunities for specialized training, such as conferences in the United States, but with recent funding cutbacks some of our external training has decreased. We go out to industry. If we want to learn about cryogenics, we'll go to a plant that makes cryogenics and ships MC 338 tank trucks. We will have people who drive the MC 338s teach us about their trucks. We have a large manufacturer of MC 306s, MC 307s, and MC 337s within the metropolitan area. We visit them for the best part of a day to learn how each of these tanks is constructed, so when one is lying on its side we know not just what's on the outside but also what things are like on the inside, what effects our actions would have if we did something to that vehicle. We concentrate on situations specific to our environment."

The original city of Toronto had a four shift system, one Haz Mat response vehicle to cover the 29 square miles within "old" Toronto, and a Haz Mat team running over 600 calls during 1997. Each shift had four people, three technicians and one captain also trained to the technician level. "We also have fire department personnel who have been Haz Mat technicians for two years but who are no longer on the Haz Mat team," according to Rem Gaade. "After the two years, they have the option to re-enlist for another two years or go back to mainstream firefighting. These people can be called upon to run the truck when we have personnel off-duty due to new time periods, vacations, workers' compensation, or whatever. We have about a 45% turnover at the end of two years. The ones we lose are eligible for recertification even though they do not go back on the truck. We see them as people who are out among the suppression firefighters, able to recognize situations where it could be beneficial for us to intervene with the Haz Mat team rather than try to muddle through with the crews at hand."

Rem Gaade was asked about concern over terrorist actions and the potential use of chemical and biological agents he has noticed in Toronto. "We share the concerns that exist in the United States. According to the intelligence services we confer with weekly, Toronto is the number one target in Canada. It's the financial and economic hub of the country and very symbolic. We recognize we are vulnerable to real estate damage but also to emotional and psychological damage to the population as a whole. We have a subway system, a very busy downtown, a stock exchange, and some other particularly tempting targets. We have been preparing ourselves with equipment for nuclear, biological, or chemical attacks. Once the NBC amendment to National Fire Protection Association Standard 472 came out, we trained our people both at the operations level and at the technician level. We are fortunate in a way that I'm on the NFPA 472 committee as well as on the task group that prepared the terrorism amendment because it gave us a bit of a head start. Probably 95% of the fire departments in Canada are not aware that they are now in the front line of NBC terrorism response.

"Our federal government hasn't awakened to the fact there is such a thing as NBC terrorism," remarks Chief Gaade. "The only people who are equipped to respond right now in any significant way are the Canadian Armed Forces. Their single NBC response team can handle only one incident at a time. In Canada, since it's the second biggest county in the world, we take a while to get to an incident because of geographic distance. If that NBC team is busy at another place, then we are out of luck. The government is not providing any guidance or any funding relative to NBC terrorism. With the new mega-city of Toronto, we're going to have 2.5 million people in a total urban area of 4 million people and we have no guidance, no training, no money, and no equipment from the federal government. We are on our own."

Are firefighters concerned about training in response to nuclear, biological, chemical agents and materials? "Well, first they are amazed, then they get concerned, and then they realize that NBC agents are just an extension of Haz Mat," responds Gaade. "NBC is just a nastier form of Haz Mat. As long as persons have confidence in their training, their equipment, and their own capabilities — and realize their limitations — they can handle incidents with confidence. In our normal day-to-day operations, we know how far we can go.

"Also, Haz Mat technicians have the right of refusal when it comes to certain situations. For instance, we have a procedure for identifying unknown chemicals in which the most dangerous part of the procedure is opening the container, taking a sample, and analyzing it. Many of the technical requirements like handling a test tube or wearing a Level A suit and three sets of gloves are never easy, but the most dangerous part is opening the container. Our Haz Mat people have, in writing, the right to refuse to perform an activity if they feel that it is unsafe to do so; and no senior officer can make them continue. This particular protocol for identifying unknown substances is based on the fact that our people are sufficiently well trained. If they think it's unsafe, you can be 100% certain that it is unsafe. Therefore, as the responsible employer, we should not be asking them to continue.

"We trust our people. Our decision making process is much more democratic than it is at an average fire. When we do our reconnaissance, the entry technicians will often discuss not just tasks and tactics, but sometimes will discuss strategy with the incident commander. This is unheard of at normal fire situations where the incident commander leads the decision making process. In Haz Mat, all our incident commanders are aware that these technicians have special skills and knowledge. If the incident commander doesn't consider these special skills, the onus for the end results may lie on the incident commander's shoulders and may be heavy. If it was an NBC problem, our Haz Mat people would recognize that the scenario is beyond their capabilities. We'd secure the area, do whatever was possible within safe limits, and activate the national defense NBC response team from its home base just a few hours outside of Toronto."

Chief Gaade considered the changes that have occurred in hazardous materials response during his time in the fire service. "The biggest change I have seen is in personal protective equipment (PPE), such as, the development of higher quality and protection levels, and the recognition that you can't use one suit or one set of gloves for everything. We carry ten different types of gloves on our heavy rescue rigs and Haz Mat truck.

"We've instituted a number of medical monitoring protocols that never existed in the past. We have been largely guided by NFPA 473 as well as NFPA 472 and NFPA 471. The standard process has been a big help to us. We in Canada are fortunate that we do not have to be full-fledged Haz Mat technicians to do work at the operations level. Under our labor laws, if you are equipped and trained to do a task, you can do it. We have people who are trained at the operations level but have also been given Level A suit training and elementary patching and plugging training and equipment. We can send them in to do a lot of the basic ground work to free up our technicians who spend an awful lot of time in the back of our truck with a computer and manuals doing the thinking and analysis. Using technicians to move several 55-gallon drums to reach the one that is leaking would be a waste of their talents. We have a limited number of Haz Mat technicians. We have far more people who are trained to what we call 'advanced operations.'"

It has been said in the United States that a firefighter in Level A personal protective equipment might have better protection than in the NBC protective clothing the military has available. "Such military protective clothing in Canada is called 'TOPP' (technical operational protective posture) gear," says Gaade, "a garment that is addressed specifically toward chemical and biological warfare. It would be absolutely no good in any run-of-the-mill Haz Mat incident that we run two or three times a day. It has limitations in that it has no oxygen supply. It may filter out the bad stuff, but it cannot make up for a lack of oxygen. With a porous, activated charcoal liner, it is totally unacceptable for Haz Mat response. It is very difficult to decontaminate because it's rough; you can't just give people a simple wash down. It is also much more expensive than a limited use, disposable chemical suit. On the other hand, it seems to be working for military personnel when they go into live chemical agent releases at Fort McClellan in Alabama.

"With Haz Mat response personnel, our everyday work rarely involves a chemical or biological terrorist attack, so we have to buy garments to match the work that we do the most. Our present clothing will provide protection against most chemical and biological agents. Even on the latest limited-use garments that are coming on the market, some PPE still has PVC face shields. Blister agents move through PVC face shields like water through a Kleenex, so there are still limitations. Blister agents also move through polycarbonates, so the Canadian military has replaced the PVC face pieces in their protective masks with lexon and/or glass. It is when you don't know such limitations and totally and utterly trust your PPE that you could get into a dangerous situation. We feel as responsible employers that we must ensure that everybody in that type of situation is fully aware of the limitations of their protective gear."

Contact: Hazardous Materials Chief, City of Toronto Fire Department, 260 Adelaide Street West, Toronto, Ontario, M5H 1X6 Canada; 416-363-9031; 416-392-0598 (Fax).

SEATTLE FIRE DEPARTMENT HAZARDOUS MATERIALS RESPONSE TEAM

The Seattle Fire Department is responsible for the stabilization and mitigation of all hazardous materials incidents that threaten public safety within the city, except for those normally resolved by the Seattle Police Department's Explosive Disposal Team. Unit 77, the Hazardous Materials Response Team, also responds to jurisdictions and organizations that have current mutual aid agreements with the City of Seattle.

Rick Picklesimer is a firefighter with the Seattle Fire Department Hazardous Materials Response Team. "The HMRT is always dispatched as Unit 77 which includes Engine 10, Ladder 1, Aid Car 5, a battalion chief, one medic unit, one air unit, the hazardous materials response van, and Safety 1. Other units can be dispatched at the discretion of the dispatcher. First-arriving units will determine the wind direction and topography at the incident scene while approaching the incident from upwind and uphill if possible. All apparatus will be stopped approximately one block (300 feet) from the incident, turned around, and parked heading away from the scene to allow for prompt withdrawal if the incident escalates.

"Captain Chikusa is in charge of the four platoons that make up the entire hazardous materials response team over all shifts. In the chain of command, pretty much everything goes through him. He is in a battalion chief's type of setting. That is, there's not a battalion chief over him since he reports to the deputy chief of Battalion 1. Every shift has an assigned line up. On this shift, Captain Chikusa is the team leader while Lieutenant Beaumont is the safety officer. The first thing we do when we go on scene is to evaluate the incident and find out what's actually going on and get a report. Usually, there's a first-in company already there that has gathered some information. A lot of times you've got workers on the scene who know exactly what the product is. At other times, we'll get on the scene where something fell off a truck; it's yellow in color but nobody knows what it is. Frequently, this type of incident will happen in a store or some other type of building where people are present. Most of the time we can figure out what the release is. A lot of the time it turns out to be pepper spray. We don't clean up the scene. We do control the leak and mitigate the situation on scene in the interest of public safety. A response contractor handles clean up."

Picklesimer says the Haz Mat team cannot be split up for multiple, sumultaneous, Haz Mat incidents because the team is required to have 11 hazardous materials technicians on a scene. "If a second incident occurred in the City of Seattle, then we would have to call for mutual aid from one of the surrounding Haz Mat unit towns or cities in the nearby area. That would happen because we must have full manpower at a hazardous materials incident, although multiple incidents have not happened yet to the best of my knowledge.

"As for general response activities and tactics, we have 'Haz Mat Mondays' where we do drills. At the beginning of each year, each platoon is assigned a number of drills for the year. Each month the drills are given to the teams so that each platoon gets the same training from the same people. Also, we do what are called 'prefire inspections' for companies that manufacture, store, use, or sell hazardous materials. The prefire plans that result from these inspections are on the computer, and are available to the battalion chief. The prefire plans also include many buildings that don't have hazardous materials present. Prefire plans are a plan of response to a specific building and include the location and a drawing of the building, storage locations, sprinkler systems, elevators, stairs, number of floors, size of the building, etc."

According to standard operational procedures strategic objectives listed in order of priority are the protection of life, the protection of the environment, and the protection of property. Incidents may include, but are not limited to, transportation accidents with the release or potential release of hazardous materials, clandestine drug labs, incidents at "fixed" sites that are known to store or process hazardous materials, incidents involving multiple victims or materials of unknown etiology, or a spill or release with obvious environmental impact.

The first arriving unit commander establishes command and operates as the incident commander until properly relieved by a higher ranking officer. Basic tactics at any Haz Mat incident include an attempt to identify the presence of hazardous materials, determine to what extent they are involved in the incident, isolate and evacuate the area and deny entry, establish control zones (hot, warm, cold), and establish an identifiable command post and communicate its location to the dispatcher. First responders also establish a staging area uphill and upgrade, locate a decontamination area, evaluate risk to response personnel, and initiate measures to rescue victims if risk is known and acceptable. If the risk is unknown, or if turnout clothing and SCBA are not adequate to such a task, first responders will wait for the arrival of Unit 77. First responders also implement defensive tactics, attempt to determine the precise name and amount of the commodity, and re-evaluate control zones.

Captain Jeff Chikusa has been a firefighter for 20 years. "My primary duties were always on an engine company, and seven years ago I was requested to take over the hazardous materials team. I oversee the budget and the training of all 60 members of our Haz Mat team. We work on a four platoon system, so we require that 11 Haz Mat technicians are on duty 24 hours a day. The Haz Mat team is funded by the Seattle Fire Department and cost recovery is done through billings that are sent to the party responsible for a spill. The team never sees any of this cost recovery money because it goes into the city general fund. The service is paid for by the taxpayers, but beyond our normal duties, if we have to use special chemical protective clothing or other specialized equipment, we try to bill the responsible party. It's mostly a charge for the materials we use. The chemical protective suits are so expensive, we really aren't able to keep buying suits all the time. As a result, we utilize limited-use suits. We determine what material we were exposed to, how much exposure time the suit has actually undergone, and then determine whether we are going to use that suit again or discard it."

Captain Chikusa is the team leader of the Haz Mat team during his shift. "I report to the incident commander who has control over the entire scene at an incident and over any support units that we require. As far as the hazardous materials scene, I as the team leader of the HMRT am responsible for that. We also have an incident safety officer and a Haz Mat team safety officer; the sole responsibility of the first is to handle safety considerations for the entire incident, while the second handles safety concerns for the Haz Mat team."

The hazardous materials response team leader assigns members of the HMRT to the following positions at roll call every shift: team safety officer, entry team(s), decontamination monitor, entry/exit monitor, product identification, back-up team(s), and manpower pool. At any incident, the team leader immediately reports to the incident commander for a briefing of all known information regarding the incident, informs the incident commander of actions necessary to correct any obviously unsafe conditions, holds a briefing with the HMRT to make all members aware of the hazards of the incident, and develops an action plan and makes recommendations to the incident commander.

The team safety officer reports to the team leader and is an adjunct to the incident safety officer. He or she assures that appropriate personal protective equipment is used, medical monitoring of the entry and back-up teams takes place as required, the decontamination area is established prior to entry, communications with the entry and back-up teams are set and working, and all personnel are briefed on what to do if the incident escalates. The team safety officer has the authority to cancel entry operations at any time if, in his or her opinion, the incident conditions have become unsafe.

The entry team should not enter the hot zone until the entry/exit point is established, the decon area is established, the back-up team is suited up in the standby position, the entry team has been briefed on procedures to initiate, on potential hazards, and on problems that may be encountered, and communications equipment has been secured and tested. The main duties performed by the entry team within the hot zone include rescue, reconnaissance, monitoring, sampling, product identification, containment, and product control. The entry team shall proceed only as far their protective equipment will allow. Minimum personal protective equipment for incidents involving known hazards shall be full structural fire fighting protective clothing and SCBA; while minimum personal protective equipment for incidents involving unknown products or hazards shall be liquid splash protective suits with SCBA. The entry team will immediately report initial observations to the team leader, and upon completion of their assigned tasks, or when told to do so by the team leader or team safety officer, shall exit the hot zone through the decontamination area. After decon, the entry team will have their vitals checked again by the paramedics, and report to the team leader for debriefing and reassignment.

The decontamination monitor is responsible for coordinating and supervising the decon area within the cold zone so he or she will not be contaminated. The person in this position will brief decon personnel on the hazards involved, supervise the selection and donning of personal protective equipment, assign at least one person to set up the decon area and mix the recommended decon solution(s), and enforce strict access control points into the hot zone and decon area with assistance from the entry/exit monitor.

The entry/exit monitor will work with the decontamination monitor to establish entry/exit points into the hot zone and decon area, monitor and document the movement of all personnel and equipment entering these zones, monitor the length of time the entry team(s) have been using air, notify the team safety officer as to the entry team's status, and relay pertinent exposure information to members working in the hot and warm zones.

The member assigned to product identification remains with the hazardous materials response vehicle and will collect and distribute information obtained from different reference sources, serve as the contact person when calls are made to chemical database services or manufacturers, relay updated information to the Haz Mat team leader and the team safety officer as it becomes available, and document the disposition of equipment taken from the team response vehicle.

"We work with the Port of Seattle and the U.S. Coast Guard on any incidents that occur in the harbor area," continues Captain Chikusa. "Probably 5% of our total Haz Mat calls are related to harbor incidents, and they are mainly spills from containers that are coming from overseas. The Seattle Fire Department contracts with a certified industrial hygienist from a private firm so that anytime we have a shipboard fire, or any type of fire that requires knowledge beyond our capabilities, we can use him or her as a reference source. In addition to incidents involving marine response activities, the industrial hygienist might deal with situations where significant quantities of hazardous materials are unknown or might be present. He or she also consults on non-routine Haz Mat incidents for which the department lacks appropriate instrumentation, or on any incident in which the incident commander determines that additional technical resources are necessary."

How does the Seattle Fire Department protect the health of firefighters who respond to hazardous materials incidents? "We have an ongoing program in the fire department where the Haz Mat team goes through a series of special tests every year including blood gases, blood draws, and lung function tests that are pretty much mandatory," responds Captain Chikusa. As far as medical monitoring is concerned, anytime we're dispatched to a large incident where we are going to be in protective suits, there is always going to be a medic unit that is dispatched along with us who will do medical monitoring for the team."

Captain Chikusa provided information on HMRT readiness for potential terrorist actions that could involve the use of chemical agents and biological substances. "At the present time, we are one of the 120 cities or major metropolitan areas to be trained by the U.S. Department of Defense in response to chemical and biological weapons. Also, we have been working with the Federal Emergency Management Agency teams starting to identify those areas where we may be deficient. We're also identifying equipment needs, and looking at areas where we may have to do mass decontamination if we have large numbers of people who become contaminated with chemical or biological agents. Right now, we just have a small capability

---

TABLE 6.1

Medical Conditions Under Which the Seattle Fire Department Haz Mat Entry Team Will Not Be Allowed to Suit Up

Systolic blood pressure:	Less than 100 mm Hg
Systolic blood pressure:	Greater than 160 mm Hg
Diastolic blood pressure:	Greater than 100 mm Hg
Pulse rate:	Greater than 120 beats/min
Oral temperature:	Greater than 103°F

Source: From Standard Operating Procedures of the City of Seattle Fire Department.

---

for a small incident but not the equipment necessary to protect a large group of people from chemical and biological warfare. We have a task force that is working on equipment and training needs specifically for the fire department, and we are working with other agencies such as the health department and a battalion of paramedics to try to identify cross-training needs so that we can work together. We have had small chemical spills at the University of Washington that were contained within rooms, so I would not call those large spills. Everything we have had has usually been contained to a small area. Probably the largest biological incident was when mustard gas was released in a residence and several people were exposed to the gas. We have not had any mass contamination involving chemical or biological substances. The Boeing Company had a couple of incidents that caused exposures and we helped out with decon of those people before they went to the hospital."

The Seattle Fire Department and its HMRT helped fine-tune the CAMEO software, a computer database for storage and retrieval of pre-planning data for on-scene use at hazardous materials incidents that was developed the U.S. Department of Commerce, NOAA (National Oceanic and Atmospheric Administration), and the U.S. Environmental Protection Agency. Now available in Macintosh and IBM formats, the software integrates a chemical database, emergency response information, air dispersion models, and local maps. Through the NOAA office in Seattle, the Seattle Fire Department HMRT members tested and commented on the brand new CAMEO system as it was perfected some years ago. The CAMEO database contains more than 3300 chemicals that can be identified by chemical response information including chemical name, formula, odor, color, synonyms, general description, NFPA hazard rating, fire and explosion hazard, physical properties, personal protective clothing, and health. It also gives tactics for firefighting, evacuation, and decontamination. "Basically, our first line of defense at a hazardous materials incident is going through the CAMEO system," explains Jeff Chikusa. "Along with the CAMEO system, we have a number of different reference manuals that we can use as a backup or in conjunction with what CAMEO indicates.

"Our job is to cover the metropolitan area of Seattle, in doing so, we have mutual aid agreements with ten different regions. Basically, we cover the whole Puget Sound area. Because of the mutual aid agreements, if we are needed, we can go outside the city to help the smaller jurisdictions when they have run out of resources. We also have had incidents where our resources have run so low that we've had to ask for help from other cities. We are probably one of the larger Haz Mat teams in the area in that we have a minimum of 11 Haz Mat technicians on each shift. We do that for safety reasons. We feel that to safely manage an incident we need 11 technicians at a time to properly respond to an incident. That is department policy.

"A telephone call to 911 is our first line of notification. Everybody is dispatched through 911, so depending on the type of call it is, the dispatcher's responsibility is to categorize the incident that we're sending the company to," continues Jeff Chikusa. "If the dispatcher feels it is a hazardous materials response, she or he dispatches the Haz Mat team, the closest engine company, and the closest battalion chief. Sometimes, if it's an unknown product that's been discarded on the street, or something such as that, just one engine company may be dispatched to try to identify the product. If they feel that it is something beyond their scope or capabilities, then they would call the HMRT. We would talk to that engine company, get some information from them, and determine whether or not to send a Haz Mat response for such an incident.

TABLE 6.2

Categories of Hazardous Materials Incidents in Seattle

Level 1 Incident	Spills, ruptures, leaks, and/or fires involving hazardous materials that can be contained, extinguished, and/or stabilized using equipment, supplies, and resources immediately available to first responders trained and equipped to the operations level.
Level 2 Incident	Any hazardous materials incident that can only be identified, contained, extinguished, and/or stabilized using the expertise of the Hazardous Materials Response Team, or that requires the evacuation of civilians beyond the perimeter of initial scene isolation.
Level 3 Incident	Incidents that require expertise and resources beyond that of the Hazardous Materials Response Team, that require evacuation of civilians from a large geographical area, or that involve multiple jurisdictions or multiple outside agencies.

Source: From Standard Operating Procedures of the City of Seattle Fire Department.

"If we have a major incident, we would try to videotape that level of incident, but to be truthful we have been pretty lucky to date, and nothing that we had is what I would call a major incident. We basically have dealt with small spills which have amounted to containment. Once we have contained a spill, we call in a commercial response contractor to handle it from there. Most of our spills have been from containers that are clearly marked. Once in while, we will come across discarded containers, but we have a HAZCAT kit which has hazardous categorization capabilities and we have team members trained to operate it. So we can break a sample of the product down to basic categories such as flammables or nonflammables. We are not able to make a perfect identification of the product, but we are able to categorize it, or identify just a basic grouping of the material. We also work with the Department of Ecology. If any further information is needed, they try to break it down further, or send it away for analysis."

How has hazardous materials response changed in Seattle? "The biggest change has been in the increase of public education. People are more aware of the types of materials they use and how to handle those materials. Our primary responsibility on the Haz Mat team has always been safety, and in the '90s safety is probably the biggest theme of all fire department operations. I think what has changed Haz Mat response more than anything else is just public knowledge of hazardous materials. That's probably due in large part to SARA Title III* and how businesses have to report, and general information the public now has about hazardous materials. In the City of Seattle, and in the rest of the county as well, there are hazardous materials stations that are set up for household goods. I think the people realize that once they have some of these products in their houses for a long time, the products could become volatile or unstable. They are thus getting rid of these materials before they become a problem, whereas before, because of being uninformed, people would hold on to these materials until they presented a problem. Basic public education has really eliminated a lot of unnecessary calls for the fire department."

TABLE 6.3

Equipment Stored within the Hazardous Materials Response Vehicle's Office

Video camera, tripod, Tasco binocular, Swift binocular, 2 clipboards, Seattle phonebook. 1 Haz Mat team coordinator vest, 2 safety monitor vests, flashlight, NEC (FAX) car battery adapter, Ethylene oxide monitor, 1 roll duct tape, 1 box extra FAX paper spools, Polaroid camera, 2 boxes Polaroid film, Weather-Pak spare keys, Weather-Pak receiver, FAX, FAX battery and charger, cellular phone, phone battery and charger, HP Deskwriter printer, Macintosh monitor, Weather-Pak receiver, Motorola Syntor transceiver, Macintosh computer hard drive, Bernouli Omega Multi-Disk, Macintosh keyboard, Windmeter, office supplies, Haz Mat Incident worksheets, Bernouli CAMEO backup, Terry towels, HIBI Stats, computer mouse cleaning kit, diskette drive cleaning kit, Macintosh reference manual, 2 MSA chargers for CGI 261, 2 MSA chargers for CGI 360/361, 2 MSA chargers for Microguard, MSA Omega Power charger for PASSPORT, masking tape, hanging folder reference file, chlorine indicator strips, Eberline manual, HAZCAT kit chemical inventory, Unit 77 response summary forms, Codebreaker 2.0 chemical list, journal card exposure list, Boeing Plant 2 radiation report, radiation emergency handbook, explosive chemical list, resource list, slide rulers, equipment replacement sheets, radiation at marine facilities, prefire and assorted Haz Mat preplans, Seattle Fire Department Haz Mat overviews, Haz Mat danger notices, catch basin eductor, waste water disposal, clandestine drug lab information, Hitachi battery charger and battery for video camera, set of tire chains, waste basket, office chair, Unit 77 team line-up card, 1/2 gallon pressurized pump can, dry chemical extinguisher, 6 Haz Mat stools, 1 portable toilet.

List provided by the City of Seattle Fire Department.

---

* Superfund Amendments and Reauthorization Act of 1986, Title III. The Emergency Planning and Community Right-to-Know Act of 1986. This law provides towns, states, and the federal government a "right-to-know" authority for nearly every firm that produces, stores, buys, or ships hazardous materials. Title III also includes provisions for emergency planning and response programs. The law mandates local emergency response planning committees, groups appointed by the state emergency response commission, to devise comprehensive emergency plans for the community.

With 60 Haz Mat technicians staffing 4 shifts of 11 technicians each and servicing 10 mutual aid agreements with other communities, does the department encounter problems with scene management, tactics, or operations when responding outside the City of Seattle? "In Seattle, we have always operated in a certain way," says Jeff Chikusa. "Because we have such a large team, we have more personnel and greater resources than the areas to which we respond, so our tactics on scene aren't always the same. However, whether we respond in an outlying area or we respond in the city, we always set up the scene in the same way because that's what we are comfortable with. We try not to deviate from that because that's how we are trained. Everything we do in the Seattle Fire Department follows the team concept; whether its 2 or 200 people working together, we are all trying to achieve the same goal. Cooperation is the primary focus of the team concept. We have different segments within the Haz Mat team such as our entry team, our backup team, or our decon team. Everything we do is in pairs or teams with each individual supporting the team effort. All have certain jobs to fulfill, so every position on the team is the responsibility of those team members working together to make sure an operation functions smoothly.

"Our training is designed so I can pull anybody from our manpower pool and put him or her in any position and feel comfortable that person will do a decent job. That's the way we train, everybody has experience doing different jobs. In case we have a large number of people off on vacation or disability, I don't have to rely on the same two people on each shift doing that job. Everybody can fill in for anybody."

Contact: Captain Jeff Chikusa, Haz Mat Team/Unit 77, Seattle Fire Department, 301 Second Avenue South, Seattle, WA 98104-2680; 206-386-1410; 206-386-1412 (Fax).
Rick Picklesimer, Haz Mat Team/Unit 77, Seattle Fire Department, 301 Second Avenue South, Seattle, WA 98104-2680; 206-386-1410; 206-386-1412 (Fax).

BRAMPTON DEPARTMENT OF FIRE AND EMERGENCY SERVICES HAZARDOUS MATERIALS RESPONSE TEAM

Captain Michael Clark has worked for the Brampton, Ontario, Canada Department of Fire and Emergency Services for 15 years doing both fire fighting and hazardous materials response. He has been on the Haz Mat team for nine years and been the Haz Mat Coordinator for six years. Before he created hazardous materials scene management computer software for the Brampton department, he had no formal training in computers.

"Our department was in the market to purchase a computer, and while I knew that we were going to use it for a database, I felt there was more the computer could do for us than just giving us a database for chemicals. I thought it could help guide us through the procedures which are consistent to every Haz Mat call and which are in written standard operating guidelines (SOG). The difficulty you have when they are written out is that people don't want to go searching for them on an emergency scene. Nobody wants to read from a whole body of literature to find specific information. The advantage of the software I developed is that it's based on a simple 'Help' program and it is easy to tunnel into necessary information, get the information, and get out. You don't need to know everything at a Haz Mat call. If you are going from step 1 to step 200 there are going to be little holes in what you can recall, or what you particularly need, for a specific incident. With computer software, the basic idea is to have all the information there but have it in a format that allows you to get just the information you need.

"I bought a book, *Developing Applications* by Christian Solomon which has a chapter dedicated to developing help programs. A help program is like any Window-based program. There's going to be a help button and if you click the help button you're going to see some highlighted text. You can jump from that into more specific information and have more highlighted text, and then jump to more specific text, and on and on until you get the information you really need. I could see the relevance of that kind of format applying to the information that is necessary at a hazardous materials incident. You can go through a consistent set of steps. Using that consistency, I began to build the software program. For instance, you go to whatever your first step is. If you clicked on that help button, it would give you more detailed information about that subject, and then maybe you'd have another five steps which are basically substeps. Then you can go to substep after substep and get into more and more detailed information. This is a good way to organize information so that somebody can tunnel down to find the information that they are looking for.

"Let's consider an incident where there is a sulfuric acid railcar that is fuming at the top. You want to send in an entry team and it is necessary to convey to the members specific information *before* the entry team enters the hot zone of what that railcar fittings are going to look like and how they operate. They may remember something from their prior training, and they may even have been up on a sulfuric acid car although that is very unlikely. You do have to have some basic knowledge of the manner in which various steps of Haz Mat response are organized. The computer program might go from 'identification,' to 'container types,' to 'railcars.' Then you could pull up the 'stenciling car' which would give you the information about where to find the railcar classification number. Knowing that, you could ascertain the type of car involved, and move to that type of car which is also in the computer program, and view various illustrations of that particular car. If you wanted a closer view of the domeway or manway, you could click for another view, or click again to see just the appliances that are on top of the manway. If you thought it was leaking from the safety vent, click again for a drawing of the safety vent including tips on how to plug or just temporarily contain a leak coming from the safety vent.

"The problem with Haz Mat calls is that they are so varied," says Captain Clark. "The problems will not be the same every time unless you're working with industry where they are constantly running into the same problems. For public safety agencies, the problems and the specialized equipment needed will be much more variable. One time it's a leaking rail car, the next time it could be a leak or spill in almost any industry. As an example, if a HMRT had a leak from a rail car, the team leader may say to the entry team, 'You are going to have get on top that rail car, find the dome cover, open it up, take a look at the devices inside, pick out the safety release valve, and this is what I want you to do with it.' Conveying that information to the entry team is very difficult. The advantage of a software program is that you can pull up a particular Department of Transportation classification of the rail car in question, take a closer look at the dome or the manway, and open the particular devices that are on top. The program may even provide tips from your own team's experience, not the regular way of handling the situation but more of a responder's way of getting the job done. One tip might be to put a rubber ball down the safety vent with some type of plug in it, the type of information that every hazardous materials technician picks up over time. A strength of the team's own software program is that every time they find a better way to do things they can enter it in the software program, and then it is shared and provides a lasting advantage to the program."

What is the starting point for a homemade, scene-management software program? "In the beginning, it was pretty much our own department's standard operational guidelines," says Captain Clark. "I've taken a look at other people's SOGs and they were pretty basic: 'come in from upwind and uphill,' and basically that's it. They don't really tell you very much. I also looked at a procedure the Canadian Fire Chiefs Association came up with a long time ago. It was vague and did not fully explain what to do if a person is wearing a Level A chemical suit. The manual treated it as if everyone was wearing only bunker gear. This wasn't practical for us and did not tell us what to do. I rewrote this procedure and went into a lot of detail. When it's a computer program, it doesn't really matter if it's a lot of detail because you are not going to read it all. You are just going to read what you think is pertinent at a specific incident.

"In the computer program, I've also done layouts, the drawings of certain procedures to support the reading material. Rather than just reading the different steps, you can click along with the layouts and get more useful information. Most of the required information for making your own computer program wasn't too hard to get just by taking courses. If you've taken enough courses, you have probably had the necessary information given to you. For example, good information and manuals on rail cars are pretty easy to obtain. You want to do a better job, so companies and government agencies don't mind giving you that type of information. SOGs are incorporated into the computer program from our general procedures all the way to the end where there is information dealing with fundamental reports. For instance, we have a general operating guideline for PPE from how to choose it to how to test it. There is also one dealing with decontamination, as well as guidelines for all the other different procedures."

How do you get started on such a large project? Do you just start stepping through your various procedures, protocols, functions, tactics, standards, medical surveillance, emergency response plans, levels of training/incidents/protective equipment, containers, team concepts, patching and plugging, or other necessary information for a HMRT? "I'll tell you how a lot of it comes," says Michael Clark. "You go to Haz Mat incidents and you get feedback after the call. You may see somebody use a Level A suit when it would be okay to use a Level B suit. You say to yourself, 'what were the criteria they used to make that decision?' and maybe there were no criteria for suit selection. If there were no criteria, then someone else could again make the confusing error. It's not enough to simply correct the individual. Under those circumstances, it's better to come up with criteria, program them into the computer, and make it simple for the responder to access that information. The logic is already there. I have gone to various industries and think to myself, 'how am I going to remember what this industrial building looks like inside two years from now?' I know now because I'm walking through it, but what am I going to do later? That kind of reason always gave me the incentive, a bit of a push. There was a need, so I started working toward filling that hole. The software is still a long way from being done. Actually, it will never be done because every time I fill another hole, there's a hole needing to be plugged."

Captain Clark confesses he devised the scene management software at home. "When some people say they don't have the necessary time allotted at work, I'm the same. I do all the software development at home. I don't have any time at work to do that. In fact, I bought a scanner for myself and upgraded my computer so I could make the software. That type of work is interesting to me since I have a feeling of ownership of the response team. I feel responsible for the team when they go out on a call even though I might not be there. When they go out and do a good job, it feels good to me. I feel I had a part in any incident and that's reward enough for the effort I put in."

As part of the software program for the City of Brampton Fire Department response to hazardous materials incidents, Michael Clark developed a risk factor concept. "I can remember that years ago we had discussions with different firefighters at both ends of the spectrum. The conservatives would say, 'never take a risk, no risk is good enough.' At one time our SOGs were to say something like 'any risk in excess of an NFPA 704 system health hazard should not be entered.* Firefighters at the other end of the spectrum would argue that 'no risk is too great if you can save a life.'

"I used to have arguments with the firefighters in that regard, telling them they didn't really understand the term 'risk,' and giving them an example. 'What would you do if your daughter was lying in a pool of Level 4 pesticide?' They all said they would go and pull her out. I thought about creating a model for the evaluation of risk. I asked myself what is the logic we use when we go to a hazardous materials incident and we consider the risk involved. I read a number of articles about how logic is used to determine risk within industry. The articles discussed the potential of the event versus the expected consequences of the intended action. As an example, a tank might release its contents. If that were to happen, what would be the consequences if the contents were released out into the atmosphere? That is how industry determines risk. I thought that this logic would work with any type of risk factor, including risks faced by a hazardous materials response team at an incident.

---

* The National Fire Protection Association in Quincy, MA some years ago developed a "Standard System for the Identification of the Fire Hazards of Materials," better known as NFPA 704. One basic purpose of this standard is to safeguard the lives of individuals who respond to fire in industrial plants or storage locations where the fire hazards of materials may not be readily apparent. The standard rates the dangers of specific materials in three categories: Health, Flammability, and Reactivity, and indicates the order of ranking in severity from "1" to "4" with "0" representing no special hazard and "4" meaning a severe hazard.

Risk is the potential of an event times the consequences of that event. These things together form the risk.

"We might think we are helping at a Haz Mat scene, but we may actually be hurting the situation. If you have a pesticide fire and are pouring a lot of water on it as firefighters have always done, common knowledge says that method is not optimal because you're producing a lot of toxic runoff. In the final analysis, we may be hurting the environment rather than helping the environment. In another example, you may have shock-sensitive, unstable chemical. While you are thinking you may be helping the situation, you may shock that chemical. Or if you have a weakened container, and people are going in there and trying to patch the container, they may cause the container to fail. So while thinking they are solving the problem, they are actually increasing the problem, and such consequences must be considered before you take action. On the other hand, motivations for dealing with a hazardous materials problem are much the same as going into a fire — to save life, property, and the environment.

"If we go into a house fire, if we take the risk to save one person, if we bring the person outside the building and do CPR on the person, all before we establish whether they have a chance at survival, we're really only choosing between an open coffin ceremony or a closed coffin ceremony. Regarding property, we might save some damaged containers and the product within them, but the owners will have to realize that our actions may not help them since our mission is to control the spill with regards to public safety. The environment is a motivating factor at Haz Mat incidents. Say a chlorine tanker was releasing its product which is going to spread out and affect miles and miles of area. All it would take to control the situation is a team member getting up on top of that tanker and shutting off a valve, a really quick fix involving very little risk.

"Another motivation is inconvenience to the public. Recently, an incident in Toronto forced a response team to shut down Highway 401, the MacDonald-Cartier Freeway, because of a transportation spill. Afterwards, people were calculating the cost to shut down that highway. The consequences don't seem severe in such a situation, but there can be great financial costs. In Brampton, we have had to shut down rail lines. The railroad companies have sometimes calculated the cost of doing that as up to a million dollars per hour. Shutting down a busy rail line for ten hours results in a tangible cost. Compare that to the number of houses you would have to burn down to equal that kind of cost. Economics is certainly a motivation. Also, one of the main motivators for us once we have evacuated a factory is to get that facility back in service. We can always say we'll wait for a clean up company to come, but they might be three or four hours down the road. We could say that we have already covered the three mandatory factors: life, property, and the environment, but when you shut down a factory for a day or more there can be a financial cost involved. We are willing to go to great risk to save that cost, so that's another motivating factor for us. No one from outside can tell a responder at a Haz Mat scene whether that responder should or should not have gone in. Nobody from the outside can do that. The only questions they can ask are, 'Did you consider the disadvantages? Did you consider the risk? Did you consider why you were going in? Was your reason life, property, the environment or inconvenience to the public?'

You must weigh all these questions at the scene to make your decision. Nobody else can make that decision for you.

"It is important to have consistent procedures for decisions like picking your level of PPE. The usual instinct is to go high. People will pick a Level A chemical suit rather than the Level B because they know it's safer. On the other hand, when persons from industry see the bill they're going to go ask, 'why am I paying for a Cadillac when a little Sprint would have done the trick?' The computer program provides some consistency in the procedures. Of course, those choices should be evaluated before responders go to the scene to make sure everybody agrees on the logic.

"One practice at an incident scene that will keep responders out of trouble more than the computer program itself is good report writing. What we do at a scene involves a lot of procedures on checklists which we check off, and we record the time that the action occurred. The hazard data sheet, all checklists, and particularly photos, are kept in a package with additional material after we leave the scene. If any response team or fire department goes into court, they should not go in empty handed; they should look like professionals prepared to defend themselves and their actions."

How do you train people to operate the computer software system? "One advantage of having a computer program is that people find it more interesting than books," responds Clark. "We have the computer at work, and just the fact that it's there makes it interesting for members of the team to play around with it. Of course, the program keeps growing and that maintains interest. I'll drop hints to a different shift or to my own shift and mention that something has just been added to the program. A couple of members start searching around and looking for things, discovering new things that were not there the last time they looked. The computer program is enticing; it draws them in. The nature of the program is so simple that it doesn't scare away anyone who's never used a computer program before. Sometimes, a computer program will scare away people who are not computer literate. When they start looking into a program, they get overwhelmed because the program may be complicated. Our program is such a simple thing, just a click-around program. Find out what you want and get out. Everybody on all shifts can use our program.

"In an ideal world where everybody is perfectly trained, and where they never run into a problem that they haven't seen before or aren't familiar with it, it wouldn't make much difference if you had a computer software program or not. The reality is that there are hills and valleys in every department across shifts. Different personnel aren't going to there all the time so the computer program allows a level of consistency. Whatever the weaknesses might be in various shifts, the gaps in knowledge can be filled. Responders may be quite well-versed in many aspects of response, but may actually run into a problem at a specific incident. They know a process should be done, but they can't remember *how* it should be done. They can use the computer program to give them that particular piece of information. They will seek out that information on the computer as opposed to looking it up in a manual because looking it up would take an inordinate amount of time. When it's on a computer program, it's right at their finger tips. All they have to do is go for it, and a second later they have the answer right in front of them. They can get in, get the information they need, and get out. That's the advantage of the computer program."

Contact: Captain Michael Clark, 43 Copperfield Drive, Cambridge, Ontario, Canada N1R 7V3; 519-740-0726 (Residence); 905-874-2714 (Work).

7 State and Regional Haz Mat Response Teams

LOS ANGELES COUNTY HAZARDOUS MATERIALS RESPONSE TEAM

In Los Angeles County, one hazardous materials incident occurs each day. Two thirds of these incidents are related to industrial production and handling while approximately one third occur in transportation. Movement of over 11 million tons of hazardous materials occurs each year. These materials are transported over 900 miles of state highways, 18,500 miles of arterial streets, and 580 miles of railroads. The Los Angeles County Fire Department protects over 2.5 million people within 50 cities and unincorporated areas of the county. Eighty-five major hazardous substance-producing facilities, plus thousands of smaller industries that use or store such materials, are inside county lines.

In order to assess the hazard, notify proper agencies, and provide necessary interim measures to minimize the effect of hazardous materials releases, the Los Angeles County Fire Department and its three HMRTs utilize the Incident Command System (ICS) which allows multi-agency or jurisdictional management of such events. A 1979 report to the county board of supervisors, developed by members of 26 state, local, and private agencies, defined a need for specialized hazardous materials response teams within the fire department.

Hazardous Materials Squad 105, located in the Carson/Dominguez area, became operational on March 1, 1982, and Squad 43, based in the City of Industry area, was instituted the same year. Hazardous Materials Squad 76 became operational on December 1, 1985 in Valencia. Each unit was staffed by a fire captain and four firefighters on a 24-hour basis. Every Haz Mat response requires two squads, leaving one unit available for a second incident. In 1991, the L.A. County Fire Department changed its Haz Mat emergency response operations to a task force system that required a 4-person engine company housed with the 5-person squads to be trained to the same level as the Haz Mat squads. The squad is mainly for mitigation while the engine company handles decontamination. The task force system ensures that three Type I or II incidents can be handled at the same time. For a significant Haz Mat release, the county fire department can immediately provide 27 Haz Mat specialists trained and equipped for Level A operations.

In a recent year, the three Haz Mat task forces responded to over 800 incidents of consequence. The majority of these were related to corrosives, flammables, toxins, or heavy metals. The jurisdictional engine companies handled the spills involving hydrocarbons, such as gasoline.

Captain Michael Kniest is the hazardous materials coordinator for the three hazardous materials response teams located in Carson/Dominguez, City of Industry, and Valencia. He spends a lot of time on the road. "My function is to coordinate the three HMRTs that are located roughly 60 miles apart within county jurisdiction. Each of these separate task forces is responsible for an area about the size of Orange County, so I probably spend about 40 hours a month just commuting among the units. There are two radios in my vehicle, a dispatch radio as well as a tactical radio. We dispatch over one frequency, but for geographical reasons, each separate squad might be dispatched over several frequencies so I monitor the respective frequencies. The tactical radio is used on the fire ground so we don't tie up the dispatch frequency.

"In my car there is an onboard computer called an MDT (mobile data terminal) where I get all the information needed by personnel on the three separate response teams. My call sign is 'Haz Mat 1.' The MDT will start beeping and will display all the information I need. It shows right now that the last incident I responded to was a train derailment last night. It shows the incident number, the time as 9:30 p.m., the location number so I can look it up on maps we carry, the grid number of the county system, and the tactical frequency that was used at this incident. The local fire department sent a first alarm, and it was reported that gas was leaking from a train. All the vehicles that went to the incident (15 vehicles in total) were recorded by this computer as responders indicated they were underway. It turned out that a small quantity of diesel fuel had leaked, and this incident was closed at 10:44 p.m.

"We no longer radio that we are going; rather, we push computer buttons for 'en route,' 'on-scene,' 'special,' 'available.' If I wanted to talk with the engines, I could type the message in. There are three cellular phones for each of the Haz Mat response squads. These are probably the best tools I have since they are almost hands-free. I also have 'cheat sheets' for my tactical frequencies. When another agency does not use our frequencies, I have a scanner so I can at least listen to them."

Captain Kniest, who has been with the fire department for 29 years, provided some history of the evolution that has taken place during his tenure. "At the present time, we have 51 contract cities for which we provide fire protection. Usually for financial reasons, they elect to continue county fire protection. Some of these 51 cities are fairly large, and the Los Angeles County Fire Department also covers unincorporated areas in the county as well as state resource lands for which the state no longer provides fire protection. For example, the last city that elected county coverage had eight fire stations, which constitutes the size of one of our battalions since we have about eight stations in every battalion. Right now, we have 17 battalions with each fire station having anywhere from a 5 to 20 mile response radius. The placement of stations in the desert area up around Lancaster is more dispersed because of reduced population. In a metropolitan area such as West Hollywood there is a dense population. Some battalions cover areas that could take you an hour to drive from one end to the other while other areas would require only several minutes if there was no traffic."

The Los Angeles County Fire Department now has well over 3000 employees. That number includes lifeguards. About three years ago 100 employees of the health department who deal strictly with Haz Mat were joined to the fire department and now respond with the Haz Mat squads. These people are the state's designated health officers who deem a situation 'safe.' The fire department mitigates an incident, and the health officers then determine the standards that should be applied to cleanup and see that they are implemented.

Captain Kniest was asked if he had any problems recruiting people for hazardous materials response duties. "It goes in cycles. Years ago we started to provide emergency medical services, basically an ambulance service offering basic first aid. Nowadays, just about every fire department in the United States provides a really good emergency medical technician system. When that system was started in the 60s and 70s, we had difficulty getting staff into the program. The training commitments and the unknown hazards scared away program staff. The same thing has happened with Haz Mat. It is similar to the early history of emergency medical services; it doesn't appeal to everyone. Some people are very good at Haz Mat response, and some are average.

"Putting on a fully encapsulated suit puts a person into a very stressful and traumatic environment. In the last four years I have seen two people who were interested in the Haz Mat squads but realized they would not do well in that duty. Our emergency medical service offers some fairly decent skill pay bonuses, about 16% of the base salary for a paramedic, but Haz Mat has not caught up to that level yet. HMRT members now get 2.75% skill pay added to their monthly salary. I think we would have people stay longer if we had higher skill pay bonuses. This sounds like you can buy good hazardous materials responders, but this is somewhat the truth. We have some pretty high standards for training in number of hours and training objectives. However, the funny thing was whenever the training standards got higher, we had more interest. More people wanted to get into hazardous materials response."

Fire Captain Dyrck McClellan is attached to the HMRT at Fire Station 43 located in the City of Industry. "I've been a firefighter for 27 years, with eight years as a Haz Mat captain. I moved into Haz Mat because I saw a need for safety. Industry and private enterprise were working with synthetic chemicals and handling a lot more hazardous materials in general. I thought that Haz Mat was a very progressive area to go into. Currently, I am the hazardous response leader for a task force that consists of a fire engine, a hazardous materials squad, and nine members who are all Haz Mat technicians to State of California standards.

"Haz Mat technicians in California must go through 240 hours of state-documented training. The state actually monitors all the training which must be done to the state standards or it is not acceptable," comments Captain McClellan. "There is also a specialist job category that fits more into health-relation Haz Mat response. People in these positions are specialists in a certain category. Members of our health Haz Mat division (industrial hygienists from the county health department) all hold bachelor of arts degrees and are Haz Mat specialists. We also fit into the specialist category because we have completed an additional 40 hours of training for the technician/specialist position.

"The health Haz Mat responders are assigned to the fire department and they respond with the Haz Mat squads. Their function at the scene consists mostly of cleanup, directing the responsible party(s) in cleanup priorities and overseeing the entire operation. They have enforcement powers in case regulations or laws have been broken or a crime has been committed, and they have their own investigation unit. They always write a citation at the scene when something has happened that needs to be corrected, and they do follow up.

"In the County of Los Angeles Fire Department we have Haz Mat incident procedures. Now those apply more or less to the first responder, usually a fire engine that gets called to a hazardous materials incident. When a person calls 911, the dispatcher makes a determination whether it's a police call or a fire call. If the reporting party says that there is a hazardous materials incident, something is flowing down the street, or there is a big cloud, the communication center will dispatch the closest hazardous materials task force which includes the Haz Mat team, two additional fire engines, a paramedic squad, and a battalion chief.

"I monitor the tactical frequency and our MDT," adds Captain McClellan. "I make contact with the first responders or battalion chief and ask them what they have, and what they are doing about it. If it's not just a static incident — perhaps a drum in a field that's not doing anything — but an actual release, they generally ask for my advice. I have to determine the situation that exists, determine if the material is an unknown or can be identified, determine whether the material can be identified without getting responders exposed, and questions such as this. I start advising them on isolation or evacuations areas necessary, and estimate our expected time of arrival. We maintain contact on a tactical radio frequency, and I will ask for radio reports every five minutes. 'Is everything okay? Do they smell or see anything? Is anything flowing down the street? Are birds falling out of the sky?' Upon arrival at the scene, I do a face-to-face briefing with the incident commander who will usually ask about methods of operations. They will need to know if they need the health department, the flood control district, the Coast Guard, public works, flood lights, or perhaps the county agricultural agency for a pesticide incident.

"I set up a hot zone. In some areas it may be referred to as an entry zone or exclusion area. Is the situation dynamic, is it going to do something very dangerous, or is it going to get worse or just 'be there'? What action do we take? I will advise the first responder on certain actions that should be accomplished. I will have him or her check the back of the area, obtain traffic control, post access control points to keep people out of the area, etc. Once an exclusion zone has been set up, we will back off 50 yards if that much area is available — or at least 100 feet — and lay out the contamination reduction zone. The second captain will generally mark the various areas with tape. We like tape because it's very effective; people don't like to cross a tape that may mean the difference between life or death. The news media are our friends and we do not want them to get contaminated, especially not their cameras. We promote the use of zoom lenses with the media.

"The first and second-in firefighters — the two people that are actually going in and doing the work — will accompany me to a briefing. We have radio contact with the Haz Mat response vehicle and a clipboard so we can write everything down. Also attending the briefing is the safety officer. We go over the situation as we have observed it, size-up the incident before taking any control actions, and discuss it as briefly as we can. Then we return to the Haz Mat squad, tell the incident commander our recommendations, and then figure out a plan of action.

"We all go back to the response vehicle. I discuss what I believe we have to do, and ask the first and second-in persons what they think. People attending this briefing may be our decontamination team, any technical experts we need, and the responsible party. The RP can advise us how to get access to the facility, how we are going to reach the spill, what other chemicals are in the area, and which systems we may have to shut down. The driver of our response vehicle records all such information, and the decontamination team members are listening since they will do all required basic work while the entry team is suiting up. We use a tarp to lay out all equipment in the order it will be used. Two decontamination team members, usually the captain and the engineer, start marking the isolation area."

The Los Angeles County Fire Department has a decontamination area in place before any Haz Mat entry is made. "The whole process of preparing for an entry is a long, drawn out defensive process," says McClellan, "and some non-Haz Mat firefighters, aggressive people who want to take action and get the problem solved right away, have a very hard time putting up with it. The best we can do is give them a job and keep them out of the area. Haz Mat response is not as forceful as response to a regular fire or EMS call would be. It is more of a drawn out process of 'let's be careful and do this correctly.' Sometimes we do nothing. If the incident could be explosive, we will call the sheriff's bomb squad because we are not trained in such a response. We have a very good bomb squad here in L.A. county.

"I go to the incident commander and explain what we want to do. Usually, my driver is with me operating a tactical radio frequency of our own, what we call a blue frequency and have to request through dispatch. We have direct communication for on-scene communication, plus repeaters on the mountains in our area of coverage. If we talk on the repeater frequency, it goes out to the entire county, so we don't do that. The repeater mode is for dispatch only. We have a dedicated tactical frequency for the incident scene that no one else may use.

"There is an incident safety officer who responds from downtown. This captain keeps in contact with the incident commander and will be reporting our actions. 'We are going in, coming out, we need this, we need that.' We are always in radio contact when the entry team goes in. Everyone else is back behind the decon area. We have monitoring equipment, video cameras, and anything we think we are going to need. A picture is really worth a 1000 words so we always take pictures, with both video and instant cameras. With videos we can show the responsible party what we have in the hot zone, and videos are very good for training. When we do our reporting or documentation, we will attach still photos to the reports. After a few years, if the reports become inactive, we will take the pictures out of the file, date them, and store them in a bag 'just in case.'"

What procedures and training do entry team members need? Captain McClellan noted an entry team member must visualize what has to be done inside the hot zone. "When an entry person gets inside the hot zone, it's always different, nothing like what you actually pictured. He has to go inside and decide if what he was told to do is actually possible. We would never ask an entry person to wade into some liquid. It just isn't worth it unless lives are in jeopardy. In a lot of cases you would be sacrificing your equipment, but our people are the most important responsibility we have. A firefighter entry person has to have basic chemical knowledge to understand any chemical processes taking place, what is being viewed, and what may happen.

"The entry person has to size up the situation, prioritize actions, and decide whether we really even need to do anything. Many chemical reactions are fairly well spent by the time a response team arrives. Often, a chlorine leak is something that has occurred and is now gone. The most important part of an entry can be saving someone's life, and the firefighter would have to be able to determine if the victim is still alive. If not, it is not necessary for us to wade in there and drag the body out of a liquid. However, rescue is always first. Our priorities are life of victims, protection of property, the environment, and protection of equipment in that order. A firefighter will take a calculated risk to save someone's life, and because of training, that risk is not anything he can't handle."

As a fire captain, Dyrck McClellan has found hazardous materials incidents to be like most fires, small and easy to handle. "Truthfully, I can handle most of them over the phone. I can picture what is described to me. Incidents are usually static, and in the eight years I've been on the HMRT, extremely dangerous situations would probably number less than 50. A dangerous situation would be a tank failure, or a compressed gas such as propane, that could explode. Personally, I probably had a half dozen situations in which materials were on fire and could explode if we didn't do something. Gasoline, although not compressed, is very flammable and dangerous. An extremely corrosive material leaking into an area could cause pipes to fail leading to a dangerous combination of two or more chemicals. But, the most common deadly poison that we get called to all the time is chlorine. Hydrochloric and sulfuric acids are potentially dangerous although we are usually able to control them. One very bad chemical in our area, ETO (ethylene oxide), is used in hospitals. It will set off an alarm in the basement of a hospital, probably one of the worst places to have an incident because you have nonambulatory patients on the floors above. You have to find out quickly how dangerous the situation is. Usually, runs I've been on are less than 3% ETO, and usually the alarms are so sensitive that a woman's perfume or a man's aftershave lotion has set them off. Ammonia is a chemical for which we receive a lot of calls, but it is more controllable than other chemicals. Ammonia is usually a vapor at a very low concentration percentage, is normally outside, and is lighter than air.

"We are required to have annual physical exams regardless of age," says Captain McClellan. "We have a baseline examination where our blood gases and other items are evaluated and compared to the previous year's results to see if we are suffering any chronic or toxic effects from any chemical we may have been exposed to. They call us 'the canaries,' but actually we're the safest because we have the knowledge required to deal with chemicals, and we are not the first to arrive at an incident which is when people tend to get exposed."

Fire Captain Martin Scott has been on the job for 26 years, and a hazardous materials captain for 10 years. Based with Fire Station 76's hazardous materials squad located in Valencia, he is a state certified instructor for technicians/specialists as well as an incident commander. "I was a firefighter, a paramedic, an apparatus engineer, and worked as a foreman running state inmate camps that taught firefighting construction work. I became an engine company captain, and when the department was putting the Station 76 Haz Mat squad into service in 1985, it sounded like a good opportunity for educational background as well as a different experience from just fighting fires. It's been a real experience. I have been to classes all over the United States, including the National Fire Academy, the American Association of Railroads, and the California Specialized Training Institute. I have had many opportunities working on the Haz Mat unit that I would not have had any place else.

"We have abundant opportunities for education, experience, and training, especially this past year when we responded to a number of big releases in Los Angeles County and elswhere. Ventura County had an incident awhile back that ended up being a significant incident of seven or eight days duration. We were there for two days in the mitigation phase, and very few other people from our fire department got to attend. The experience was fantastic, especially working with federal, state, and local agencies."

Los Angeles County is geographically divided into thirds for hazardous materials response since there are three separate Haz Mat task forces; one in the City of Industry, another located in Carson in the South Bay area, and a third at Valencia in the northern area. "In Task Force 76 at Valencia, we cover all of Malibu, Lancaster, Antelope Valley, Newhall, Saugus, etc. Our jurisdiction runs north to the Ventura County border, west to the San Bernadino County line, and east to include the whole Antelope Valley area," says Captain Scott. "We cover and assist in some of the 'front country' in the Burbank/Pasadena/Glendale area. These cities work as a triad and have their own HMRT. Our coverage area is quite large, and some big transportation corridors run through it, including I-5 which goes to Oregon and Washington.

"We do have a number of highway incidents. At the California Highway Patrol scales where they pull trucks over for inspection, we have sporadic incidents. We have a number of illegal drug labs, and a lot of illegal dumping in the remote areas of the desert. We are finding that with PCP labs, instead of making the drug in their homes or apartments, the people involved are going out to canyon areas, cooking their product, taking their finished product with them, and leaving everything else. Recently, just three or four miles behind this fire station, we had a very costly drug cleanup."

Some years ago in East Los Angeles all three Haz Mat task forces faced a chlorine release that required the evacuation of about 30,000 people. The county board of supervisors thought that the county health department assisting at the incident did not have as good an emergency response program as the fire department. The supervisors determined that the county health department responders would come under the control of the fire department. "At the present time we have about 90 industrial hygienists called 'county fire health officers' who are managed by the fire department," continues Martin Scott. "Now we have all the same communications, including MDTs and dispatch.

"We have a better working relationship with the health department now than we did before. The relationship has always been good, but probably not as effective as it is today. Nowadays, our incidents resolve much more quickly because we are here to take a dynamic incident and make it static. We don't normally do cleanup unless the situation presents an immediate threat to life. If a non-threatening cleanup is necessary, generally the responsible party contacts the contractor and gets the cleanup completed. Cleanup is usually under the supervision of health Haz Mat, our name for the industrial hygienists. These health officers are the only ones who can deem an area 'clean' and therefore safe to let the public in. The fire department does not have the ability to do that, and we don't take on that responsibility. We find that incidents go smoothly and quickly with them. If they have to enter with us, they use our equipment. We may be unique in the United States in that our health agency is part of the fire department.

"All the state highways in California are under the control of the state department of transportation. They have an excellent response system with private contractors preassigned to certain areas of the highway system. In the event of a release, a contractor can be called immediately. Also, many CalTrans (state highway department) people are trained Haz Mat responders with 40 hours of training. On a highway, if we deem a spill nonhazardous they can clean it up. They can also pick up diesel, gasoline, or oil.

"We have a HAZCAT system to identify various chemicals that we find at an incident site. It helps us to learn if a compound is nonhazardous or to determine its hazard class. It has been an asset to us. We can resolve incidents quicker and get a highway open sooner."

In one incident at a truck scale, there was a spill of green, granulated material. It reportedly looked like a copper compound, so the Los Angeles County Fire Department Haz Mat task force was called. Responders used the HAZCAT for some analytical chemistry. The system works like a decision tree and has a series of tests and charts. Perhaps one test will give a positive color change or become effervescent. This may lead to another test which may take off in another direction. Eventually, you learn the hazard class of a material. There is also the ability to rule out substances that are not hazardous. The material in this case was completely inert and nontoxic. It was green-dyed aquarium rock that people put in fish tanks. An expensive cleanup job was thus avoided.

"Northridge is a large, local industrial park right off I-5. When an earthquake hit there, we had a number of releases, so we, along with Task Forces 43 and 105, worked that area for two days. There are some big film studios there, a lot of Hollywood productions, and a number of sound stages. These do not normally present Haz Mat threats, but they produce a lot of small commercial films with electrical processing and silicone-type processing, so a lot of acids and bases are in use. People came back to their facilities after the weekend of the earthquake to find that product had seeped into the next occupancy. That's when we started having problems. As a matter of fact, we put our reserve squad into operation and had it there fully staffed for eight or nine days.

"During the recovery work after the earthquake we had problems with highways, power outages, and many fires, particularly gas fires," adds Captain Scott. "While the Haz Mat squad was working on these problems, they were called to a major release at a packager of chlorine and acids for pool supplies. After that we went door-to-door in the industrial park. In the future, we would like to see our disclosure system target certain geographical areas and have a computer list of all the handlers within the area. We could go to our hard files, which would be marked as to which are handlers and which are not, and be able to pull the ones we need. We eventually did go to the hard files and went to check those properties. We checked every commercial property in the area, prioritized each as to sprinkler systems that were inoperative or operative, what problems the hazardous materials handlers had, and what mitigation was needed. County health even had to bring state health agency representatives to this area to work on mitigation because we were working with so many releases."

Fire Captain Ben W. Wolfe III has worked for the Los Angeles County Fire Department for 14 years, spending 3 years as a hazardous materials responder in a firefighter position, and 4 years as a captain supervising a hazardous materials response team of five individuals. He talked about the response functions of the Valencia team.

"When we respond to Haz Mat emergencies we operate basically as a technical resource within the ICS. We are there to handle the hazardous materials incidents. Some of the services we do supply to the overall department are identifying unknowns, taking samples, doing mitigation, and patching and plugging. We drill holes in overturned tanker trucks to off-load product. The main capability we have is to enter hazardous materials environments to do any kind of work. We have specialized personal, protective equipment, and the monitors and meters to do this work safely.

"We provide support to the incident commander by way of information and technical expertise about chemicals. An engine company comes with us as part of a task force. They do our decontamination and we handle the entrywork. We have specialized tools and equipment we carry with us. For example, we use a pneumatic patching system. We have big patches, small patches, plugs to go inside pipes, non-sparking tools we use when working in flammable atmospheres, and we carry monitors and meters.

"I tend to have a lot of confidence in our procedures and our protective clothing. I'm vigilant against suits failing, injuries, or falling debris. But, the only bad experience I have ever had was a chlorine leak in Ventura County. Chlorine gas was escaping and expanded as it cooled. It cooled the Level A suits that Ventura County Haz Mat personnel were wearing. The suits got so cold they just cracked, shredded, and shattered off the bodies of entry personnel. All the Ventura County entry personnel went to the hospital and the team called us in for assistance.

"Typically, a Haz Mat incident is not dangerous to us because we park a long way away and have the knowledge, training, and equipment to be able to handle it. Our chance for exposure is really not too high provided we follow the correct procedures. Still, you never know what is going to go wrong. Usually, in a fire situation our mentality tells us to go in and take care of the people. We want to do it quickly, and we want to be aggressive. In Haz Mat response you really have to stop yourself to be safe. You have to approach an incident more slowly and cautiously, and do a little more planning and preparation. We have to be very methodical in Haz Mat response. When we get new people on the squad, it takes them a bit of time to get their minds set on how we do it."

Station 76 at Valencia, Station 43 at the City of Industry, and Station 105 at Carson are all "dedicated" HMRTs. "All of us on the squad are full-time Haz Mat people, but the engine company is dual-purpose," explains Captain Wolfe "The engine will respond to fire and rescue calls as well as Haz Mat. The squad will do some rescue work infrequently. If our engine is busy, we will take rescue. All the people on the squad are trained firefighters and EMTs first, but we are the only ones who have the necessary training and equipment to do Haz Mat. We are kept available in case an incident occurs and are not sent for first-alarm structure fires or brush fires. We do go on second-alarm structure fires for manpower when the department has few personnel, but rarely."

Ben Wolfe was asked what qualities he seeks in a candidate for a squad member. "Actually, we don't look for anything," he says. "Our union negotiates that all the firefighter spots in our program are open for bids based on seniority. We do require candidates to be fully trained before they can work on the squad. That is, they need 80 hours of Haz Mat training before they are allowed to work.

"Since we cover a very wide area — Malibu, the San Gabriel Valley, Palmdale, Lancaster — our longest response time might be 1 hour and 20 minutes. In Los Angeles, response depends on the time of day and the traffic. One of the big problems we have is the Antelope Valley. In the summer, temperatures are usually 105° to 110°F. Inside a Level A suit, all your exhaled air stays in the suit, and you end up in a 'shake and bake' bag. Heat stress can be a big problem so we really have to limit the time people can be in such suits under these conditions. We recently obtained cooling vests that have helped a lot. We have had problems with people starting to show signs of heat stress out there on the desert after only 15 minutes in a Level A suit.

"We've had difficulties with the organizational structure at an incident. The other agencies involved in hazardous materials incidents don't have the same type of training and equipment as we have. They tend to go on incidents so infrequently that oftentimes events do not run as smoothly as they should. For instance, a fire chief will have the same mind set he always has, 'Now let's get this job over and done with so I can get my district covered.' It takes a lot more time, involvement, and support to actually field the team to go in and deal with a Haz Mat incident. The problem typically involves several other agencies. Everyone comes from a different background. We don't usually have a lot of problems dealing with our own people, but coordinating different agencies, getting everybody thinking the same way, and getting the organization down is really a big problem.

"We have had some interesting incidents. Last year a truck full of blasting agent — basically the same material that was involved in the Oklahoma City federal building bombing, a mixture of ammonium nitrate and fuel oil that is used in quarries for blasting — overturned and spilled its cargo on the ground. The truck driver told us what the product was. The real problem here was that the explosive was one with which we did not have much experience. Explosives in our county, including bombs, are handled by the sheriff's bomb squad. This product was spilled and not a 'bomb.' It did not meet the sheriff's office criteria, and we were called. The product was basically a hazardous material. Even though we knew what the explosive, hazardous material was, we didn't know *how* hazardous it was in its present situation. We were able to call the manufacturer to learn how hazardous it might be, how to approach it, how to clean it up and stop the leak. We have both a cellular phone and a FAX machine in our response vehicle, two of the most useful items we have.

"In another incident, we encountered an unidentified product at our local prison. It was a cleaning product, and some inmates had sprayed it on a hot oven. The product had vaporized and four people were complaining of difficult breathing after inhaling the fumes. When we got to the prison, we asked the staff for a material safety data sheet (a document that contains information regarding the specific identity of hazardous chemicals, including information on health effects, first aid, chemical and physical properties, and emergency phone numbers) they are supposed to keep on file, but it could not be located. We suited up the team and sent them in to retrieve the bucket and test the atmosphere. They came back out with the can, but it had no real hazardous materials warning label. However, it did have the name and address of the manufacturer.

"We treated the victims, and sent them off to a hospital. I got on the cellular phone to call the manufacturer, said I was with the L.A. County Fire Department, and asked to speak to somebody with chemical expertise. Worried about product liability, the manufacturers were more than happy to comply, and connected me to one of their chemists. The chemist was able to tell me immediately what kind of hazardous materials were in this product and their composition. I asked him if he had an MSDS, he said he did, and I told him to FAX it to me at the scene. I hooked the FAX machine to the cellular phone and received the FAX copy of the MSDS which provides instructions for treating victims. I then sent a copy of the MSDS from the Haz Mat squad to the hospital so the physicians would have some idea what the product was and how to treat the victims. The cellular phone and FAX machine have been valuable tools for us."

Captain Wolfe said the team had a temporary problem using cellular phones during the earthquake recovery when everybody seemed to be on phone. "Immediately after the earthquake hit, there was a bit of a problem using the cellular phone at certain times, but it seemed to clear up very well," he remembers. "However, we were totally isolated in our coverage area. We were flying firefighters and water into Magic Mountain (a recreational park near Fire Station 76 in Valencia) because I-5 and Highway 14 were completely closed down due to damage to the interchange. The transportation corridors were really disrupted, but communication with cellular phones seemed to work fairly well.

"We had a tremendous amount of damage in the canyon area. Keebler (the cookie company) virtually came down. We had a brand new regional post office, that hadn't even opened yet, that sustained a lot of damage. People saw on television the remains of the freeway interchange that collapsed killing a police officer on a motorcycle. Many fires ignited in mobile homes which did not have any earthquake retention devices to hold them in place. When the trailers started shaking, they broke away from their natural gas lines and the gas found an ignition source. Many water supply lines were down so some mobile home parks had no water."

Ben Wolfe believes first responders are integral to the success of hazardous materials response teams. "All firefighters should have basic background knowledge obtained in a first responder's course. We, as a HMRT, always get to an incident after it has developed somewhat and after the first responders have already arrived on the scene. First responders who have some good basic knowledge are invaluable.

They can deny entry to people not required to be there, handle some notification requirements, and perform mitigation within containment areas. First responder courses are very important."

The Valencia HMRT has an onboard computer that carries information on 20,000 chemicals as well as a complete library of chemical manuals and other publications. "We come across so many chemicals that frequently we find some that are mixtures of chemicals, are named in a way that is different from what is normally used, or have a trade name that is a proprietary secret of the manufacturer. In such cases, we rely on our cellular phone to call the company.

"Because of our chemistry background we can sometimes look at the formula or the chemical name of a product and break it down to learn its constituent components," adds Ben Wolfe. "It happens frequently with pesticides. We can call the agricultural commission or the farm outfit, and they will have information on some these trade name pesticides.

"When the space shuttle project had a problem with tiles coming off the space capsule, many manufacturers tried to make a product that would hold these tiles to the outside surface of the capsule. An outfit in the City of Industry produced a glue for that purpose. No chemical formula was involved, and the product overheated and started an actual reaction. Task Force 43 was called to mitigate this reaction. They studied it for a long time before they packed it in dry ice to reduce the reaction. When people have problems with chemicals, the end result is they call the fire service to solve the problem. Sometimes, we're short on answers, especially when dealing with mixtures."

Contact: Haz Mat Coordinator, HMRTs, Fire Operations, County of Los Angeles, 1320 North Eastern Avenue, Los Angeles, CA 90063-3294; 213-881-2485; 213-780-0307 (Fax).

SOUTHTOWNS HAZ MAT AND COUNTY OF ERIE HAZ MAT TEAMS

Dean A. Messing works two volunteer jobs, one as team leader for the Southtowns Haz Mat Team in upstate New York, and one as commander of the County of Erie Haz Mat Team. He also has a paid job as Deputy Commissioner with the County of Erie Department of Emergency Services. He has 19 years in the fire service.

"Southtowns Haz Mat Team formed back in 1986, a result of a grant from a New York state senator," remembers Dean Messing. "We wrote a grant proposal for $350,000. The first part of the grant we received was for $25,000 which we used to buy training and equipment for the Southtowns team. We got an additional grant for $55,000 that helped us to continue building the team as well as building our training resources. Then we got additional funding, $271,000, to continue to put the team together. The Southtowns team went on-line October 1, 1990 with full Level 2 response capabilities.

"My paid job is as disaster coordinator for Erie County, NY located in Buffalo where I do disaster planning and review disaster plans of communities. Our county is one of the only counties in the state to have a disaster advisory board which is made up of industrial representatives, disaster coordinators, and other community representatives. As a result of my work with the board, I became one of the co-founders of the county team. That team started as a result of a fire on July 4, 1988 when medical personnel were decontaminating firefighters in the parking lot of a hospital. Firefighters going through decon were exposed, naked, to the public. A task force was formed to solve the problem, and the county purchased a portable decontamination unit for this team. The county team started out as just a decon unit."

Dean Messing was asked if he thought being on two Haz-Mat teams was beneficial. "I find a benefit in being on both teams because any training I don't pick up at the county team I pick up on the Southtowns team," he says. "The county team was funded strictly for the purchase of a decontamination unit, a vehicle, and a command post. The county team is not funded by the county any more. Now that team gets it funding through settlements of lawsuits and donations. The funding for the Southtowns team is a little different. We take the total assessments of the communities we cover, add it all together, and then come up with a percentage each year. During four years, the money totals $5,000 for each town or village. We cover eight towns and four villages.

"Our funding comes from sources other than the municipalities. The towns of Hamburg, Eden, and Evans fund us through their respective fire districts. Alma is funded through a legislative resolution from the county. We have a board of directors that governs our monies and how they are spent. We submit a budget, they say 'yes' or 'no.' If there is a need to change the budget during the year, we have to submit budget resolutions."

The Southtowns Hazardous Materials Response Team has 45 members, 4 vehicles, and 2 buildings, a rough total of three-quarters of a million dollars worth of equipment. "However, we don't get out of funding. With the volunteer fire companies we cover, we ask them for their old or surplus equipment," says Messing. "If we can't use the equipment on our team, we swap it or sell it. Our command post is a 1987 school bus that has high and low radio capabilities, our own licensed FCC frequency used for entry team members, four cellular phones one of which is strictly dedicated to a FAX machine, and communications with controls that we cover. We have a complete reference library, and two computers on board the command post. When we arrive at an incident, we punch the chemical in question into CAMEO and check our reference library. If we can come up with two references in addition to what CAMEO says, we recommend to the incident commander the action we should take and get an approval before we do it. We never take over an incident.

"Besides the four vehicles we use, we also have two buildings. The first building is to store our equipment and supplies. The second is a 1200-square-foot office building that we also use for training. The Southtowns HMRT covers about 300 square miles of Erie County in New York State. Our 45 members have to be active members of a volunteer fire company in order to belong to the Haz Mat team. We are now in the process of putting a proposition into our by-laws that will allow us to accept personnel other than volunteer firefighters as long as they are involved in some sort of emergency activity such as EMS, disaster coordination, or chemistry or have some hazardous materials background. We have a fairly young membership. Our average member is 25 to 30 years of age."

The Southtowns HMRT requires an annual physical examination including a test of blood gases in accordance with the CFR 1910 guidelines. Every other year, chest X-rays are required. An ambulance is required at every incident in case a firefighter is exposed to a hazardous material and requires transport to a hospital. We don't transport anyone until the person has been decontaminated. Each team member has a medical file. If they leave the team, we retain the file for 30 years to comply with CFR 1910 guidelines.

"The county team consists of 40 to 50 members, and you don't have to be a firefighter to belong to this team," continues Messing. "A member has to be a technician level responder by the completion of one year of membership (this requirement holds true for the Southtowns team as well).

"Both teams have sectors — EMS, decon, command, or security. Personnel are assigned to a particular sector in various ways. If they have an EMS background, they are assigned to the emergency medical service sector. A person who has a background in communications and is good at handling paper, might be assigned to the command section of the bus. Also, we have our own security sector who goes along with the HMRT. They secure the area for us. If you don't have an identification card, you don't get on the incident site. They also secure the command bus to deter anyone who does not belong on the vehicle. If a candidate does not have a particular assignment during the first year, we will assign that person to the decon sector and let him learn decontamination first. A person has to start in another sector before he or she can join the entry team.

"I mentioned settlements of lawsuits as a funding source for the county team. The environmental group, Atlantic States Legal Foundation, is a group of lawyers who sue companies that don't report under SARA Title III and IV requirements the quantities of extremely hazardous substance and non-extremely hazardous substance that the firms release. The lawyers also sue companies who don't report or who are lax in their reporting. They try to get the companies to settle, and they give the money recovered to the Haz Mat teams who are going to respond to emergencies.

"The county team received two settlements in this manner, one for $17,000 and another for $20,000. They also received a third settlement, but off-hand I don't know the amount of this third settlement. The Southtowns team also received two settlements in the same fashion, one for $17,000 and one for $20,000. With these settlements we purchased equipment we needed but normally could not get from our budget.

"Anytime we have an incident, we bill the spiller unless the responsible party is one of the municipalities that fund us", says Messing. "When we know who the spiller is, we have about a 95% recovery rate. We bill the spiller for maintenance of equipment at the rate of $25 per vehicle. We also bill them for the use of our cellular phone. When we send out a bill, we attach the federal legislation that requires us to recover costs. If the spiller won't pay, their insurance company usually pays.

"We don't respond to that many incidents. Last year we responded to five incidents. The county team had about three. The year before that we had about eight incidents. One of the bigger incidents we had last year was a train derailment in the Town of Hamburg which involved nine rail cars. Five of the cars contained butane while another four cars contained #4 fuel oil. The cars were twisted and mangled, and we thought one of the butane cars was leaking. One fuel oil car lost its full contents of 28,000 gallons. The fire department had arrived very quickly and put out the fires that resulted from the derailment. Still, we had a bomb sitting there waiting to go off. One spark could have set off the butane in five cars. We stopped the fuel oil from leeching off-site. A commercial response firm was called in and cleaned up the fuel tank cars and then the butane cars. The incident lasted about 18 hours.

"Under the incident command system, the Haz Mat team leader reports to the incident commander and advises what actions the team should be taking. The incident commander says 'yes' or 'no' to the advice. Ninety-nine percent of the time the incident commander will say, 'Yes, go ahead.' My operations officer works with the incident operations officer.

"The Incident Commander will probably ask what the team is going to do with this incident, how we are going to handle it, what additional resources or equipment do we need, and what can be done to make the job go safely. Any time a news reporter arrives, I refer him to the incident commander because that is where the answers are supposed to come from."

The Southtowns HMRT has a requirement that after every incident a critique has to be done within a week. Every agency that was involved with an incident is invited to the review. "We go over each incident," continues Messing. "We want to ensure that we follow procedures. Each community has a Haz Mat plan that requires a critique. The incident commander picks the site where the critique is to be held and starts the meeting. We file complete reports of the critique. The one for the train derailment was 70 pages long. We include all the log sheets, the diagrams of how the train was wrecked, the layout of the wreck, what cargoes were involved, the material safety data sheets of the cargoes, CAMEO printouts of the cargoes, the comments that were made at the critique, etc."

The Southtowns and county HMRTs are volunteer organizations. "We give them nothing but their identification passes and a baseball cap when they join the team(s)," says Commissioner Messing. "We are a specialized unit and we take pride in being Haz Mat team members. If they want a team shirt, they have to go out and buy it. Once they are a member of the team, they'll receive a lot of praise from the people who recognize the time and effort team members expend. It is similar to being a volunteer firefighter; they enjoy it and take pride in it."

Contact: Dean A. Messing, Deputy Commissioner CD/ODP, County of Erie, Department of Emergency Services, Room 1351, 95 Franklin Street, Buffalo, NY, 14202; 716-858-8477; 716-858-8072 (Fax).

HARFORD COUNTY HAZARDOUS MATERIALS RESPONSE TEAM

James W. Terrell is a chief officer with the Harford County Department of Emergency Services located in Forest Hill, MD. One of his responsibilities is the operation of the county hazardous materials response team. Terrell joined Harford County in 1968 as a dispatcher, was promoted to lieutenant in 1974 and then chief in 1985.

He joined the fire service in 1963 with a volunteer fire company in the northern area of the country and there came up through the ranks of lieutenant, captain, assistant chief, and chief. After 11 years as chief in the volunteer fire department, he was appointed chief in the county and had to give up the volunteer position as a conflict of interest.

"Harford County is a large area with almost 250,000 people and 454 square miles of area serviced completely by volunteer fire departments," notes Chief Terrell. "The volunteer departments run 21,000 calls a year for a very small dollar figure compared to what it would cost with paid departments. We try to help the volunteer departments as much as possible because they provide this public service. Any bureaucratic burden that we can take from an incident commander we try to relieve.

"The Harford County Hazardous Materials Response Team was formed in 1988, and the county began seriously funding the team in 1989," remembers Terrell. "We have always tried to put our funding toward equipment that will protect team member's lives rather than toward vehicles. We would be fortunate to get one large, fancy vehicle, but that has not been the situation. The vehicles we have are surplus school buses and Army ambulances. These vehicles have served us very well and have been efficient and functional. We do all the work on them ourselves including the work on the command bus generator.

"The team runs all chemicals spills and petroleum spills within the county that are 50 gallons or more. We average anywhere from 200 to 300 calls a year. We also handle investigations for the county sheriff's department who has a corporal assigned to me full-time to do investigations. Hazardous materials laws in the county allow us to investigate environmental crimes and problems. I can cite people civilly, and the corporal can cite them criminally. Overall, we can fine such people or take some other penalty action if necessary. We are also responsible for SARA Title III with about 200 companies reporting to us the hazardous materials they store at the present time.

"Our goal in hazardous materials response is to assist the fire department incident commander. When we arrive on scene, we provide technical assistance and support. Any entry work is done by the HMRT. By law, the incident commander of the volunteer fire department can turn the incident over to me if he or she desires to do so. Such action has to be done formally and officially. The county becomes responsible for the scene. The volunteer fire department then provides assistance to the team rather than the team assisting the fire department."

The team is allowed to bill for services such as time and materials in order to fund the efforts and recover costs. "We bill the spiller with one bill and reimburse the fire department for expenses. We receive payment on 62 to 68% of the bills we send out," says Chief Terrell. "One of the challenges we face is that I-95 runs right through Harford County as do Route 40 and U.S.-1. A certain percentage of our spills involve out-of-state trucking companies. Sometimes such bills can be a challenge to recover. After 30 days, our law department takes steps to recover such funds. Also, part of our budget is dedicated to hazardous materials response.

"The hazardous materials response team is staffed by seven of us who are full-time members of the Department of Emergency Services, while other members are not full-time employees of the department but are paid for calls and training. Once their pagers are activated, they become employees of county government. They sign a contract each year saying that they will respond, train, and keep their certifications current. In return, we pay them for responses and training.

"We generally have an easy time keeping up with our technician training requirements. My deputy chief is training officer, and we have two crew chiefs who assist with training. We bring in outside trainers, do our own 'hands-on' training, send people to conferences, and participate in the training provided by the U.S. Environment Protection Agency each year. We have some local industries that help us with training and equipment. The Philadelphia Electric Company that operates the Peach Bottom nuclear power plant, and the Aberdeen Proving Grounds, have been helpful. Systec, which used to be called American Cyanamid, has been very kind to us with equipment and donations. Our guys built a response trailer, and Systec paid for some of it.

"Only one vehicle we use is a new, custom built unit — the Harford County command bus. The catalyst for that purchase was the night that three police officers were shot. One was killed and the other two wounded. All the police departments in the county were involved that night and we had no way to coordinate all our communications. With the county command bus now available, the HMRT, sheriff's department, and hostage negotiation team might work out of it at different events. It can also be used at any big public safety event such as a walkathon, police, fire or ambulance event.

"The team often uses an older bus as a command center for hazardous materials responses. It carries all the reference materials, computer, FAX machine, Level A and B equipment, as well as meters and monitoring gear. That bus is used for spill control requiring gear such as diking equipment and overpack drums. We use a surplus Army ambulance for small spills and investigations. It pulls a pump-off trailer that we use for transferring product from one vehicle to another, or from a vehicle to a drum. There is a Dodge ambulance that we leave empty all the time so we can bring whatever extra equipment we need to a scene."

For medical surveillance the hazardous materials team follows regular OSHA standards. Each district's fire company provides a medical unit who will be responsible for medical command. HMRT members hand the medical command unit a kit that has all the information, including the necessary forms, to do all the medical monitoring. All recordkeeping is turned back over to the Haz Mat team for their records after the incident.

"Although the team does all first stage decontamination, fire departments do second and third stage decon," points out Chief Terrell. "We go out to each fire department yearly to train or recertify them in decon procedures. We also provide specialized training to the fire departments on specific types of incidents they may want to know about. However, decon is the area where fire departments assist us on scene because, although we have a 23-member Haz Mat team, there are not 23 members available at a time. We operate like a volunteer fire department; if you are available, you come. We have always operated with an adequate number of people. I've never been on a scene where we did not have the people we needed."

The county HMRT is dispatched along with the fire department that is called to the scene. In addition to Haz Mat response, the team does scene cleanup in the event of suicides, automobile accidents, blood, or body fluids. "There seemed to be a need for that service," adds Terrell. "We work very closely, 24-hours a day, with the sheriff's department. Early on, if a crime scene was a mess, we had nobody to clean the scene. Our team started doing it. Now some private contractors can handle it, but they charge and we don't. We use a medical waste contractor who weekly picks up manifested waste from an area that is OSHA-approved. This program takes a lot of responsibility off a volunteer fire chief. Even if he cleans the medical waste, he can't dispose of it legally. We can.

"Harford County has a great deal of industry. In the beginning we figured such facilities would be our nightmare. However, fixed facilities have actually comprised a very small percentage of our calls while transportation accidents accounted for our largest number of calls. We have faced a good variety of chemicals, plus an awful lot of petroleum products, particularly on I-95. We use a tiered response. If dispatch gets an 'unknown chemical,' we respond with a full Haz Mat team. If petroleum is involved, we gear our response to the amount involved.

"We had a recent incident that we handled as a Level A alarm. A guy bought a farm in a rural area of the county. He was cleaning up behind the barn and found a drum that obviously had product in it and was marked, 'Mustard Agent, APG,' a blister agent, a good bit of which is stored at Aberdeen Proving Grounds. The guy called us. At the scene, we did a little research and learned that the farm's previous owner had been indicted for mishandling hazardous waste. We called the fire department for assistance and called the Aberdeen Proving Grounds to send a battalion chief to give us some advice on what he thought might be the best procedure. We ended up overpacking the drum which was then transported by the Army back to the Aberdeen Proving Grounds' chemical transfer facility to be analyzed. The contents of the drum turned out to be used motor oil.

"We tend to have a lot of variety. We had a fuming nitric acid spill at the H.P. White Laboratories where they do weapons testing (they got a lot of publicity years ago when they tested the rifle that was used to assassinate President Kennedy). They had a small storage area where fuming nitric acid got out-of-hand, and we were called in to handle the emergency. Then we had the saddle tank punctures on I-95. Truck drivers can stop in the county and fill-up, so when there is an incident we find a lot of guys driving with 200 gallons of diesel fuel."

Chief Terrell was asked about the issues an incident commander faces on scene. "We face different questions every time," he says. "You have to be very precise when dealing with hazardous materials, but Haz Mat is not always a precise subject. Many of the people transporting chemicals don't know what they are carrying. The driver hooks a trailer to a tractor but doesn't really know the characteristics of the load. We ran into an incident on I-95 where we saw a liquid dripping, but the driver had a MSDS that defined the chemical as a powder. A secretary had typed the information incorrectly and given it to the driver. This type of mistake happens all the time. You have to check the spelling and be very careful. We don't assume that what the shipper says is correct. We make phone calls to knowledgeable people at the company to confirm information on the paper provided to us, and try to keep surprises to a minimum.

"The lieutenant who is in charge of the SARA Title III program for the county also runs with us and does all of our research. She is often on the first unit out because she lives close to headquarters, and will check the computer for applicable information. By the time we get to the scene, she hands me all kinds of pertinent data. She works in the bus/command center doing whatever research is required. We have a pretty good library on the vehicle as well as several computer programs that give us characteristics of the chemical(s) and proper protective equipment. We also get information from shipping papers and material safety data sheets. Our computer can give us immediate access to information we keep on fixed facilities."

Harford County has a contingency plan and a manual of standard operating procedures for personnel. We also have a hazardous materials and emergency operations plan that is updated annually. "All team members get a copy of the standard operating procedures," adds Terrell. "We have pretty comprehensive procedures that are realistic and not meant to sit on the shelf. It's a living document. Everyone has a manual. We want our manual to be useful, but we also give the incident commander the flexibility to deviate from the plan if necessary and justify such action later. Hazardous materials response is not an exact science.

"When we come on the scene, the fire department, without exception, will always be there before we are. First, we get a briefing from the fire department as to what they think they have and what assistance they think they are going to need from us. Whoever is commanding the HMRT will work closely with the incident commander for the fire department. We size-up the situation based on what they have told us, and after we take a look around, we can add or subtract from that evaluation. A lot of people in the various fire departments within Harford County have taken a lot of Haz Mat training, so we have a high level of confidence in what they are telling us.

"We also check with other people involved, such as a truck driver or a facility manager. The Haz Mat equipment will remain away from the scene until the Haz Mat commander arrives and decides which equipment will be needed at the scene. On the Interstate, we are particularly careful and don't take extraneous equipment to the scene because it could cause a dangerous situation. That is, we may stage the equipment away from the scene. Basically, the HMRT will do the entry, the recon, and necessary chemical research. The fire department will help us with decon.

"We always try to critique incidents because you can always learn from watching other people," adds Terrell. "Depending on how critical the incident was, the incident commander may have the critique done immediately or done at the next training session."

Chief Terrell talked about lessons learned over time. "One of the lessons I have learned is that you can't have tunnel vision on a Haz Mat scene. You have to look at the entire picture. What appears to happen is not always what actually happened. I think we have probably had some incidents where we assumed we had one situation when in fact we had another. After being in this business for seven or eight years, we have learned to stop, take a look at the big picture, and proceed very cautiously, never assuming that what we think we have is what really exists.

"An example would be the way in which we choose our PPE. We secure a great deal of backup information beforehand. We want to have four information sources about protective clothing and characteristics before we make any decisions. A lot of conflicting information exists. One source may tell you Level A clothing, another source may tell you Level B. For safety, we always go with the higher level of protective clothing. We have respirators, but almost always use SCBA.

"Another lesson I have learned is that industries sometimes downplay calls to avoid public embarrassment. You may be told you don't have anything to worry about when in fact you do. Most industry tends to be honest and very helpful, but some industry is not. To stay out of trouble, we don't always listen to what they say. This has paid off."

Contact: Chief James W. Terrell, Harford County Department of Emergency Services, 2220 Ady Road, Forest Hill, MD 02150; 410-838-5800; 410-878-5091 (Fax).

NORTH CAROLINA REGIONAL HAZ MAT RESPONSE TEAMS

The State of North Carolina takes good care of its contracted regional response teams. Six state-supported regional response teams have each obtained state-of-the art equipment and supplies carried in a specially-designed tractor-trailer truck, complete with a communications center work area in the rear of the trailer unit. Each truck is valued at more than $148,000 and is equipped with more than $250,000 worth of hazardous materials tools and equipment.

A Regional Response Team Advisory Committee in August of 1993 recognized a central problem in providing statewide response to hazardous materials incidents; most local emergency response agencies cannot afford the required training and specialized equipment necessary for an organized and effective response. The North Carolina General Assembly in 1994 directed the Secretary of Crime Control and Public Safety to establish a means of responding to Haz Mat emergencies on a regional basis.

Ultimately, six Regional Response Teams (RRTs) were contracted, funded, equipped and trained to a technician-level capability. The resultant teams are RRT-I in Williamston, Martin County; RRT-II in Wilmington, New Hanover County; RRT-III in Fayetteville, Cumberland County; RRT-IV in Durham, Durham County; RRT-V in Mount Airy, Surry County; and RRT-VI in Asheville, Buncombe County. The purpose of the RRTs is to protect citizens and responders alike, and to provide all communities regardless of size or population an effective, professional response to Haz Mat incidents in a safe, expedient, and cost effective manner.

RRTs supplement the efforts of local government agencies at incidents requiring a high level of training and more sophisticated equipment. The six teams became operational after being certified to OSHA standards by the North Carolina Division of Emergency Management within the Department of Crime Control and Public Safety. Members of a RRT must have a Firefighter #1 certification plus more than 100 hours of Haz Mat training. Funds for equipping, training, and managing the six teams are provided by the state General Assembly. Operating costs are recouped from responsible parties.

In August of 1995, a first bid award was made to the Williamston Fire and Rescue Department for North Carolina RRT-I. In April of 1996, Williamston became the second operational RRT for the state. Under a two-year contract with the state of North Carolina, the response team provides hazardous materials response services to 24 counties in northeastern North Carolina stretching from the Virginia border to the north, Nash County west of Rocky Mount, through Craven County to the south, all the way to the Outer Banks on the Atlantic Ocean.

Williamston Fire and Rescue is a municipal department governed by the Town of Williamston that has a population of 5500 to 6000 residents. The hazardous materials response team is limited to 36 members. These members come from within and outside the Williamston area, hold a minimum of Level I Haz Mat Responder certification, and are affiliated with a fire and rescue department. Currently, the Williamston HMRT has 34 Haz Mat Responders Level II (technicians).

Members of the North Carolina RRT-I may be career or volunteer members of the Williamston Fire and Rescue Department, career or volunteer members of other fire or rescue departments, employed by private industry, or private citizens. Personnel not affiliated with a fire or rescue department as a career are eligible for volunteer membership in the Williamston Fire and Rescue Department to ensure workers' compensation coverage. When activated by the State of North Carolina, all members of RRT-I represent not only their own employer, either public or private, but the state and the Town of Williamston.

Chief James B. Peele started with the Williamston Fire and Rescue Department as a firefighter in 1977 and has been there ever since. Six years ago, he became chief of the department. "With a population of 6000 or less, Williamston is not large enough to have a full-time Haz Mat team, but in northeastern North Carolina you can't hide from hazardous materials," says Chief Peele. (Of the 24 counties RRT-I covers for the state, about half have I-95 running through them.)

"In the early '90s when plans for a regional response team were being discussed we became interested. I had enough people working here who liked Haz Mat. The program gave us an opportunity to do more hazardous materials response. At the time, we did not have an established team at all. The legislators realized that many areas in the state did not have the capabilities or resources for Haz Mat response. They formed a committee that looked at other states such as Virginia and Pennsylvania. The committee reviewed standard operating guidelines and policies and agreed that North Carolina would pattern its regional response teams program after one that was developed by the State of Oregon.

"Eventually, the program provided money to six regional teams to pay for equipment, vehicles, training, administrative costs, medical monitoring, and physical examinations, and to supplement required workers' compensation coverage. Information about the program was sent to every county manager, every mayor, and every emergency management coordinator. The officials really saturated local government with an opportunity to at least look at the program and bid accordingly."

The area that Williamston eventually contracted to provide emergency hazardous materials response services for did not have a single fire department large enough to take on the task. "In the proposal we provided to the state, we had to have a minimum of 12 people ready to respond 24-hours a day," continues Peel. "I was talking to a lot of friends in the fire service in these counties, had good backing from our emergency management people, and got permission to send a flyer to fire chiefs asking for any firefighters who would like to be on a Haz Mat team. If other fire departments are like this fire department, they have people who like Haz Mat and people who don't like it. Those who do like it will do a good job, and I began receiving calls from all over the area. I set up a meeting for the chiefs and told them what the state RRT program could offer and what would be expected if they wanted to become involved with our team.

"Ultimately, we developed a regional response team, the only state team that is really a true, regional team. The other five state RRTs are based in single fire departments where shift personnel handle calls as needed. After Williamston bid on a contract, the state increased our area of coverage. Another meeting was scheduled with chiefs from the overall area, and we brought in three or four additional members. As of January 1998, we have 34 members on the team; 16 or 17 are from Williamston and the rest are from the 24 county area.

"We signed a two-year contract with the state, which was renewed for a second two years during 1997. We do not have a rollover renewal process. That is, in the first part of a contract-ending year, the state again requests proposals so everybody has an opportunity to go after the RRT program. In our area, nobody submitted a bid for it other than the established team.

"The state developed a budget of $18,000 a year for administrative costs, $14,000 for training, $5,000 for physical examinations, and $5,000 for workers' compensation. We keep all the administrative money up to $18,000, but what is not spent of the other monies goes back to the state at the end of the year. We do have to be accountable for what we spend, and we do a spending report once every quarter. We get reimbursed for each quarterly report, but what money we do not spend during the year we turn back to the state.

"We have a two-year proposed training budget. Our first contract was done a bit differently. We had to spend it first, and apply for reimbursement. We had to go through a request for loans just like the state employees do, get a two-week advance, request training, get the training approved, then submit a request for the money which had to be approved. We would go to the training and spend our money. Once we came back, we had to fill out the state's travel vouchers to be reimbursed based on the state's per diem rate. But, we are not state employees; we are contractors. The teams didn't like this practice, had it changed, and now we look after our own money. We are accountable for the money, and don't mind being accountable for it.

"When we signed the first contract with the state in 1995, we were not yet an established hazardous materials team. I was not willing to establish a 30-plus member team without assurances of a contract from the state. It just was not feasible for us. Basically, we could not afford it. We did say that if we received a contract we would put together this type of team to meet the training standards for the State of North Carolina. Our standard operating guides require the team to be certified by the North Carolina Fire and Rescue Commission as a Level 2 team. That type of designation requires 80 hours of technician training and 80 hours of chemistry.

Prior to that, members would have to be Level I firefighters and be trained to the operations level in incident command.

"Our firefighters were Level I firefighters and we had the operations level requirement. We met all the criteria other than having 80 hours of chemistry and 80 hours of technician training. We signed a contract in September of 1995 to become a regional response team with the State of North Carolina. In October, we scheduled an 80 hour chemistry class followed by an 80 hour technician class in December. The North Carolina Division of Emergency Management schedules such classes through local community colleges. When we want training, all we have to do is ask for it.

"From September of 1995 through April of 1996, the state had not yet received appropriate equipment. For example, each team chose the air packs they wanted to use. The state waited to order air packs until they were confident that the items ordered were those we were familiar with. In the interim, we didn't feel we could officially do a call. Eventually, we received all our equipment, we got our response vehicle, we got equipment on the truck as we desired it, we got our training completed, and we went on line in April of 1996 and have since been available for calls statewide."

Chief Peele recognized a good opportunity and went after it. He received a good deal of support from the Town of Williamston and fire chiefs in northeastern North Carolina. "Early on, I went to Raleigh several times to learn about the possible program, knowing that the Williamston Fire and Rescue Department could not afford a Haz Mat team and knowing at the same time that we had as much potential for an incident here as anybody does. I knew the town would benefit from this program. For example, the equipment that is stationed here is available to us for local use. We were given a ton of equipment, a lot of training, and the physical examinations required to do hazardous materials response. Our board, the town administrator, and our commissioners believe the team is an asset and have supported it 100%.

"If a local government or a local fire department has a Haz Mat incident that they cannot handle, they can request a state RRT. Ideally, they would go through their local emergency management coordinator who would call the state emergency operations center (EOC) and request a response. The operations center has a list of questions they ask. If the incident is within our area of response, the operations center might give us a phone number to call. We will call the incident site to find out what is actually occurring. If the incident meets certain criteria, an official in Raleigh makes the decision as to whether we go. We can't go on our own. Someone from the state EOC must authorize us to respond. Once we receive authorization to respond, all the expenses incurred from that point on will be billed to the state. If we went out on an authorized response today with 12 people, and our payroll ended up being $2000, I would bill the state for $2000 and the supplies we used. The state will pay us, I will pay my people, and the state will go after the spiller for these expenses."

How do the various RRTs protect the health of their personnel? "We have a standard operational guideline that spells out what types of physical exams are required," says Peele. "I called a number of hospitals and sent them a copy of our SOG to get estimates. We found our local hospital could give us a lower price in meeting the criteria set in our SOG. One fact any RRT should recognize is that $5000 a year isn't going to cover initial physical exams. We had 36 people at $400 each (for a total of $14,400). The state allows the physical exam fee to vary. They told us to go ahead and get our exams done and they would pay the bill. The $5000 would cover maintenance follow-up each year. Last year was the first year we did maintenance exams, and they ran about $4900. We have a physician here who really looks after our RRT members, both in Williamston and the surrounding area."

Who keeps the confidential health records of team members? "The hospital does keep copies," relates Chief Peele, "but we also keep a medical file on every person on the team. On the vehicle we have a small safe for papers which we carry with us every place we go. These papers are in a vault because the Department of Labor oversees the confidentiality of and access to medical records. Only I and one other person, in case I am not there, have access to these medical records.

"We write an exposure report for any incident where we enter into the hot zone. We don't fear that an individual was exposed, but we do need to document if he or she entered an environment that may be hazardous. For those who actually made an entry, we write an exposure report for their medical record. That way we know that the person did go into a hazardous environment, although wearing protective clothing, and we have that record for future reference.

"We don't do cleanup. We are not a clean up crew. Once we have mitigated the incident, the spiller decides who should clean it up. We want to ensure the cleanup company he chooses is a reputable firm, and we will often leave someone at the scene to oversee that the job is done properly. In addition, the State of North Carolina through the Emergency Management Division will have an area coordinator on-scene any time we are dispatched as a state regional response team. The state will also call to the site any representative we may need, such as a water quality person or someone from the environmental agency.

"The intent of the state is to dispatch a fully trained and equipped regional response team quickly anywhere in the state. Every part of the state needs to be equally covered. If you look at the map, the Wilmington RRT-II based in New Hanover County has a total of eight counties to cover. Here in the Williamston RRT-I, we have a total of 24 counties, mostly rural, to cover. In the Wilmington RRT-II they have two state ports, one of the largest ammunition ports in the country, a major Interstate highway (I-40), plus a number of four-lane highways, and major industry. Wilmington's RRT-II risk in eight counties is probably far greater than Williamston's RRT-I risk with 24 counties. Each of the towns or cities that was granted a contract was assessed according to risk. Here in rural, eastern North Carolina where we cover 24 counties, the risks are low. However, there is risk. We cover I-95 for over half the total distance as it crosses the state from north to south.

"We are a state resource. If someone in a county within our RRT area of coverage is at a loss at a hazardous materials incident and doesn't know what to do, the state wants to send someone to assist him or her. That is what we do. We don't take over their incidents in any way. The local incident commander is still the local incident commander. We will assist and make recommendations, but we will not take over the show."

> **Contact:** James B. Peele, Chief, Williamston Fire and Rescue Department, P.O. Box 602, Williamston, NC 27892; 919-792-3521.

WILMINGTON HAZARDOUS MATERIALS RESPONSE TEAM

The City of Wilmington with 55,530 residents is a port city in the southeastern corner of North Carolina. Captain John Forestell is the coordinator of special teams (Haz Mat Team, Confined Space Team, and the Scuba Water Rescue Team) with the fire department. He is a 13-year veteran with the fire service who was originally a paramedic in the state of Maine. After moving to Wilmington, he joined the fire service and worked on the trucks. He trained in hazardous materials and worked as a fire inspector before a position opened up for coordinator of special teams in hazardous materials. He was already certified as a hazardous materials technician and started to learn confined space entry procedures.

"Presently, we not only maintain a local Haz Mat team, but we also run a regional response team in eight counties. For our regional Haz Mat response duties that meet certain regional response teams requirements, the State of North Carolina provides a trailer truck loaded with specialized equipment, pays our team compensation for training costs, covers charges for our annual physical examinations, and covers certain other benefits. The city of Wilmington and the Wilmington Fire Department contracted with the state of North Carolina to respond as a regional response team in eight counties within this area. So, really, I am with two groups: I am the coordinator for the Wilmington Fire Department Haz Mat team, but I am also the coordinator for the regional response team.

"There is a lot of industry in the eight counties as well as pipelines. The Wilmington Fire Department maintains RRTs at the Wilmington state port and at the Moorehead City state port, and covers the Sunnypoint Terminal which is the largest military ammunition shipping depot on the east coast. We have a nuclear power plant less than 20 miles away, and a General Electric plant that makes nuclear fuel rods filled with uranium dioxide.

"We have Du Pont, International Paper, Takeda Chemicals U.S.A. and other large industries in the area. Most of the industrial facilities are really cooperative with us. We are familiar with their facilities, and we may even train with them. Takeda Chemical is outside city limits, but the city contracts with the company to provide fire protection. One of my assignments is to maintain the fire prevention inspections for Takeda and for two hospitals in the area. I also do all the sprinkler plan reviews for all facilities in the city. I maintain my inspection certification and maintain my hazardous materials training so I can establish rapport with industries in both areas.

"The organization of all regional Haz Mat response teams in the state is the responsibility of the Secretary of Crime Control who is in charge of the Highway Patrol and the Department of Emergency Management at the state level. An advisory committee that includes team members, industry personnel, fire chiefs, firefighters, chemical industry members, and various state departments develop the standard operating guidelines and the criteria for the RRTs. We have a matrix we use for a regional response. If we get an alarm, it goes to the operations duty officer at the state level. The state will then call the team leader in that particular region. Together, they go through a matrix to decide what the level of response is going to be. The levels are broken down into subgroups so in some cases you may only have a telephone advisory. In other more serious cases, you might have a multi-agency regional response activation. In a serious incident, the final criterion is getting the approval of the state secretary of Crime Control, or the state director of Emergency Management.

"The state bought the specialized equipment for each regional response team. In the Wilmington area, we have a brand new 46-foot tractor trailer truck, with an International cab, five seats in the front, and a resource center with computers. Each team receives nearly $500,000 worth of equipment and can use this equipment in their own locale even if the alarm is not a regional response call. However, for a local Haz Mat incident, we have to reimburse or replace equipment that is used. If we use or damage a suit, we have to replace that suit. The benefit though is that we did not have to buy all that equipment up-front.

"All the Haz Mat responders on the Wilmington Fire Department and the regional response team are paid, career personnel. We do not use any volunteers. The regional response team is activated by people on-duty on a particular day, and we can call in off-duty personnel as necessary. There is a medical surveillance program for our Haz Mat responders. They go annually to a physician for blood work, a bicycle test, and a stress test.

"The Wilmington HMRT members do all fire department duties. Some of our people are cross-trained in all three special team skills (Haz Mat, Confined Space Entry, and Scuba Rescue Team). The HMRT is stationed at one location, but the team is comprised of about 36 people with 12 technicians for each shift.

"Besides having computer programs, we also have many printed reference materials in case the computer program is unavailable at any time. But, normally, we use a computer program to get specific response information. One of these programs is called 'TOMES.'* Much of the available reference information is built into the TOMES program. For instance, if you pull up the chemical, methanol, you can get the following data bases: the U.S. Coast Guard CHRIS manual, RTECS (Registry of Toxic Effects of Chemical Substances), OHMTADS (Oil and Haz Mat Technical Assistance), HSDB (Haz Substances Data Bank from the National Library of Medicine), DOT Emergency Action Guides, IRIS (Integrated Risk Management System), and NJ Fact Sheets (New Jersey Hazardous Substances Fact Sheets contain employee-oriented exposure risk information for over 700 hazardous substances). We can print such data right onto a hard copy so we don't have to waste time looking it up in printed data manuals.

"What is really nice about the TOMES program is that it has a medical section. If we had an incident where people were injured, we could download all the medical information such as drug therapy, respiratory therapy, and all other therapy information that a physician might use at a hospital.

---

* Micromedix, Inc. 600 Grant Street, Denver, CO 80203-3527; 800-525-9083; 303-837-1717 (Fax).

"We do work under an incident management system, so we do a lot of work developing SOGs which we follow. The SOGs, the manuals themselves, are broken up into 14 different chapters covering 40 SOGs that range anywhere from response criteria and medical surveillance to decontamination and specialized equipment. For example, we use the HAZCAT system to identify unknown chemicals at the scene so we have a pretty good idea of what hazardous materials we may be dealing with.

"Our personnel who work on the scene are constantly pulling our equipment and maintaining it. We also have check lists that they go over on a daily, weekly, or monthly basis. Some equipment we will visually check every day. A weekly check item, for example, is ensuring that all battery are charged. Every three months, we pressurize the suits to inspect for any holes or leaks to ensure our safety.

The weather is relatively moderate in Wilmington except for one area. "We are in a hurricane belt and that is a concern with preplanning of hazardous materials incidents," says Captain Forestell. "The LEPC (Local Emergency Planning Committee) and industry got together to look at how we could handle the situation during a hurricane. People would have difficulty getting out and working in the area during a storm, and after a hurricane is over, we would have quite a clean up to do if there were a number of Haz Mat incidents. We work quite heavily with industry.

"Weather isn't regularly a serious problem, and the temperature rarely falls below 35°F. A freak storm here a couple of years ago dumped 17 inches of snow on us. Since the city has only two snowplows, that storm really tied up the city. Another minor problem we face is that salt air and water from the ocean sometimes cause problems with ocean transfer containers. A couple of weeks ago a ship arrived after battling 30-foot seas. Hazardous materials were stored down in the hold, and when workers opened the hold they got a face full of what they thought was a hazardous materials odor. They became nauseous and sick and called the HMRT. Since it was midnight and dark, we could not make an entry until daylight. In the meantime, we monitored the ship and cordoned off the area. When we made our entry, we could find nothing wrong. We had to take all the necessary precautions, and entered in fully encapsulated suits. It was an 18-hour ordeal that accumulated about 300 man-hours, so it was a relatively expensive response.

"We have a good rapport with the U.S. Coast Guard and get a lot of training from the local shipping area, but you can learn a lot about one specific ship and the next one may be entirely different. One problem I find in shipboard firefighting is that some ships have super firefighting equipment while others may not have anything aboard. We try to get all possible information on board the specific ships where we have an incident. The best information we can get is from those who are qualified to give it.

"Since we use the incident management system, we always hold a briefing before and a critique after every entry. In a briefing we will try to learn as much about the situation as possible. We may contact representatives from the manufacturer or handler of the chemical. After an incident, we hold a review meeting to critique our response. If we come across any situations that were not anticipated, we review those. We write this information down so we can learn from our mistakes.

"We maintain a printed form procedure so each sector officer has criteria taken from a printed form that tell him or her what to do at the incident. The forms help ensure that everything is running smoothly during the operation. Also, when we have a critique, we can review the forms to ensure that all the various tasks were completed."

> **Contact:** Captain John Forestell, Coordinator of Special Teams, Wilmington Fire Department, Station 5, P.O. Box 1810, Wilmington, NC 28402; 910-799-7174.

ALLEGANY COUNTY HAZARDOUS INCIDENT RESPONSE TEAM (HIRT)

Al Ward is the Public Information Officer of the Allegany County Hazardous Incident Response Team located in the northwestern section of Maryland. The county has a population of 75,000 and covers 421 square miles. The county seat is the City of Cumberland.

"We are basically a volunteer organization funded through the county government. Ultimately, we answer to the county commissioners through the Office of Emergency Management. The forerunner of our team started in 1990 as a response to SARA legislation. The commissioners and some other public officials recognized that they were responsible for some sort of emergency response, and they had no one in the area trained or capable of doing that function. Our county has 1 career fire department and 23 volunteer fire departments, none of which had financial resources or manpower to get into the Haz Mat business. The forerunner of our team was created by drawing people from the different fire departments and training them in incident command, Haz Mat tactics, research, and similar skills. This effort went on-line in 1990.

"Then came the realization that this effort was not sufficient, and we needed to establish our own, full-fledged Haz Mat response team. The present team went on-line in 1993. It has about 25 members drawn from local fire departments and ambulance companies, some volunteer and some career. We have two funding sources: one is a line item in the county budget and the other is a collections process. Under Maryland law we are allowed to bill the responsible party for any transportation incident, and we have had a fairly high success rate in collecting such bills. The county government and their accounting department handle all the billing for us. If necessary, the legal department will pursue responsible parties who do not pay. One case went to court; it took us three years to finally collect, but we did get paid."

How does HIRT draw up a bill, and what do they charge? "We have an itemized list that includes all our equipment, vehicles, and manpower. I don't remember the entire list, but I know we charge $100 an hour for our vehicles, $50 per hour for each of our team officers on scene, and $25 for each technician. Most of the equipment we use is charged at replacement cost plus 25%.

"Anything we get through the county government has to be a line budget item, and we have specific training and equipment needs for all this money. However, the money we get from billing is our own fund. We have total discretion over what we do with it. By law, it cannot go to the county's general fund and is used strictly for our team operations. The county still administers this money; we need a county purchase order to spend it but basically we can spend it as we wish."

What did the team do to recruit inductees? "We went around to the different fire departments looking for prospective members," remembers Al Ward. "We went to the county fire department convention, we put a nice article in the newsletter the county sends out, and we recruited a few people. A lot of our recruiting, however, was done on the side of the road during an incident. Whenever we go to a call, the first in are the firefighters. Especially in a long, drawn-out incident we have some time to talk. The firefighters start looking around our rig, see what we can do, and some become interested. We bring them to a meeting or a drill, and let them get a little more hands-on close-up of what we actually do and what kind of equipment we have.

"For insurance purposes, all potential candidates for the hazardous incident response team must be members of one of the fire departments or ambulance companies in the county. That indicates to us how much training they already have. Within our team, we have two divisions: people trained to the technician level or the specialist level and support personnel who do medical work and research. We have one older person who's no longer physically capable of putting on protective suits or that type of activity, but is excellent at keeping records. He sits in the command post and take notes that are a big help to us. We appreciate it greatly. He has been to a technician class so he knows what type of records he needs to keep. He does keep us straight. He knows what has to be in his log and helps remind us of what we have not yet done. When he doesn't make a call, we really miss him because someone else has to do that function. Being an all-volunteer outfit, we need all the people we can get to do the suit work and other necessary tasks. When we have to commit one of these people to recordkeeping, that becomes a major bind in our resources.

"There is a duty officer for the team available 24-hours a day, 7 days a week. That person pulls a duty shift like everyone else but also has the responsibilities of logging some of the papers, attending meetings, and similar tasks. These seven duty officers have several other officers available to handle vehicle maintenance, medical recordkeeping and other jobs. As stated before, our team members are divided between technicians and non-technicians."

What happens when the Allegany County HIRT goes on-scene? "We are alerted by pager through our county office of emergency management," says Ward. "Usually, the duty officer on call responds directly to the scene. He can use his private vehicle or a team vehicle. Whatever members are closest respond to our station to pick up our response vehicles. For most of the other team members, transportation depends on where they are and where the incident is. On the initial alert, or at least shortly thereafter, we try to identify for everyone responding where the staging area is going to be so private vehicles do not inundate the scene.

"When the duty officer arrives on-scene, he will immediately do a consultation with the incident commander from the first-in fire company. He wants to know what has been going on, what's been done, and what we are going to need to do. The duty officer with HIRT will probably have 10 or 15 minutes between his arrival and the arrival of the rest of the team to start getting some plan of action in his head.

The duty officer and the incident commander will organize a plan, determine where the vehicles are to be set up, and establish where the decon area will be located. By the time the vehicles roll in, the plan is pretty well firmed up, and we start following whatever plan they have in mind.

"One of the key factors about our operation is that the first-in fire department is still in charge. It's their incident. Their officer remains in charge of the incident. We are there strictly to handle the Haz Mat aspects of the incident. We don't do anything without consulting incident command, and they have the option of rejecting our proposals if they so desire. We will give them the best educated guess we can based on our knowledge, research, and our contacts. We will give them our best suggestion of how to handle the situation. If they were to suggest some other solution that we really disagreed with, we would say, 'it's your ball game. We are going to go home.' This has never happened. Usually, they are quite willing to accept our recommendations.

"We have worked with fire chiefs outside the county. As an example, the county immediately next to us has no Haz Mat team, so we do mutual aid with them. We have actually gone into Pennsylvania and West Virginia because incidents that occurred there had no response team nearby. We were physically a lot closer than other teams even though we were outside their political boundary. We will assist the first-in fire department to the best of our ability until whatever authority in these states can provide a trained and equipped Haz Mat team. We will mitigate the emergency until their team can arrive.

"We have certain written SOPs and protocols. For example, SARA Title III locations are listed in what we call the 'red book,' which is our Local Emergency Planning Committee master plan that covers our operation, SARA Title III locations, evacuation routes, and similar matters. Our team also has SOPs that cover our training, procedures on scene, and the checklists for the safety officer, the research officer, and some other people.

"The Local Emergency Planning Committee is staffed by community leaders, representatives of the fire service, and the local state police barracks. The head of our organization also has a seat on the LEPC, which operates to meet existing regulations. It also serves as a funding source for some grants, and this committee signs off on our grant applications. They have very little to do with our day-to-day operations, but when we write our budget, we usually run it past them for their review and comment. Some of the people on the LEPC are a lot more knowledgeable in this sort of thing than we are, and we appreciate their expertise. Also, whenever anyone joins the team all applications are run through the LEPC. Usually, they will accept the recommendations from our team officers. Since the LEPC, and ultimately the county commissioners, are paying insurance on team members, they want and are entitled to some say as to who is on the team, who is driving their vehicles, and who is going out and exercising authority."

Why do people join the Allegany County HIRT? "Primarily, people join because it is interesting, challenging, and different from their fire department routine," says Al Ward. "Many of the volunteer fire companies in our county are fairly small. They may handle one or two major structural fires a year. Our candidates are looking for a little more to keep them busy, active, and challenged. They understand the need for Haz Mat and the risk that is out there. We're the ones that are doing it, and they want to help.

"We keep our personnel trained by having a drill once a month which usually involves two or three hours during an evening. Once a year, we have a major, day-long drill where we may shut down a highway and roll over a truck. Sometimes we have someone like the DuPont company come and assist us. This type of training exercise has been very well received, and we can involve the local fire departments. Also, as part of their accreditation process, some hospitals in the county are required to test their emergency disaster plan annually, so we might schedule a drill involving a major highway Haz Mat incident where we have 15 medical emergencies that need hospital care. The hospital will run their drill along with our scenario and everybody gets to work their little piece of it.

"For medical surveillance, team members get a physical examination when they join the team, and a cursory exam the following year. Every two or three years, depending on the level of activity, we do another thorough physical exam on members. At an incident and during exercises, we do pre-entry and post-entry medical monitoring of vital signs, EKG, and other tests. We maintain a computer database of this information.

"Our local hospital has been very helpful to us with our medical surveillance program. They donate their manpower, and the fees to us for items like X-ray film and blood work are minimal. The hospital performs those tasks at cost which has been a big help to us. We do have to coordinate scheduling. We stagger physical exams so the hospital only has two or three of us a month. They have been pretty cooperative in maintaining our files.

"We stage a briefing and a critique on any incident where we use Level A protective clothing, most incidents where we wear Level B, and any long, protracted incident that involves a lot of mutual aid with local fire departments or other Haz Mat response units. We get all the records that we can from the incident and go through the incident step-by-step. When there is a problem, a discussion, or another opinion about how a decision should have been made, we sit there and thrash it out to see what we did do, what we didn't do, what we could have done. We try to invite someone knowledgeable in hazardous materials response who was not at the incident to come in as sort of a mediator, a neutral party to focus the discussion and ensure that personal ties don't get involved.

"With my position as public information officer, I often act as the moderator because I am fairly comfortable getting in front of the group and talking. I call the meeting to order, explain what we are going to do and why we are here, identify the players, and start the ball rolling in a free-wheeling discussion about the incident. My job then is to keep it rolling and make sure we cover the whole incident in some sort of reasonable time. Records are kept of the critique and become part of the incident file. If necessary, we go back to our SOPs and add changes.

"Currently, our team operates three vehicles; two of them were surplus, acquired from other government agencies and modified to suit our needs. Being volunteers, we have people with all sorts of expertise in mechanical endeavors, so we do a lot of the necessary work ourselves. One rig was bought second-hand from a commercial business, and we fitted it out ourselves after we looked at trade magazines and attended conferences to see what other teams have done with their vehicles. Sometimes, out on the side of the road at incidents where we are helping other teams, we will look at their vehicles and ask a lot of questions. We've had many good ideas come from networking, and some of the modifications were just trial and error.

"We have arranged the bins in the vehicle three or four times to make items more accessible. The items we use frequently are close at hand, the items we use infrequently may be buried. We have tried to give the truck some organization. If you look at it closely, most of the things on one side are what we call 'clean' — the suits and monitoring equipment. Most of the things on the other side are 'dirty' — absorbents and shovels. Some of the items are organized simply for convenience sake, for location, or for safety; we don't want to have a Level A suit full of absorbent. Mostly, the organization is trial and error and includes ideas we've seen at various places.

"Our hazardous incident response team has grown over the years, and we have acquired operating funds," concludes Al Ward. "We have been able to buy more equipment and monitoring tools. Our level of experience has increased, and we know how to do a lot more. We are more comfortable with what we are doing, and our SOPs are more streamlined to reflect that fact. We have realized that certain procedures are not necessary on every incident, so we have removed such items from our SOPs and made them optional at the bottom of the page."

Contact: Al Ward, Public Information Officer, Allegany Hazardous Incident Response Team, c/o Office of Emergency Management, P.O. Box 1340, Cumberland, MD 21502; 301-777-5908.

8 Industry and Commercial Response Teams

PCR CHEMICALS, GAINESVILLE, FLORIDA

John E. Hudson is Manager of Safety and Emergency Preparedness at PCR Chemicals, a small chemical manufacturing company that focuses on organosilicone and organofluorine chemistry. "We make a lot of silicone intermediates such as the products in shampoo that make your hair shine, silicones that are beta blockers in pharmaceuticals, and a silicone that coats every Intel processor chip. We've developed a drug called '5 Flurouracell' that is a basic chemotherapy building block, and we manufacture specialty heavy fluids used in gyroscopes for the U.S. Department of Defense. Basically, we run a diversified batch operation.

"Most of our production is market driven. People come to us and say, 'Can you do this?' Our R&D department develops the product and tests it. We put it in production if it is a viable product. Other companies come to us and ask if we can manufacture a chemical product cheaper than they can do it for themselves. Very often, we can because we already have the infrastructure that would support the product. If they were to make it themselves, they would have to invest a great deal of capital where we have already spent the necessary capital and can easily manufacture these products.

"We have had incidents here, one in 1991 and one in 1994, which prompted us to coordinate an emergency response team as a first response team for incidents at the plant. Our biggest problem can be a runaway reaction that results in a fire. We can now make the initial response and keep the situation under control until the Gainesville Fire and Rescue gets here to help us out. As a result of the '94 incident, we purchased an industrial foam pumper with 1000 gallons of foam aboard which gives us a capability not only to fight a fire but to suppress vapors. Many of our products are so toxic that this unit has really helped us with vapor suppression."

On June 17, 1994 a release of trichlorosilane at PCR resulted in the evacuation of 600 residents near the plant. Hundreds more were advised to shelter in place, and 148 were treated and released from local hospitals. The nearby airport was closed for four hours.

"The emergency response team is purely voluntary and currently staffed by 42 persons with specific duties within our plant operations," continues John Hudson. "Right now, we are in the process of training 17 persons to a Level 3 technician level. We began training with the Florida State Fire College located in Ocala about 35 miles south of here. Half of the team will be firefighters and half will be Haz Mat technicians, but all will be cross-trained so they can do any of the activities that are needed.

"Once a year, we open up the emergency response team for people who are interested in becoming volunteer members. They are usually brought in through the decon team which is a good place to get them involved. We will train them immediately through Level 1 and Level 2. Once a year, we offer a Level 3 course. As they work with decon, they participate as support people for the Haz Mat group. If they are interested in firefighting, we get them involved in some of the courses at the Fire College. It is purely a voluntary program. We don't force anyone to go into any particular field. If they want to stay with the decon team, that is no problem. We continually have spills that the decon team, actually called 'Decon and Spill Response,' responds to as well as decontamination drills.

"We have found it very advantageous to have a response team. We've had several near-misses that through quick response remained near misses. We are a 7-day, 3-shifts-a-day operation, and all our evening and night shifts are trained in either the fire program or the Haz Mat program. We have developed what we feel is an excellent relationship with the Gainesville Fire and Rescue Department.

"When I came to work at PCR in 1993, I brought with me the good relationships I had with the fire department. I used to work in a teaching hospital at the University of Florida, and we had so many fire alarms at the hospital that firefighters would respond several times a month. When I joined PCR, our first column went bad shortly after my arrival. When the firefighters responded, they were afraid to come in the front gate. Rick Lust, who was Haz Mat Engineer with the fire department at the time, and I got together and decided that we should do some joint drills, preplans, and activities so the fire department was comfortable in coming out here. At the same time, we started working on our emergency response team here at PCR.

"We had a group of five or six people who would respond prior to 1993, but they were not well trained. They were mostly chemists and persons who operated the vessels. We started a program to train everybody in the facility to the OSHA Level 1, the awareness level. Next, we trained everyone who was interested to the operations level so they could take defensive actions in their own work areas. At that point, we developed a training program that could be presented internally to train them for the Level 3, or technician level, under 29 CFR 1910.120 (9)(2).

"We trained our personnel as responders with 160 hours of training, and we purchased about 45 sets of bunker gear, SCBA, and more support equipment so we could handle anything on a first response basis. We have some good fire training, but I don't classify our personnel as firefighters. All we can do is hold the line until the Gainesville Fire and Rescue gets here. We make our equipment available to the fire department, and we train together at least four times a year so they are familiar with our equipment and we with theirs.

"The PCR emergency response team can provide 40 people who are Level 3 Haz Mats who work with hazardous materials on a daily basis. We have nine Ph.D. chemists on the team so risk assessment is one of the things we do on the fly which really stymies most Haz Mat teams. The expertise we have in dealing with chemicals gives us an advantage and gives the Gainesville Fire and Rescue a big advantage because they have our pagers and home phone numbers so they can call us any time of the day or night. If they need us to respond, we will. If they need information, we will supply that information. It's been a great relationship for the two of us, and we do a lot of training together. I would certainly recommend that relationship, and I do every time I get an opportunity.

"PCR's Level 3 responders have four, 4-hour refresher courses a year to keep them up to speed. We have about six Level 4 persons, the specialist level, 'plug and patch' people, and we are continuously doing exercises for them. Sometimes, damaged containers will come in and we will treat them as an exercise for the specialists.

"We have people assigned to roles and trained to fill those particular roles. For instance, a safety officer role in a fire department response can be filled by any command officer. We have a Ph.D. chemist to fill our safety officer role, and his designee is a Ph.D. chemist as well. They can analyze what is happening at an incident, and we have trained them through the incident command system so they recognize how this system works. But they also bring Ph.D.-level knowledge of adverse effects as in a BLEVE (boiling liquid expanding vapor explosion) or a flashover. We use the same system for decon. While the fire department will assign an officer to the decon team, we permanently appoint a deputy chief for decon. He or his designee will be available 24 hours a day."

The PCR emergency response team has been available for off-site incidents in the past. "It's good for the community to have an asset like our team," says Hudson, "and we have responded off-site a few times in the past. In October of 1996 a smaller chemical company here in Gainesville burned. They had a lot of cyanides so we supplied our truck with foam and about eight team members, four or five to operate the truck and the rest to provide oversight on the chemicals involved. I was the incident commander of our unit and worked closely and communicated with the incident commander for the other chemical company.

"My management gave me permission to take the team off-site for that incident and turned the response team over to me. If a public emergency type of incident proves to be something we can't handle, we won't do it. We've talked about a mutual aid agreement, but decided that was not particularly necessary. We also talked about doing a memorandum of understanding, and I believe we are going to do that to address the issues of cost recovery and liabilities. We don't ask for any pay or liability coverage if we pump 600 gallons of foam, but we would like someone to reimburse us for the cost of the foam. We make an incident-by-incident decision for off-site response. For instance, a pentaborane disposal issue arose recently. I was asked whether we could provide some people to operate as a backup for the effort. I said absolutely not because my people have no training in new toxins like pentaborane. We can help in some ways, but we are not going to respond to a pentaborane incident.

"For a response at PCR we have material safety data sheets for all the raw materials and intermediates," relates Hudson. "Some chemicals go through several steps to move from a raw material to a finished good. Depending on where it is in the process, we can pull information about that chemical. We also have response information on our local area network. Most of our people are pretty familiar with the chemicals in their areas; and we have people from every area of the plant involved

with the response team including people from our laboratories and the R&D department. We have plant-wide coverage so that in every area we have personnel familiar
with what's going on. That expertise certainly helps a response. That's one of the
reasons I have such great respect for the fire department. When they leave the station,
they never know what they are going to be running into so they have to be prepared
for anything and everything.

"Another PCR staff member and I are members of the Local Emergency Planning Committee and are familiar with the enhanced hazard analysis that's been done
in the community. We have used the LEPC as sort of an information dissemination
program. Chief Williams was the LEPC chair for a couple of years. I succeeded
him as the chair and have held the job for five years. The LEPC covers seven counties
in the northeast fire district, about 7000 square miles in total, and keeps all of our
fire community aware of current issues. The LEPC has been active in making training
available to the rural and volunteer departments.

"As a company, we've done exercises and participated in rural areas up to 65
to 70 miles from Gainesville, and we feel it is very important that we share our
expertise. We have leadership in the company that will allow us to do this, and we
have people in the community structure like myself chairing local LEPC. I am also
a member of the state emergency response commission, and a member of the Florida
Fire Chiefs Association Emergency Services Coalition as well as a member of the
Florida Transcaer Committee (Transportation Community Awareness Emergency
Response sponsored by the Chemical Manufacturers Association and other national
associations).

"Our LEPC has 35 members who range from persons who have been there since
the committee began in 1987 to folks who came in the last year. We have little
turnover in our LEPC. Once somebody comes on, he tends to stay for five to six
years. Other LEPCs in Florida have a pretty high turnover rate. The most active
committee is their membership committee, while with our LEPC, the public information committee is the most active group. As an example, a week in late February
was designated 'Hazardous Materials Awareness Week' in Florida. Employers using
extremely hazardous substances from the EPA list had to submit their Tier II reports
to the state capital in Tallahassee. They must report to the state what chemicals they
have on hand, where the chemicals are stored, volumes they have on hand, and the
average daily volumes on hand. The governor published a proclamation. We sent it
to 44 communities and municipalities and they published it in their areas.

"An incident occurred at the male correctional facility a year or so ago. The
Governor's staff appointed me as an investigator through the LEPC. Well, I learned
a lot doing it. I learned I never want to do that again. The difficulties of doing
hazardous materials work are compounded by the difficulties of working in an
environment such as a prison. Something had permeated the air and made a lot of
people sick. We discovered in the investigation that one of the guards had released
prisoner control gas. It permeates clothing and stays with a person. Prison officials
did not do any decon. They took all the injured to a hospital and contaminated all
the ambulances and the hospital emergency room. They wanted to do what was
right, but they just didn't know what was the right thing to do. Of course, that

information didn't endear us to the prison superintendent until he learned that we could help him with good information and training. We have done Level 1 and Level 2 training for their staff, and made arrangements for them to take the 'Hazwoper' course. The head of the corrections department has now mandated such training for all his staff in Florida.

"We are continuously working on making our chemical processes safer. We had a trichlorosilane incident in June of 1994. A hose ruptured and the product, which is water reactive, ignited when the humidity that day reached about 95%. If we had an initial response capability for fire fighting at that time, we could have gone in there and resolved the incident in less than an hour. As it turned out, the incident took about seven hours to get under control because the fire department waited until they could make a safe response before going in. Presently, we use remote-activated valves. Every set of valving and pumps in the plant can be shut down from another location. That system was instituted because of that particular incident. We have also poured a lot of concrete containment way, far beyond what the regulations require, so that if something should break we can keep it from going into the soil or to the waterways.

"We also have a number of suppression systems so that if anything that is not water reactive does give way, we can immediately flood it with water. If it is water reactive, we have five foam systems with which we can lay a blanket of foam. We use foam as much for vapor suppression as we do for fire suppression. Many of our chemicals will liberate HCl (hydrochloric acid) vapor which can be very damaging to mucous membranes and respiratory systems. It important that we are able to suppress it."

A fire department Haz Mat team is now required to have medical surveillance for response personnel. How does that work within industry? "We require it," continues Hudson. "It's part of our response plan. Before you go into any activity, you have your blood pressure and heart rate checked. They are checked again when you come out. Every month when we have an emergency response team meeting, we do checks. If we have an exposure during an incident, we send that person or persons to our company doctor. Only one person has had to have blood gases drawn. I think it was such a bad experience that no one will ever again say that they were exposed. We have had one of our response team members go to a community college to become an emergency medical technician. We hope to send another one this summer as well as one in the fall, so we will have at least three EMTs. One person on site is a paramedic and can make the initial medical response. We are working with the Gainesville Fire/Rescue Department on a first responder course which covers basic life support. Twenty people will go through that course. We are trying to get ourselves in a position where we can take care of ourselves and then help the fire department if they need help.

"We keep about 20 Level A suits and 35 to 40 Level B suits on hand. We only use them for a specified period before we paint a big 'T' on the backs of them and they become training suits. We supply a lot of support to the fire department because their resources are usually limited to half a dozen or so suits. One of the areas we cover in our Level 3 course is how to dress. It sounds rather funny, but it is very important to know how to get in and out of protective clothing without damaging it. If we have an event that requires a Level A dress out, we use disposable suits because it makes more sense than trying to decon permanent Level A clothing.

"We have quite a few chemical computer programs such as Sax's Dangerous Properties of Industrial Materials. For most of our emergency response we use CAMEO because that is what the fire department uses. Using the same program prevents a lot of confusion. We had looked at the possibility of ordering the CHARM module which is supposedly one of the more sophisticated programs, but the more we thought about it the more we realized that it was not a good choice since the fire department doesn't have it. If we are going to interface with them, we need to do what they are doing. We have several laptops that have CAMEO on them so we can go off-site to incident command and help the fire department with modeling and tasks like that. The Gainesville Fire and Rescue Department is getting ready to put a cellular FAX machine on their Haz Mat response vehicle which will give us the ability to get information at the command post.

"We have also used ALOHA (a segment of CAMEO) and it was pretty accurate. As a matter of fact, we were so pleased with it that we are going to use CAMEO and ALOHA for our dispersion modeling under the Clean Air Act 112R. I was at the Chemical Process Safety meeting in New Orleans recently, and the companies were all talking about what program they were going to use such as SAFER, FAST, and CHARM. These are $15,000 to $35,000 proprietary software systems. I mentioned CAMEO and ALOHA. Others responded that CAMEO and ALOHA are not very sophisticated programs. I replied that may be, but our responders are using CAMEO and ALOHA and it is very important that we are able to interface with them on a basis they can understand. Some of the people there said their first responder groups used CAMEO and maybe that's what they needed to use. I urged them to think about it and talk to their responders to help them make a decision. That's one of the reasons we started using the same terminology as fire departments. It makes sense for us to converse with them in a way they can understand and we can understand. We feel very comfortable with our relationship to the fire department. It is a positive thing and I think it is good for the community. Within the seven county area, the Gainesville Fire and Rescue Department, PCR, Inc., PSC in White Spring, the Hamilton County Fire Department, the former Proctor and Gamble pulp mill in Buckeye, and the Perry Fire Department, all use the same terminology."

Contact: John E. Hudson, Manager of Safety and Emergency Preparedness, PCR Chemical Company, P.O. Box 1466, Gainesville, FL 32602-1466; 904-376-8246, Ext. 284; 904-373-7503 (Fax).

TEAM-1 ENVIRONMENTAL SERVICES, HAMILTON, ONTARIO, CANADA

The July 1997 fire and toxic cloud at Plastimet Recycling in Hamilton, where at least 400 tons of plastic burned and released gases raised serious questions about the dangers to people who live or work near the 630 waste transfer sites in the province of Ontario. The fire burned for 77 hours, and an estimated 5,548,000 gallons of water were poured on the incident site. A commercial response contractor was on site recovering contaminated water for 144 hours after the start of the blaze.

The toxic substances that were released into the air included dioxin, any of a group of compounds known as dibenzo-p-dioxins. When tested on laboratory animals, dioxins were found to be among the more toxic synthetic chemicals having an oral LD_{50} of 0.022 mg/kg in male rats and 0.045 mg/kg in female rats. Dioxin became widely known as a potential danger after the herbicide 2,4,5-T exploded at a manufacturing plant in Seveso, Italy years ago. Many workers were exposed, and vegetation in the town was destroyed. The workers developed chloracne, a disfiguring skin condition characterized by the appearance of blackheads, cysts, folliculitis, and scars.

The small Missouri town of Times Beach was evacuated and eventually abandoned by the EPA in 1983 when high levels of dioxin were found on unpaved city streets. The accumulation of contaminated oil which had been spread on the streets to control the dust caused Times Beach to become a ghost town. Dioxin is known to be a carcinogen. *The Firefighters Handbook of Hazardous Materials* lists its toxicity for lungs and toxicity for skin as high, potentially causing permanent injury or death. For the Disaster-Atmosphere category, dioxin is judged a level 4, with a cautionary statement that "many factors influence this point, such as degree and area of confinement, air and wind currents, type and scope of involvement, etc. and is a relative value only."

One sample, taken from a stream of wet ash flowing from the main part of the Plastimet fire, contained 25,000 ppt toxic equivalent of dioxin. Another sample, sooty residue in a stream of water running across a nearby street, contained 7600 ppt. Any area with dioxin concentrations higher than 1000 ppt is considered unfit for industrial use according to provincial guidelines. In high amounts, dioxins may be linked to medical problems including cancer, suppression of the immune system, and reproductive problems.

About 200 firefighters fought the fire at Plastimet. About half have reported health problems including respiratory difficulties, nasal and throat irritation, skin rashes, eye infections and fatigue. Despite their protective gear, many had skin peel off their hands and feet. The International Association of Fire Fighters demanded a provincial investigation stating, "More than two months after the fire, many of the firefighters at the scene continue to experience the ill effects of toxic exposure … the Plastimet fire has raised serious safety concerns not only for the Hamilton fire fighters, but also for the citizens of the community … exposure to burning polyvinyl chlorides (PVCs) raises the real possibility that the fire fighters and citizens will be stricken with serious illness in the future."

Firefighters are really "burned," in more ways than one, with the politics that have gone on since the fire. They are worried that chemicals such as dioxin and benzene, both of which cause cancer, could cause drastic health problems as years go by and the Plastimet fire is all but forgotten. The fire department has refused to pay for liver and kidney tests to ensure that vital organs have not been damaged, and the Ontario Health Insurance Plan usually does not pay the $350 required for such exams. The Occupational Health Clinics for Ontario Workers informed

Hamilton-area physicians that such tests would be appropriate only if firefighters were showing visible signs of organ problems. The fire department has said it will pay only for tests sought by doctors. The firefighters are left in a Catch-22 position. They have no baseline medical data after the Plastimet fire to judge whether the fire damaged certain organs. So far, no medical follow-up has been authorized for the firefighters.

Other pollutants at the Plastimet fire with levels that rose sharply and then fell off were carbon monoxide, nitrogen oxides, hydrogen chloride (which mixes with moisture in the air or in a person's lungs to form hydrochloric acid), vinyl chloride and benzene. At its peak, hydrogen chloride in the air was almost six times the accepted government standard. The vinyl chloride peaked at 2.5 times the provincial standard for air quality, but the standard is set based on long-term exposure, and temporary, short-term exposure to higher levels is not a risk according to a report released by the Ministry of Environment and Energy. The readings for benzene peaked at 250 ppb; the normal range in the Hamiliton area for benzene is 1 to 8 ppb.

Mitchell Gibbs is manager of emergency services at TEAM-1 Environmental Services Inc. located in Hamilton, Ontario. This commercial spill response contractor provides Canada-wide incident response, contingency planning, and Haz Mat and confined space training. In Ontario, TEAM-1 has dispatch centers located in Hamilton, Burlington, Toronto, and London. About 30 minutes after the start of the Plastimet fire, TEAM-1 was asked by the Hamilton-Wentworth Region to respond.

"Our main goal was to recover the water runoff that was deemed toxic," says Gibbs. "Responders put 210 million liters of water on the fire. It was estimated that 10% of that was lost through vaporization leaving a balance of 189 million liters. Fifty percent of that went into the storm sewer system and allowed for direct discharge into the harbor. The other 50% went into the sanitary sewer system and to the sewer plant. The amount recovered compared to the amount put on the fire wasn't very good. The excess was a concern for a lot of people. Some of the streets were under two feet of water; the railroad track lines were covered. Obviously, the runoff water carried a lot of toxins: dioxins, lead, zinc, chrome, magnesium — a lot of heavy metals.

"The site was an industrial location. A smelter had been onsite for the past 50 or 60 years and had been abandoned for ten years. The location was deemed a highly contaminated site even before it was rented to a person who recycled polyvinyl chloride. The exact amount of polyvinyl chloride on site is still under question. The site had numerous violations under the fire code, and the company that operated the PVC recycling was approximately $850,000 in arrears on taxes. A lot of issues relate to that fire. A lot of people claim it never should have happened. A lot of questions surfaced about how the fire was handled and how much hydrochloric acid fallout occurred. The big question is, 'shouldn't there have been an evacuation?'

"The alarm was called in by a fire department tactical unit out doing building inspections. They came around a corner and discovered the fire, which was called in as a 'still' alarm (common term for when an active fire is found, not when a fire is called into the station). Immediately, trucks were dispatched from a station two kilometers away. Response time would have been less than two minutes. Within 10 to 12 minutes, however, the entire facility was fully engulfed in flames. Responders suspect a carrying agent accelerated the fire; adapters and beams could have been contaminated with a carrying agent such as zinc dust that could have caused the fire to travel so quickly. Off-duty fire department personnel were called in to respond. At that point, it was declared a major fire with environmental impact. The ministry of environment, the ministry of health, all regulatory agencies, the mayor, and all politicians responded.

"Advice was given on how to handle the fire," remembers Gibbs. "At that point, the fire department made an external attack, no personnel were allowed inside the building. It became a 'surround and drown fire,' a common term familiar to all the agencies involved.

"All two acres of the site were involved. Firefighters placed a number of aerial trucks around the area and flooded the site. The building on the site collapsed on the PVC which led to a lot of problems. The firefighters were not able to direct water at the piles of PVC. Heavy equipment was eventually brought in to remove pieces of the roof from tightly wound vinyl and other PVC products. The fire was very intense and stubborn.

"The fire service was taking advice and recommendations from numerous agencies such as the ministry of the environment, the directors of health, and the director of emergency preparedness. They determined that evacuation wasn't necessary the night of the fire. However, three days after the start of the fire, an evacuation was ordered because of a temperature inversion which would not allow the toxic plume to escape and because of the amount of toxic flooding. Over 77 hours, the plume migrated approximately 30 kilometers affecting many areas. Initially, the wind kept switching. First, the plume headed south, but during that first night, we experienced at least 10 to 15 wind switches at a minimum of 180° each including some that changed 360°. Such movement caused the plume to flood the command post with fallout. You could never get a safe position.

"TEAM 1 was on site to recover the toxic waste. It was a massive task utilizing the services of numerous carriers, but Laidlaw Environmental was the main responder. At one point we had 26 large-scale industrial vacuum trucks on site trying to recover water at different points. Luckily, the main area collection point was a low-lying area on the street. It probably covered a good acre. Had it been on a hillside, we would never have been able to recover the water. The decision to pick up the contaminated water was made by the regional sewer people because it was obvious that any water in this known toxic site would have elevated levels of heavy metals and hydrocarbons. Knowing that polyvinyl chloride was involved, officials decided right away that some form of environmental recovery would be necessary.

"The Regional Municipality of Hamilton does not clean up spills in their area. They contract with private industry and oversee the cleanup done by private industry. TEAM 1 Environmental has a contract with the city of Hamilton to respond to chemical spills and similar incidents. We knew immediately that hydrochloric acid and dioxin would be the products of PVC combustion. These were the two main concerns at the Plastimet Recycling fire. Dioxins, of course, were a huge concern. The problem that we ran into was that dioxin test results were not available for 48 hours from the point of sampling whereas hydrochloric acid results were immediately available. The fire started at 7:30 p.m. and through on-site analysis of the waste water, we knew by 2:00 a.m. that the water was toxic. The Ministry of Environment, for whatever reason, could not respond with certain high-end equipment until 5:30 a.m."

Mitchell Gibbs was asked if TEAM 1 members have been called on to respond to a chemical agent like a poison gas or a biological substance such as anthrax or typhus. "We have a couple of pathological biological laboratories in Ontario that retain our services. A spill there would be treated no differently from a chemical spill such as chlorine or ammonia. The same type of suits would be worn with the same types of precautions, same manpower, same chemical setup, and the same system. The suits we have would protect us against most of these agents, and we have a careful monitoring program for the specific agent we might be working with. In 14 years of doing this type of work, I've only had two occurrences involving an unknown. We have a workplace hazardous materials information system in Canada. Anything that's in an industrial plant has to have a material safety data sheet attached to it or in an accessible place. Very rarely do you get called to a drum that's in the middle of the road with nobody around."

When called to the scene of a hazardous materials incident, TEAM 1 personnel make use of a standard operational guidance that was designed after the National Fire Protection Association Standards 471 (Recommended Practice for Responding To Hazardous Materials Incidents) and 472 (Standards of Professional Competence of Responders To Hazardous Materials Incidents). Mitchell Gibbs functions in the role of the incident commander, and Debbie Vanderlip is the health and safety coordinator.

"Upon arrival, we report to a pre-assigned command post to gather information. At the point that the two of us agree we have sufficient information, we would establish a plan of attack for the situation. Debbie would be responsible for selecting personal protective equipment and protocol. From that point we would enter the site to confirm what has been reported to us, exit the site, and report the findings to the overall incident commander. From that I would either readjust my protocol or form a new protocol, depending on the nature of the incident. We subsequently re-enter and mitigate the situation.

"At any time during that process, we have provisions to alter our original plans based on our findings. If we get in there and find an additional agent involved, we can remove ourselves. At the same time, if we find that the air quality or the risk level is less, we can degrade our suiting level. Downgrading our suits can be very important. The very last thing you want to do is to work in an encapsulated suit needlessly. If you can get the incident response down to a hard hat and safety glasses, it is much easier to do work at such a site, but that can only be arranged when you have accurate findings and accurate air quality readings."

TEAM-1 Environmental Services, Inc. provides Canada-wide emergency response to government, industry, and transportation agencies and businesses. Specifically, they handle tractor trailer/tanker truck roll-overs, Haz Mat spill response, Level A–D chemical handling, hydrocarbon spill cleanup, high hazard tank cleanout, radioactive incidents, explosive materials control, high angle/confined space work, specialized heavy equipment, municipal fire/police department support teams, illegal drug lab dismantling, low profile rapid response units, abandoned property mitigation, contingency spill plan consulting, and training facility services, with fully trained, certified, and insured personnel.

Contact: Mitchell Gibbs, Manager of Emergency Services, TEAM-1 Environmental Services, South Ontario Division, 1650 Upper Ottawa Street, Hamilton, Ontario, L8W 3P2 Canada; 905-383-5550; 905-574-0492 (Fax).

THE BOEING COMPANY, SEATTLE, WASHINGTON

Gary D. Gordon is a toxicologist and a captain with Boeing's Security and Fire Protection staff in Seattle, WA. His official title is Environmental Programs Administrator, while his departmental title with the Boeing fire department is Hazardous Materials Specialist. "I ended up in hazardous materials basically by accident," says Gordon. "I was going to community college pursuing a degree in engineering. At the same time, I was interested in emergency medical services since I was working what they call 'part-time per diem fill-in' with Medic One. That sort of sparked my interest in the fire service. I continued to work for Medic One through my college days until I graduated from Western Washington University with an undergraduate degree in environmental toxicology and health. At that time, the City of Renton was hiring a hazardous materials specialist to work in the fire marshal's office as a liaison and technical resource for incidents involving hazardous materials. I spent a little over two years with the City of Renton fire department in a cooperative effort with the Boeing Company and some of the other industries that were located in Renton. I joined Boeing in 1990 and have basically worked in hazardous materials response and training since that time.

"In 1992 we integrated the hazardous materials initial training and refresher training into one core group. Joseph Richards, Mary Hinds, and I comprise the Puget Sound area hazardous materials training group. To date, we have trained about 13,000 student-hours per year. Participants include Boeing employees as well as municipal and county firefighters. We try to train on an interagency cooperative effort. Anytime we can do that it's not only to our best interest but to our municipal and county counterparts' as well."

The Boeing Commercial Airplane Group Haz Mat student training manual, *Hazardous Materials Emergency Response Training*, a copyrighted publication prepared by the Boeing Fire Department, is one of the most complete, thorough, well-researched and useful training documents the author has ever read. Clear details, examples, and complete coverage of basic Haz Mat response are the norm. Among items covered are competency checklist, regulatory law, hazardous substances, exposure guidelines, reference sources, SCBA, incident command system, chemistry/basic survival, monitoring/detection/alarm instruments, protective clothing, explosives, containers, containment, health and safety plan, heat stress, decontamination, incident termination, confined space entry, radiation, and bloodborne pathogens.

"We currently have approximately 350 certified hazardous materials technicians in the Puget Sound area staffing seven Haz Mat response teams for Boeing," continues Gary Gordon. "The company has its own internal fire department, and 150 of these technicians are firefighters who staff five company stations. The other technicians on the seven response teams are trade group people: plumbers, electricians, and millwrights. Our facilities people bring with them specific trade skills to create a cross functional hazardous materials response team. We had about 30 or so incidents this year at Boeing facilities throughout Puget Sound.

"Our hazardous materials technicians initially attend 40 hours of training in which we cover personal protective equipment, respiratory protective equipment, containment control techniques, the incident command system, and chemistry. During the final two days, we integrate all these subjects with a hands-on scenario implementing the incident command system, recognition and identification of the problem, mitigation, termination and demobilization. We feel confident that when they finish training they have the skills necessary to perform in the Boeing system. Our annual training update is an additional 16 hours. We do that on a quarterly basis, four hours each quarter so we see our Haz Mat technicians four times a year. We mix classroom and hands-on training and work on specific areas that are critical to maintaining the competency of the individual technicians. Each of our plant sites has an emergency response coordinator. For example, Bill Christie is the fabric division response coordinator. These people form an interface between the company fire department and the facilities group to ensure that training, activities, and equipment purchases are coordinated.

"For our response vehicles, we've chosen one-ton vans as our basic chassis with some modification of the interiors to accommodate different types of equipment. The interiors were adapted on site by Boeing personnel. Obviously, when you have a company this large, you have many internal resources that other companies, cities, or municipalities don't have access to. We use both MSA and Scott air packs (SCBA) because in the implementation of our first team efforts in 1989 we did not have a centralized focus or a standardization process. Some teams opted to use MSAs and some chose Scott. We have cross-trained all team personnel, both in initial training and ongoing training, in both types of air packs."

The Puget Sound area Boeing Haz Mat teams use Responder CSM personal protective suits, 8 to 10 Level A suits, and 8 to 10 Level B suits. For decontamination equipment they keep it simple: tarps and off-the-shelf wading pools. If these become contaminated, they dispose of them as hazardous waste and don't try to decontaminate them. "We have a pretty wide range of air monitoring equipment," says Gordon. "We have an infrared spectrophotometer that gives us a pretty wide range of capabilities for detection of organic materials. We also have a laptop integrator that has a library of 400 chemicals. We can take an unknown sample, create an infrared 'fingerprint,' compare the 'fingerprint' with the library, and the instrument tells us the highest probability of what the sample is. Also, each one of our Haz Mat vans is equipped with combustible gas/oxygen meters. Most of our vans have or will have photo ionization detectors. We also have some specific electrochemical sensor cells for use in areas where we have a significant amount of anhydrous ammonia. The electrochemical cell is a small, hand-held unit about three inches by four inches, battery operated with a LCD display, that is a sensor designed specifically for a particular chemical. We also have these units for xylene, arsine, and chlorine. We often will rely most heavily on colorimetric tubes for initial categorization of an unknown atmosphere. At the present time, we have in the neighborhood of 25,000 active MSDS, so it's sometimes difficult with all these chemicals to pinpoint exactly what we are dealing with.

"We have an internal emergency reporting number that connects with our central security and fire dispatch center. We have EMS-trained dispatchers who take the calls and dispatch the appropriate units. Depending on the type of call, the location of the incident, and the magnitude of the call, we might make a second call, or hotline transfer, to the county to ask for additional help, or to notify them of a hazardous materials incident. This is not mutual aid although it could be perceived as mutual aid. Because we are a private fire department, we cannot legally execute mutual aid agreements because they are government to government agreements. However, there are certain operational parameters where we immediately notify cities in the area of a Boeing plant that has an incident. For example, if there is a hazardous materials release at the plant in Auburn where we have exposed people and we need to go a Level A suit up to evacuate potentially affected individuals, we automatically contact the City of Auburn. They respond, and we form essentially one team so we have sufficient resources in equipment and personnel.

"We have a relationship with all the regular fire departments in the Puget Sound area where Boeing plants are located. Examples include the City of Everett where we just finished a series of hazardous materials chemistry classes at their station, the City of Kent, the City of Seattle, the City of Tukwilla, the City of Auburn, and Graham Fire District 21 in central Pierce County. Again, we advise the local fire department when we have an active release, when we are evacuating personnel, when we have potential exposed victims, and when we are utilizing Level A clothing. We like to bring them in early because it helps to establish a unified command system. We all follow the incident command system that the National Fire Academy has published. Here at Boeing, our captains, crew chiefs, chief officers, and inspection officers have all been through the hazardous materials incident command training class. We try to make the fire departments and Boeing teams as seamless as possible.

"Our response system is guided by the greatest concern for the health and welfare of our response personnel, both firefighters and facility representatives. Response teams in our facilities strictly adhere to the nationally recognized system, or the system adopted by the city or municipality, so we have no confusion or conflict in terms of understanding each other's systems. That is, our incident management structure at this facility looks the same as the City of Auburn's. Our passport accountability system looks identical to the City of Auburn's. When any Boeing facility in the Puget Sound area has an incident and functions jointly with a municipal fire department, we have essentially adopted the fire department's approach.

"Under certain circumstances the Boeing Company has provided resources off site. At the Walden fires in eastern Washington in 1994 we sent 2 engine companies for a period of 21 days. For Haz Mat, we evaluate our ability to assist, and I and Joe Richards, a chemist, are on the central dispatch availability list. For instance, if the City of Auburn has an incident on the freeway and they need some technical information or resources, the company freely allows us to support their responders. We've worked with city and county agencies, police departments, Seattle-Tacoma International Airport, the Federal Bureau of Investigation, the Bureau of Alcohol, Tobacco and Firearms, and the Secret Service over the years pertaining specifically to chemicals, chemical impact, and chemical potentials."

Does Boeing have membership on the SERC (State Emergency Response Commission) for the State of Washington, or an LEPC (Local Emergency Planning Committee)? "I am a voting member of the SERC," responds Captain Gordon, "and Chief Bob Johnson of the Auburn Fire Department is the chair of that committee. Collectively, wherever the company has an operation, we always like to have a Boeing representative active in the local LEPC; this would include in the Puget Sound area King County, Pierce County, and Snohomish County."

What are the response procedures and tactics involved when there is a Haz Mat incident at a Boeing plant within Puget Sound? "First and foremost, all of our personnel have three tactical priorities during a hazardous materials incident: protection of life, environment, and property, in that order," stresses Gordon. "With that as your guiding tactical mission, everything you do is in support of those priorities. First arriving units will typically assess the situation and look for any visible signs that they have incident in progress, your typical recognition and identification assessment. They will confer with the reporting party based on his or her evaluation and the on-scene crew's evaluation. They may or may not evacuate the area. Obviously, if no one is in the area there is no need to evacuate. Now that we have identified we have protected life, we consider environmental aspects. Over the years the company has made a concerted effort to install safety devices such as secondary containment systems to prevent a release from leaving the building. They assess whether the site is a contained area. If it is, a spill is not going anywhere, and it won't be able to impact the environment. Next, we consider protection of property. Based upon the material concentration of airborne contaminants, we have to select the proper level of protective clothing, either Level A or Level B. If the material is anything less than Level B and can be cleaned up using Level C, we do not consider it to be a hazardous materials emergency. It is still a 'situation,' but not an emergency.

"We follow fundamental guidelines found in NFPA 471 and NFPA 472. We believe that for every person you put in the hot zone you must have a counterpart in back-up. You also must have a decontamination corridor set up and ready to go before you send any personnel into the hot zone. We delineate zones early on in the incident where nobody goes without proper protective clothing: the hot zone, and the warm zone where we are going to set up our decon corridor. We also use a green zone where our support, staging, operations, and command will be contained.

"Because our firefighters are here for the protection of people, the environment, and the assets of their respective locations, we deal heavily in fire prevention. Our goal is that fire apparatus sitting there right now will never turn a wheel. Because of that goal, we have an active, fixed fire suppression system. For instance, take the sprinkler system of this office we are sitting in right now. We check and maintain these sprinkler systems, and issue cutting and welding permits for all operations on company premises. This familiarity with Boeing processes and the knowledge of Boeing firefighters really are the best resources for fire departments from outside the facility. They don't have the familiarity that we do with our systems. An example would be the overhead crane systems. Every one of our firefighters understands the overhead crane systems, how they work, and where they are located. They also understand aircraft, one of our key businesses.

"The Boeing fire department also maintains emergency medical services. Over 80% of our firefighters are emergency medical technicians with AED (automatic external defibrillation units). Currently, the King County save-rate is in the neighborhood of 10 to 12% for cardiac arrest. At the Boeing Company, if you have a cardiac arrest that is witnessed, your save-rate is in excess of 80%. Our average response time is three minutes or less to anywhere in this plant since this site is rather large but fairly compact. We know where the people are and processes they are doing. An understanding of how the company operates gives our firefighters a much better handle on what is going on than someone walking in off the street. They can really focus on the true hazard without first having to understand what's going on in the process. The firefighters who are hazardous materials technicians perform many functions. Highly trained and skilled, all 150 of them maintain five areas of competencies. They are Haz Mat technicians, EMTs, structural firefighters, confined space rescuers, and airplane crash rescue personnel. This particular site does not have an airport that supports flight operations, but we have three others that do, and all our firefighters are trained in responding to that type of incident."

How are the Boeing facilities in the Puget Sound area funded for fire service, particularly for Haz Mat service? The fire department resides in a group called Information Support Services, or shared services group. According to Captain Gordon, "Our funding comes from the operating divisions on an annual basis and is meant to provide services to our 'customers.' A customer for us is any operating division in the company. We present them an annual operating plan, they review it and decide those are the services they want at the cost they want, or they decide that's not the service or cost they want. Essentially, it's a negotiation process. Emergency response coordinators represent each of the operating groups at the specific plant sites. If we are talking about chemical protective clothing, for example, they would purchase the suits. The individual sites have agreed to purchase the Haz Mat equipment on the response van, and we have agreed to maintain the equipment and the van.

"Several years ago we made an effort to install CAMEO in our hazardous materials response vehicles. Unfortunately, with the amount of jarring, heat, cold, dampness, and rough handling in the vehicles the fixed-type systems that you would purchase at your local computer store did not stand up very well. The reliability that we could count on just wasn't there. We have since placed our information on a laptop computer, and that information is available by calling Joe Richards or me at anytime. We are also looking at installing the CAMEO software in our central dispatch centers so that our dispatchers can also access that information and provide it to the incident commander at the time of the call. CAMEO is the primary software we use in a response mode, but we need a total of three resources that give the same data before we put together our final tactical operating plan. We might use CAMEO, the MSDS for the chemical, or *Dangerous Properties of Industrial Materials* by N. Irving Sax and R.J. Lewis, Sr.

"The Haz Mat technicians at Boeing know we have a contingency plan. In that contingency plan, we have resources, special containment control equipment, and information about people who function in certain capacities. The information is listed at each of the various areas where we use hazardous materials. The contingency plan is a resource primarily for the incident commander to double check what resources we have available at each site. A number of our hazardous materials technicians were selected from the trades such as plumbing or hazardous waste; they know where the equipment is and how to use it."

Hazardous materials technicians from the trades represented at various Boeing facilities are volunteers for such duty. If few Haz Mat technicians are available, the company asks for volunteers from the trades as well as from various shifts. The fire department is at the facility 24 hours a day, but the company also needs site-knowledgeable facilities technicians across all three shifts. According to Captain Gordon, they have no trouble recruiting people from the various trade groups. "We never had a problem with recruitment. Interest does not seem to be affected by gender because we have a substantial number of female hazardous materials technicians who make outstanding response personnel. We have some personnel, both male and female, who plainly state that Haz Mat response is not for them. One of the first things we do in our 40-hour training course is tell everyone ahead of time what they are going to have to do. 'We are going to put you in Level A clothing. It's going to feel tightly enclosed when you are zipped into the suit. If you have claustrophobia, this may not be for you.' One of the reasons we emphasize suit time and air path time in their initial training is to allow them to identify adverse reactions now rather than later on when an incident occurs."

Boeing uses an incident management passport accountability system to track the movements of people so there is no freelancing, and so that supervisors do not lose track of individuals who might be in the staging areas or operations area. They follow a procedure developed by the City of Seattle. "For a Boeing employee, the name tag, has his or her last name and the last four digits of his or her social security number which we call a 'clock number,'" explains Gordon. "The name tags identify you when you are given an assignment at a hazardous materials incident. Your name tag goes onto a 'passport' which is nothing more than a piece of Velcro with a label or division across the top. If the staging officer assigns two technicians as entry team members, that officer takes the name tags from the two technicians, puts them on the passport, hands them back to the two entry team members, and says, 'I want you to report to the operations officer at the northwest corner of building 2460.' They do, and because of the passport, the staging officer knows he has allocated these two technicians to operations, and how many people are left to assign. When the two entry team technicians report to the operations officer and give him the passport, the staging officer knows that they are now under the direct control of the operations officer. If something goes wrong, if there are any circumstances where technicians cannot be located, such as an unforeseen explosion, the incident commander will know how many people are missing, who they were, and what they were assigned to do. That's why the practice is called the passport accountability system. It's just like traveling from one country to another, if you don't have a passport you don't get by the checkpoint.

"In terms of our Haz Mat technicians' long-term health and safety, the federal regulation makes it clear that if you are a member of a hazardous materials response team, you are required to be part of a medical surveillance program which includes a complete physical examination upon entry, an annual exam while a Haz Mat technician or Haz Mat team member, and a final exam upon exiting the program," continues Gordon. "Also, anytime you exhibit signs or symptoms of chemical exposure, you need to be evaluated. Within the company we have a safety health and environmental affairs organization known as 'SHEA.' This organization is part of Boeing medical and is staffed by occupational physicians who perform our physicals based upon a standard that is recognized throughout the occupational medicine community for the type of work we do.

"On the incident ground, we follow recommendations put out by the National Fire Protection Association," relates Gordon. "We do pre-entry vitals for all person-nel in Level A or Level B chemical protective clothing. Blood pressure, heart rate, respiratory rate and body temperature are the four key indicators. We try to establish a baseline so that when a technician comes out of the hot zone, we can compare 'before' and 'after.' If we notice any abnormalities, we immediately send the person to one of our medical clinics which has a full range of diagnostic equipment, or to one of the local area's hospitals.

"We use the incident control system to maintain order on the incident scene. You function in your assigned role based on training, and the system is there to protect you by providing resources and personnel. The incident control system by its very nature establishes order and discipline. Working with Boeing, and working with outside agencies as we do on a regular basis, we find the system works best when the incident control system structure and guidelines are adhered to from the time the call is made until the incident is done. We can get any type of call — it does not have to be a hazardous materials call, any emergency response would do — where we think we have done this many times before and expect this incident will proceed like the previous ones. We might become lax in our adherence to the incident control structure, at least in the initial stages. We must maintain a disciplined approach when we become lax in the initial approach and integration, or we may have to go back to the beginning and start all over again. That costs us time and frustrates the personnel at the incident scene. We strive to teach that on every call you respond as you trained. You institute that discipline and management structure no matter how seemingly small the incident. Small incidents can escalate to large incidents with just a few minor changes in conditions."

"The incident command system is disciplined in structure but it still relies heavily on people. People have personalities, and people are human. We find the best way we can overcome the people and personality issues is to train with our municipal firefighter counterparts and other response agency counterparts who work with us on major incidents. If we train with them and work with them on a regular basis, many of these potential conflicts or misunderstandings go away. When you're walking down the street and you see somebody you recognize, you have a certain comfort level immediately; you go up and talk with him. But how many times do you walk up to an absolute stranger? Severe hazardous materials incidents have the potential to become life threatening situations. In such situations, whom are you more willing to trust, someone you do not know, or someone whose capabilities and limitations you are familiar with? It is important to train together so you know your counterparts' training, equipment, capabilities, and limitations. When complete strangers come on the incident ground, the human factor has a tendency to get in the way.

"Hazardous materials response, and your education as it pertains to hazardous materials, is a journey not a destination. If you talk to someone who says, 'I'm it, I'm there, I have arrived,' it's time for him to quit the business. With that in mind, we change our training approach based on everything we learn. The three of us who comprise the training group work on the operations site. We're there to see the mistakes, things that could have gone better, and we take that information, turn right around, and incorporate it into training. After all incidents, we hold a follow-up debriefing and critique. A debriefing is essentially getting everybody who worked on the incident scene together to go over the facts so we know what had occurred. That way, everybody walks away knowing his or her significance in the incident response. The critique, which happens about ten days after an incident, is where we start to analyze. What things were correctly done, what things didn't go quite as well? Is there anything we can do to correct our training, operation, or tactics? After each of our incidents, both the debriefing and the critique will take place."

Gary Gordon notes that not much has changed in principles of fire science. "However, look at hazardous materials," he says. "The Chemical Abstract Service is growing by leaps and bounds every day. The last figure I heard was 27 million chemicals and they were adding 1800 a month. That's dynamic. This field will change as we learn about how to build different processes and move away from different metals to composites. What does this mean for the hazardous materials technician? It means he will face more organic materials as opposed to metals which pose little to no hazard. We should be wary of the individual who claims to be a Haz Mat expert because I truly believe there is no *one* Haz Mat expert. We all have our areas of specialty, our niches with which are intimately familiar, but I find it very hard to believe that in a field which requires so much diverse knowledge that you could have one person who says, 'This is it, I understand everything all the time.' In this particular business of hazardous materials response, when you have reached the point where you are complacent about incidents, then it's probably time to do something else because complacency breeds accidents. You *can't* be complacent.

When you deal with hazardous materials, sometimes you have to deal with "fight" or "no fight" decisions. "Oftentimes in the fire service or any emergency services situation the public believes that we are here to do 'something'," responds Gordon. "When people call the fire department because of a house fire, firefighters put out the fire. When they call the fire department for an emergency medical condition, the department will transport the patient and provide the best medical care within their capability. In hazardous materials response, on the other hand, you have to understand that there are fight/no fight situations. A statement by my predecessor who has retired from the company comes to mind, 'all incidents are self-mitigating over time.' In other words, you always have to go back to those three technical objectives: protection of life, protection of the environment, and protection of property. Sometimes you will place other people at risk. To save another's life, people in the response field are willing to accept risk to their own lives. The important consideration is that we are not risking others' lives to protect something that is of less value. You must ask yourself, 'Does the benefit of what I am doing outweigh the risks of what I am committing?' At some point, all incidents solve themselves. The tanker full of gasoline will stop leaking because it's empty. The fire in the house will stop burning; unfortunately, no contents are left, but it will stop burning. In the Haz Mat arena, we understand that there is definitely a time to do battle and a time when we don't do battle. We need to make sure in our risk/benefit analysis that we are weighing an appropriate amount of risk against an appropriate amount of benefit."

Has the Boeing Company considered the possibility of a terrorist or another form of attack against their seven facilities in the Puget Sound area? "That's always an issue," answers Gordon. "We consider any type of attack, not just a terrorist act, a direct effort to interrupt or do harm to our personnel. The company has a security force in Puget Sound of approximately 250 officers. We have a badging process in effect. People at specific areas prevent unauthorized entry, and our parameters are distanced from our buildings. We have the ability to recognize the potential threat and, as necessary, to take actions. This plan is constantly evaluated. A new thrust is related to chemical/biological terrorism in light of the Tokyo, Japan incident. Speaking as a hazardous materials individual, not as a Boeing Company representative, I look at a chemical terrorist act as essentially a multi-dimensional incident. You are going to have to deploy two separate divisions. One of these would be your medical services division because you have the potential for an enormous number of people to be affected. In the Puget Sound area, we have to implement what's called a county-wide MCI (multiple casualty incident) which sets up a system for finding, locating, triaging, treating, and transporting patients as quickly and efficiently as possible to multiple area hospitals so that we don't overwhelm any one location's resources. On the Haz Mat side, once a chemical agent is dispersed, you are dealing with a hazardous materials release and you treat it as such. There are three caveats in such a situation that you have to be able to consider: one, this incident was intentionally done; two, if possible, try to preserve evidence; three, look out for secondary devices — devices designed to injure, harm, maim, or kill the responders who are coming in to take care of the problem. In a number of ways, however, a chemical is a chemical whether presented in a terrorist attack or in some other arena."

Contact: Gary D. Gordon, Toxicologist, Haz Mat Response, Security and Fire Protection, The Boeing Company, P.O. Box 3707, MS 34-67, Seattle, WA 98124-2207; 253-657-8657; 253-657-9988 (Fax); 206-949-2529 (Cellular).

9 Urban Search and Rescue Teams and a Mass Casualty Incident in Washington

THE PUGET SOUND URBAN SEARCH AND RESCUE TEAM

The U.S. National Urban Search and Rescue Response System (USAR) was developed by the Federal Emergency Management Agency (FEMA) as a means of providing existing response personnel from the local level for integration into mobile response task forces trained, equipped, and available to counter and control a national emergency. Such task forces would be dispatched only when a natural or technological emergency overwhelms state and local response capabilities, the state government requests federal assistance, and the president formally declares that a disaster has occurred (as required by the Robert T. Stafford Disaster Relief and Emergency Act, P.L. 93-288, as amended by P.L. 100-707).

USAR task forces can provide the following capabilities: physical search and rescue operations in damaged or collapsed structures; emergency medical care to disaster response personnel; emergency medical care to injured persons; reconnaissance to assess damage and needs and provide feedback to local, state, and federal officials; assessment and shut-off of utilities to houses and buildings; hazardous materials surveys and evaluations; structural and hazard evaluations of government and municipal buildings needed for immediate occupancy to support disaster relief operations; and stabilizing damaged structures, including shoring and cribbing operations. A USAR task force is meant to support and enhance local government efforts, and local government should have ruling authority. A task force must respond from their location within six hours of notification, and while at the incident site must be able to maintain operations without support for 72 hours.

There are 27 USAR task forces throughout the country which are composed of local emergency personnel trained and equipped to handle structural collapse. A USAR task force has 62 specialists and is divided into four major functional elements: search, rescue, technical, and medical. Members include structural engineers and specialists in hazardous materials, heavy rigging, search (including highly trained search dogs), logistics, rescue, and medicine. Each task force has a comprehensive equipment cache weighing 29 tons including communication gear and equipment for locating, roping, rigging, hauling, lifting, pulling, sensing, extricating, cutting, and drilling. A medical team has four medical specialists, many of whom are both paramedics and firefighters, and two physicians, who are generally specialists in emergency medicine. The medical teams carry advanced life support equipment.

The Puget Sound Urban Search and Rescue Force (WA-TF-1) was established in 1991, one of the first 12 task forces to be deemed deployable by FEMA. Sponsored by the Pierce County Department of Emergency Management, the overall task force is drawn from fire departments, local hospitals, law enforcement agencies, public works departments and the military. The largest contributor to the task force is the Seattle Fire Department, with other significant support coming from the King County Fire Chief's Association, the Pierce County Fire Chief's Association, and the Pierce County Sheriff's Department.

TABLE 9.1

Agencies Participating in the Puget Sound USAR Task Force (WA-TF-1)

Pierce County Emergency Management	FEMA - Region X
Pierce County Sheriff's Department	Seattle Fire Department
Seattle Emergency Management	Tacoma Fire Department
Pierce County Fire District #2	Tacoma Public Works Department
Pierce County Fire District #3	Central Pierce Fire/Rescue
Pierce County Fire District #5	Western WA State Hospital
Puyallup Fire Department	Madigan Army Medical Center
Good Samaritan Hospital	King County Emergency Management
Tacoma Mountain Rescue	Mercer Island Fire Department
Tukwila Fire Department	King County Fire District #4
Renton Fire Department	King County Fire District #26
Woodinville Fire/Life Safety	Federal Way Fire Department (District #39)
WA Structural Engineers Association	Northwest Disaster Search Dogs
WA State Emergency Management	Pierce County Planning/Land Services

---

On April 19, 1995 an estimated two tons of explosives inside a Ryder rental truck were used to collapse the north side of the Murrah Federal Building in Oklahoma City killing 169 people and injuring 475 more. Eleven USAR task force teams from throughout the country were rotated during the days that followed. Puget Sound USAR was alerted on April 23, mobilized on April 24, arrived in Oklahoma City at 7:39 p.m. on April 24, and worked 12 to 14 hours a day for 8 days.

The executive summary of the Puget Sound USAR Task Force Oklahoma City Deployment has this report of the deployment. "Our operation on site primarily consisted of shoring partially collapsed concrete floor sections, breaking and removing concrete, hauling debris from the shelter, and assisting with extrication and removal of victims. The search component of our group gained access to the upper floors on the east side of the structure by entering a nine-foot diameter HVAC tower on the southeast corner and using ropes to ascend inside the tower to the ninth floor. At this point we established a rope hoist system enabling access to all floors on the east side of the building. Rescue squads gained invaluable knowledge of concrete breaking methods with various tools, rebar cutting techniques, heavy object movement, and the shoring of unstable, partially collapsed floor and wall sections. More visible among our accomplishments were those achieved by the combined efforts of our riggers with the crane crews, responsible for movement of the massive sections of reinforced concrete.

"Search team managers reorganized their groups into rescue squads employing technical search specialists, canine search specialists, and Haz Mat specialists. Both safety officer positions were utilized to monitor the number of squads working at a given time. A paramedic filled the role of safety team manager as well as assistant safety team manager. This allowed both doctors the freedom to assist any squad involved with victim recovery operations. Medical monitoring of our personnel under this system proved to be very successful; WA-TF-1 members experienced the lowest injury rate of the task forces. The work load for logistics personnel remained high for the entire time in Oklahoma City. They provided for on-site equipment, tool repair, and procurement of new equipment. Logistics, technical information, and communications personnel worked well beyond the scheduled shift periods on a regular basis. After installation of a repeater on a nearby high-rise, communications were uninterrupted and clear."

Captain Bryan Hastings has been with the Seattle Fire Department for 12 years and has worked in urban search and rescue since 1991. "All the USAR task forces were federally funded by FEMA with an initial grant to develop a team and buy equipment. Then they were given annual grants to train the team and care for the equipment. On any deployment we have been sent on, we have been reimbursed for both wages and equipment, if we could make the argument that such equipment was expended on this particular training evolution. All the task forces around the country are staffed with 62 members as an operating team. Most of the task forces have three to four times that number of people. We can break our team up into three sections so we are never going to run short on any deployment.

"In the Puget Sound USAR Task Force, the Seattle Fire Department has assumed a leadership role and has provided the organizational base for equipment and training. We have 62 members ourselves, while other organizations in this joint task force have about 130 or 150 members (on a FEMA mandated alert to a distant location, the Seattle Fire Department would provide 36% or 22 members of the 62-person responding task force). When we are dispatched, FEMA does not mix teams, but most teams are made up of mixed organizations due to the types of skills and training required. The destruction of the federal building in Oklahoma City could serve as an example. The standing protocol is that after 72 hours officials evaluate the fatigue factor of the team on duty and determine when they will rotate teams in and out. Usually, at the five to six day mark they've already developed a rotation schedule.

"At Oklahoma City, we were probably the seventh or eighth team to be rotated in, but I don't remember how many days that was after the explosion. FEMA had alerted the Pierce County Department of Emergency Management which receives the orders for the deployment of our task force. The military transports the task force and its equipment to the incident site. We have team members who are trained in setting up the pallets that the military will accept for the C-130 aircraft that is used. The U.S. Department of Defense has close liaison with all the task forces so we can have all the transportation needs met. When we arrive at the incident site, defense department and FEMA representatives meet with our task force leaders and provide a briefing. We unload our equipment, set up our quarters, and within two to four hours they put us into action and establish a work rotation for us."

TABLE 9.2

The Specialized Skills Seattle Fire Department Provides to Puget Sound's USAR Task Force

- Technical search and rescue to locate trapped victims
- Canine Search
- Structural collapse extrication and rescue
- Collapse shoring and stabilization
- Collapse cutting, shoring, breaching, and void penetration
- Heavy equipment and rigging
- Emergency medical field care for collapse/confined space medicine
- Collapse operations when exposed to hazardous materials
- High and low angle rope rescue and rigging confined space rescue
- Below grade rescue
- Incident command system coordination and command structure

"We have approximately 65,000 to 68,000 pounds of equipment that includes approximately 2500 to 3000 different items from hand tools to generators to big wrecking tools," relates Captain Hastings. "The inventory list is about the size of a small telephone book. Our job is to locate trapped victims, to provide an avenue for escape, to detect any hazardous materials, to provide emergency medical care to team members, search and rescue canines, and victims, and provide a support network to the local government. We can administratively support local government, aid in any of their ongoing operations, and fit into any command structure. We work very closely with the local jurisdiction since they have overall control, except in incidents such as the Oklahoma City bombing where the Federal Bureau of Investigation controls security.

"Urban search and rescue at Oklahoma City was for one building, but generally it takes place in a wider area like that affected by an earthquake or a hurricane, probably measured in blocks. We are usually looking for injured or dead people. Of the 62 members sent to any catastrophic incident, probably 50 to 52 will be firefighters with some law enforcement personnel included even though most of the search and rescue canines (K9s) are handled by firefighters. We usually take two to four canines on a trip with us.

"The Summer Olympic Games in Atlanta during August 1996 was a different type of activation for us. We were pre-staged to reconnoiter the area, develop maps, gather intelligence, and practice before the games began. Base camp was 8 to 12 miles outside the site with military helicopters ready to fly us in. We could get to the Olympic Village as quickly as the Atlanta Fire Department. Although we had received training related to NBC tools used by terrorists, we were there as an urban search and rescue team. We do have some law enforcement personnel on the team, but they are trained in search and rescue. We do not have any element on the team for security. In fact, we are there in case some component of security fails. Officials pre-staged search and rescue teams at both the Atlanta Olympics and the Democratic National Convention and restaged them during the events.

"So far, we have been deployed six times to a federally declared disaster: to a typhoon in Guam in 1991, the Northridge Earthquake in Los Angeles in January 1994, the Oklahoma City bombing in April 1995, Hurricane Hugo on the east coast, the Summer Olympic Games in Atlanta in August 1996, and Hurricane Opal in Florida. We have gone into action at two of these incidents. If not assigned, we take time to train with the equipment we brought since most of the team has only a limited ability to train with the specialized equipment we bring to a federal disaster scene. Our people are well versed in these tools, but if we are not assigned on a deployment, we turn the event into a big training exercise and start having demonstrations and exercises. On deployments, everything is reimbursed by the federal government. At other times, we get a small stipend of federal money, either $30,000 or $40,000 on an annual basis. When you consider how many people we have to train and the amount of equipment we maintain, that really isn't very much. What normally happens is that the supporting entities like Seattle fund the USAR team with money from their own training budgets. That is, during non-deployment assignments, the team gets supported on 30 to 40% federal funding and 60 to 70% local funding.

"Our team is required to have one deployment annually in case we don't go on one that is federally sponsored. Typically, we set our own deployment up for Alaska, but every single year we have been federally deployed, so we have to cancel our inspections and training. We send all of our training records and reports to the federal government. FEMA sets up annually the minimum requirements to belong to the team that include minimum hour requirements."

What about local disasters? "The urban search and rescue equipment purchased by the federal government is kept in a Seattle Fire Department USAR semi-trailer which has been pulled out for three incidents, all building collapses," responds Hastings. "You have to write a report to the feds indicating what the use of the equipment entailed. It was purchased by FEMA and if Seattle uses this particular equipment for a local incident, a memorandum of understanding that all entities have signed with the federal government outlines the use of such equipment for dire emergencies only. FEMA has approved the use of this special equipment in each of the three local emergencies in which we used it. Basically, the equipment supports FEMA's efforts in national emergencies, and use by local entities is tightly controlled.

"With regard to the general health and welfare of task force members, good training should eliminate most of the hazards. Some of the hazards are just inherent to the type of work we do. They are no different from the hazards we face in the fire service. The hours of training we have done should prepare us for most catastrophes. Some of the jobs and some of the equipment are just a little tricky, so we do have a small civilian component on the task force including structural engineers and hazardous materials people. We receive a physical examination prior to the end of a deployment. Also, we have very strict guidelines. If anyone does not meet the standards applied by the doctors giving the physical exams, he or she cannot be employed at the building or geographical location where the incident has occurred.

"We maintain a lot of equipment including computers, databases, resource materials, and manuals. We carry six computers with the team, and when we go on deployment, we leave a contingent of the team here in Seattle with two laptop computers. Wherever we go, we link our laptops so we have access to all the databases located in both areas. As examples, we have a personal medical history of each member, as well as the training levels reached by each member. Oftentimes on a deployment, we have to move team members around to different functions. With 62 members on the team, the Seattle Fire Department has a great deal of flexibility in adapting to different situations. If we had to convert the entire team to search, and officials decided that everyone was to do a really quick search of the entire city of Seattle, and then follow with rescue elements, we have enough flexibility including equipment, scene management and evaluation, staging area, team concept, and tactics to accomplish the task. We have certain tactics that the team is always trying to refine. Since the Atlanta Olympics, we've developed a recon team, basically an advanced scout team of six members. They bring in lots of recording equipment: digital cameras, video cameras, still cameras, and compasses and maps. After they reconnoiter, they call back and tell the task force what type of equipment we will need, the areas on which we might want to concentrate, what type of clothing might be required, etc."

While on a deployment, the Puget Sound USAR Task Force provides a daily briefing and, at the conclusion of the event, turns in an after-action report to FEMA. The records must be complete. According to Captain Hastings, "Our financial spread sheets on both wages and equipment, our scene management description, an analysis of the event, and our daily activities are all tied into the after-action report. If something doesn't jibe, then you certainly get the report back. Every USAR Task Force also eventually gets audited. Three days from now will be the first time in six years we have been audited by a FEMA team. They will spend up to nine days checking every nut and bolt. They audit for every penny that's ever given to a team.

"I think one of the most important things we have learned with the urban search and rescue task forces is that the level of success you have is in direct proportion to the level of training you receive. Some teams sit on their hands and don't seek out the training that FEMA offers, and FEMA offers a lot of training. It shows up on deployments. Some teams are well trained, and some teams have gone into it for the compensation rather than for the training. Most of these teams are made up of people who volunteer for this type of duty, and generally 30% of their time is not compensated. They are donating 30% of their time to better their skills within the department and to become an asset to the federal government. A team's level of training is obvious. We can tell in a short time which team components are well trained and which are marginal.

"What is unusual about the present USAR Task Forces is that the whole organization came about as the military was paring down. The federal government needed some element out there to take up the slack and develop a little more diversity than the military was capable of providing. The military basically used their own manpower, some equipment, and some local people. Now, with local fire department personnel from 27 areas around the country the federal government is getting experienced people who can take their various skills and move them around to have the USAR Task Force conform to the incident and the incident conform to the team. If I go to a fire with nothing but fire engines when I need a lot of ladders, I'm stuck. We are unique in that we can do many types of functions. We have the necessary equipment and the training to use the equipment. Everyone is trained in the various disciplines required for urban search and rescue. So wherever we go, we can adjust our tactics to meet the needs of federal and local jurisdiction officials. Most organizations can't do this. They can get a specialist to drive a big piece of machinery or a specialist who can do rigging. Most of our people can do most anything related to urban search and rescue."

Contact: Captain Bryan W. Hastings, Seattle Fire Department, 301 2nd Avenue South, Seattle, WA 98104; 206-386-1420.

THE TRAINING AND HANDLING OF URBAN SEARCH AND RESCUE CANINES

A new breed of dogs has come to live at the firehouse. Years ago, the only firehouse dogs were Dalmations. Their calming effect on horses and their ability to navigate quickly among the axles of the horse-drawn fire engines made them exceptional coach dogs and assets to early fire departments. Nowadays, the breeds may have changed but the firehouse dogs still do duty that requires a special talent.

On January 19, 1997, the Seattle Fire Department received a request for assistance from the Bainbridge Fire Department. A mudslide had buried a home with a family of four inside. Bainbridge needed technical assistance with search equipment and K9 capability. Because the victims would only be able to survive a short while under these conditions, the Seattle team had to locate and extricate the victims quickly. The team was airlifted in two helicopters across Puget Sound to Bainbridge Island. The following information regarding search and rescue dogs, plus other reporting requirements, is contained in the Seattle Fire Department after-action report for this incident. "The first K9 Team was sent in at 1145 (military time) and within 10 minutes a 'hit' was made. The second K9 Team was sent in for confirmation and an identical 'hit' confirmed the first body, a 3-year-old child. We continued in the Rescue Mode due to the presence of small survivable voids within the debris ... The location of the mother and father in the same area as the 3-year-old was confirmed with both K9s and then with visual confirmation after debris removal ... All 4 members of the family were in the master bedroom on the ground floor with the parents and the 3-year-old in bed and the infant in a crib at the foot of the bed. They were instantly pinned by the back wall of the structure collapsing onto them and perished immediately ... The use of the K9 capability significantly improved our ability to assist in the effort. The pinpointing of victims was essential to the appropriate use of resources ... This was clearly a mission where lives could have been saved due to the ability to use specially trained K9 and Technical Search Equipment."

On January 23, 1997, the Mayor of the City of Bellevue wrote a letter to the Chief of the Seattle Fire Department. "On Sunday, January 19, 1997, several of your firefighters, trained in urban search and rescue, were airlifted to Bainbridge Island to assist in the rescue attempt at the residence of the (name withheld) family. On behalf of the employees of the City of Bellevue, I would like to express our sincere gratitude to your staff for the assistance rendered. As you may know, (name withheld) was a long term employee with the City of Bellevue. We understand that your Technical Search Specialists armed with search cameras and several rescue dogs were airlifted to assist the Bainbridge Island Fire Department in the search for survivors. During these rare and difficult rescue challenges, it's comforting to know the Seattle Fire Department has the technical resources, as well as the willingness to offer them when needed. There are no words adequate to express the gratitude and appreciation for your department's assistance; this has meant a great deal to the employees of the City of Bellevue. Once again, please extend our sincere appreciation to all members that assisted in the rescue attempt."

On February 10, 1997, the roof of the International Travel Building in Seattle collapsed, and the Seattle Fire Department was dispatched to the scene. The following statements concerning rescue dogs were contained in the after-report among other factors related to this incident. "The incident commander called the Seattle Fire Department search manager at home to assemble K9 search capability with a minimum of 4 teams. Seattle Fire Department on-duty personnel with K9 capability were relieved to get their animal and report to the scene ... Searches were conducted, using K9 resources, from the east and west sides. On the east side, chain saws were used to provide triangular access points for the dogs. Search markings per FEMA standards were used at the entrance openings. Trained spotters were used to assist the K9 handlers ... the spotters were on the roof tops. This was done because the handlers were unable to see their dogs in the extensive rubble pile. The thermal imaging camera was used to track the K9s and assist in confirmation if a patient was located."

Urban and search and rescue dogs are not the same as police dogs. Police dogs are guard dogs. They are defensive. USAR dogs are "lovers." They are defenders. USAR dogs must be extremely bright, but do not need impeccable breeding. Several were recruited from dog pounds.

A.J. Frank is a young, career firefighter who serves at Station 2 with the Seattle Fire Department. Because of a strong personal interest, he is also an urban search and rescue dog handler and trainer. His co-worker is "O" (Ohlin), a very friendly and beautiful brown Labrador retriever who has some outstanding talents. For example, on a one word command, he will climb a ladder paw-over-paw; at another one word command, he will descend a ladder paw-under-paw.

"I got 'O' almost six years ago as a pet since Labrador retrievers are supposed to be easy to train, lovable, and good companions," remembers A.J. Frank. "I wanted to get the dog into what the American Kennel Club (AKC) calls 'Obedience Training for Companion Dogs.' To get that title, your dog has to pass a basic obedience test three times — he comes when you call him, sits down for a specified time, lies down for a specified time, can be with other dogs and/or people, can walk off-leash, can walk on-leash next to your side — just basic obedience. 'O' and I passed that test four times and then started training at hunter certification.

"I understand that most search and rescue dogs should start training the day you get them. I didn't start that type of training until he was about 18 months old, so it took awhile to get Ohlin up to speed. One of the Seattle Fire Department offices sent out a memorandum saying if anyone would like to come and learn about search and rescue, show up at this fire station at a certain time. I got the dog involved, and we have been working for about four-and-a-half years now. It has been a lot of fun, but it is time consuming. Most of the training is done on my own time, and most of the work is done with a group of people outside the fire department. I started training 'O' for search and rescue duties by working on a number of obstacle courses: climbing up and down stairs and ladders, walking on planks that wobble under his feet, handling slippery inclines and declines, and traveling through what is called a 'metrobox.' A metrobox is like a small-scale, collapsed apartment building. It is dark inside, with small holes cut to see if your dog can get through a combination of obstacles to learn if he will be okay in that type of situation. 'O' weighs 75 pounds. I guess I could carry him up a ladder, but it's a lot easier to just set a ladder where you want it and give the dog a one-word command. Most of the time we have a 100-foot aerial ladder up against a building or rubble. I have him climb up and then climb back down. I really don't want to carry him.

"You teach any canine how to do things in small steps. They know where their front feet are but not their hind feet. You just have to instill in them that the hind feet follow the front feet. You stand the dog against a ladder, place the right front paw up on a rung, place the left front paw on a higher rung. Now that the dog is stretched out, you have to get him to follow through with the hind legs. This takes more time. You follow him along; as he moves his front legs, you bring his rear legs up. Pretty soon, an intelligent canine will pick it up, and they will work on their balance and get that mastered. Sometimes instead of climbing a ladder 'O' must wear a lifting harness; we put him in a harness and lower or raise him as required.

"The Swiss have been doing canine search and rescue for years. I belong to a group in California called CSSDA (California Swiss Search Dog Association). If a Swiss dog doesn't do well in obstacle courses, he or she won't do well in urban search and rescue. Swiss trainers come to California once in awhile, and I have gone to drill with them a number of times to learn more about canine search and rescue training. They have a lot of expertise, and I have learned a lot from the Swiss and CSSDA."

A.J. Frank was asked if he gets extra pay from the fire service as a dog handler and trainer. "No, I'm not getting paid extra for such duty, but the FEMA team has been gracious enough to pay for some of my drills every month. I do get a small amount of money for drills. I got into dog training and handling just to see what my dog could do. I could not go to Oklahoma City because the dog and I were in California training with the Swiss. We went to Georgia for the Summer Olympic Games as back-up. In the Seattle area, an airlift medical helicopter crashed in the bay, and officials couldn't find the location of the crash. Just for training, a female dog handler and I went out in a small boat with both our dogs. Both her dog and 'O' alerted at a certain spot, and we were later told that was where the helicopter crashed. About three years ago an arson fire in a warehouse killed four firefighters. Rescuers found two bodies but could not find the other two bodies. 'O' and I were able to look for the two missing firefighters. Searchers were really close to finding the bodies but walked right over them. 'O' found the two bodies. At another incident, a building collapse (International Travel Building), 'O' did a search for potential victims and cleared the building. He found no victims.

"The way a canine searches during rubble work is via his or her nose. They only see in black and white. Their eyesight is not that good, but it is reliable enough to get them through an obstacle course or a collapsed building. Canines smell through their olfactory nerves. If you took a person's olfactory nerves out and lined them up on the ground, they would cover an area the size of a postage stamp. A canine's olfactory nerves would cover one square yard of fabric, so dogs can smell a lot better than we can. They smell like we see. If I told a human to go find a guy in a blue shirt while other guys all wore red shirts, it would be a piece of cake for him or her to do that. 'O's' job is to find live human beings. If you were injured and stuck amid debris, and another person was dead in another part of a collapsed building, we will need two hours to dig out either one of you. Would you rather have us spend that first two hours on you or on the dead person? On a body recovery, it doesn't matter how long we take, whereas you probably need help at once. That's what the dogs are for. The FEMA teams have other means of locating live humans. They have listening devices, heat cameras, motion detectors, and all sorts of gadgets, but they take time to set up. Everything has to be quiet to use the listening devices, and everybody has to be out of the area for the heat cameras to function properly. A perfect scenario for the dog would have all responders away from the rubble pile so the dog can concentrate on the scent of trapped people. I don't think you're ever going to get that situation. You're always going to have a lot of people around the dog. The dog has to figure out the human scent he is after. 'O' does not have to sniff a piece of clothing from a trapped individual to find him or her. He's searching for the person rather than their clothing, and in training USAR dogs are tested on that. If many people are alive and working in a room, the dog can smell the difference between them. The dog thinks, 'I smell a lot of guys in blue shirts. Oops, I now smell another live person but I can't see or get to that person. This must be the person I'm looking for.' Dogs smell the way humans see, so they can determine who they are smelling.

"If I tell 'O' to find human beings who are alive, he knows what to smell for. Live humans slough off skin that is deteriorating. That's what the dog is after. Your skin is always falling off you in very small particles. 'O' is just like a tracking dog that tracks your skin falling off as you walk down the road, but his main duty is to find living people in building rubble or to clear other areas where disasters have occurred. If a human is buried in a pile of rubble, a scent is coming off that person. If the scent heads downwind, the dog only needs to pick up that scent and follow it to its strongest point. A good search and rescue dog should give a barking alert while digging and trying to penetrate into the scent. Urban search and rescue workers see the dog up there making a lot of noise over a pile of rubble and know that something is there. Where the dog is digging and trying to penetrate is probably the closest spot the dog can get to the scent. There may be a shaft or two, or something may be moving the air scent from one spot to another, but for the dog, that is as close as he can get, and the search team will need to remove some of the rubble."

A.J. Frank was asked if Ohlin is bothered by flying in airplanes or helicopters. "When we fly by air, he goes in a crate down below. Some people prefer to have their dogs with them, and some people say that dogs always have to worry about other people being around them when they fly. 'O' wants to play with everybody, lick everybody, kiss everybody so when he travels in the passenger compartment he doesn't get any rest, and I don't get any rest. I put him below in his kennel where he can relax and do whatever he has to do. He is not bothered by flying. We've gone to California five or six times, and in Atlanta at the Summer Olympics we flew in a helicopter with crazy pilots and the dog seemed to love it. 'O' has flown more air miles and goes to more places in the world than many humans. The hard thing about a search and rescue dog's life is that the work is tough on his joints, jumping up and down, learning to climb up and down ladders, falling off ladders, and overcoming obstacles in training and on deployments. I'm sure flying isn't easy on human bodies, so I don't think its easy on a dog's body. The stress of being a USAR dog probably takes a toll on his life, but I think he enjoys it. When he stops enjoying it, then we'll stop the searches. Urban search and rescue dogs can probably do searches until they are about eight years old depending on problems with their hips, backs, or joints. Other than normal wear and tear, 'O' has been lucky. He has not had a sore pad, ripped off a nail, or fallen on anything recently. Most of the time, the handlers step on nails, get holes in their hands, or suffer from the results of bad footing. The dogs do really well.

"Ohlin is probably the most spoiled dog out there now. We just moved up to five acres, so he gets to live and play on five acres. One point where he is not like a normal dog is that he does not have house privileges. Other than that, he is just a normal dog. I keep him in a kennel and I make sure I know where he is more than someone else might do. I guess anyone who has a dog that they spent a lot of time and money on would do the same. He eats packaged, dry dog food rather than canned, and I have five and ten pound packages at several locations. I can't say that I never mix canned food as something special for him during training. Most dogs are prey driven, and 'O' is a hunting dog. Years ago Labrador retrievers had to survive by catching birds. 'O' is a good retriever, but he loves birds. If you can get that prey drive working during his training, that drive that his breed has to hunt birds, you will do okay. Some people use a toy while others use food as a reward. If you are hidden someplace and the dog finds you, he gets a reward. When I show up to work, they give me a reward once in awhile; it's the same for 'O.' When he shows up for work or training he get his 'pay check' once in awhile. The worse he does the less pay he receives; the better he does the more pay he receives.

"The only bad habit 'O' has is a selective hearing loss. My parents have told me I have the same bad habit. Other than that, he is a great dog. He's lovable, wants to play, and likes having a good time. When he is on-site, people like to pet him because he is particularly friendly, but this can be a problem for me. I worry about who they are and what they may or may not do. Most of the time, it is not a problem. When we are searching, it's best to just let the dogs do their own thing and not talk to them. However, when we are standing around somewhere, he's just a Lab. He loves everybody."

How is 'O' activated to do a specific function? "One word equals one command," responds A.J. Frank. "When I ask him to go search, I remove his collar and give him a one word command. I may talk to him for motivation saying, 'Do you really want to go search?' maybe get him to bark a couple of times, and then let him go.

He's just smelling for live human scent, doesn't care who it is, doesn't care if he has smelled that person before. The neat thing about a dog is that if a two-story building a block long and a block wide collapses into a pile of rubble, the dog can cover that pile very easily and quickly because he walks differently than humans do. The dog walks on all fours thus spreading his weight out over all four pads and achieving a more stable base. While two footed humans would be standing upright, getting blown around by wind, and inching along to secure good footholds, the dog is low, has great balance, and can jump over debris. He is just sniffing, sniffing, sniffing for live human scent. Canines used to chase mice and other small animals for a living; either they captured their prey or they did not. An urban search and rescue canine must be able to find a hidden, live human in a pile of rubble. If he cannot do this, you have to send him down the road and get another dog.

"FEMA sanctions a test for USAR dogs. A FEMA team comes in, and an experienced handler and the dog go through obstacles, directionals, drop and recall, and a time-limited search for a person within a pile of rubble. The handler cannot see the dog work, and the dog can't look to the handler for help. The advanced test is three piles of rubble with six live humans somewhere in the area. Overall, the dog has to find five persons out of the six hidden 'victims'. The trainer and dog have 20 minutes work time on each pile. At the first pile, you can go anywhere you wish; at the second pile, you have to stay within a perimeter; at the third pile you have to stand in one spot and send your dog. You hope the dog barks and does not 'false alert' because the examiners also put food, clothing, cats, or mice in the area of the pile to ensure that the dog knows what he is doing. We had such a test in California during 1997, and expect to go to Boston in October for another. In Switzerland, where I think they have some of the best dogs, teams have two days and one night of searching. I've never taken or seen that test. On the 27 urban search and rescue teams funded by FEMA in the United States, I think we have 4 to 6 dogs who have passed the Swiss test, perhaps 10 to 20 Advanced dogs, and probably 50 to 60 Basic dogs.

"A dog for urban search and rescue work should be one who loves to learn, loves to please, is agile and very durable. A fragile dog would break easily; a durable dog should be able to climb over rubble, rocks, and everything else. The dog must also be obedient; when you ask a dog to do something, you would like it done right now instead of tomorrow. When you have to leave your dog while you go someplace else, you command him 'down' and 'stay' and the dog cannot go running off. You expect him there when you get back. You would like the dog to listen all the time. Granted, that talent comes with age. Trainers should allow their dogs to have access to all sorts of environments. For instance, 'O' practices on his balance when going into small areas. Like most dogs, 'O' gets used to an environment and does well in a number of environments. At the Summer Olympics in Atlanta, we were exposed to a number of negative G forces in a Black Hawk helicopter while tied to the floor. The dog didn't puke, but the humans did."

Contact: A. J. Frank, Seattle Fire Department, Station 20, 2334 4th Avenue, Seattle, WA 98212; 206-448-5234 (Work); 425-334-5793 (Residence).

A MASS CASUALTY INCIDENT IN WASHINGTON

Robert K. Johnson is the Fire Chief and the Coordinator of Emergency Management for the City of Auburn, WA, located about 20 miles south of Seattle, home to 34,000 residents. He is also a member of the Washington Emergency Management Council and chairman of one of its sub-committees representing state agencies, local fire chiefs, and local emergency planners who respond to hazardous materials incidents. The emergency management council consists of up to 17 members who may be city and county personnel, sheriffs, police chiefs, state patrols, military personnel, department of ecology staff, state and local fire chiefs, seismic safety experts, state and local emergency management directors, search and rescue volunteers, medical professionals, building officials, private industry personnel and others. The emergency management council advises the state governor on state and local emergency management including specific progress on hazard mitigation and reduction efforts, implementation of seismic improvements, reduction of flood hazards, coordination of hazardous materials planning and response activities, and all related matters.

"We are trying at the state level to promote the regionalization of hazardous materials response teams, and we try to help such teams get equipment and supplies and recover funds they spend on responding to spills and in getting equipment and supplies," says Chief Johnson. "We are also attempting to fund local emergency planning committees (LEPCs) to support their local community operations. I had been a fire chief in Oregon before I became fire chief in Auburn 14 years ago. After a year on this job, the mayor appointed me as Director of Emergency Management. I started going to classes held by FEMA and attended the National Emergency Institute at Emmitsburg, MD and found out that disaster management is a whole different profession than fire service. Because of my service as the fire chief and the director of emergency management for the City of Auburn, I was selected by the governor to represent the state fire chiefs' association on the Washington Emergency Management Council.

"The Federal SARA Title III Act required the governor to appoint a State Emergency Response Commission (SERC) that has certain responsibilities. Our state chose to become involved in other issues involving hazardous materials not necessarily mandated by law but of interest to our state. We quickly became overloaded, and this newly formed state emergency response commission said the first order of business was to do what is required of a SERC before we branched into other areas. One of the universal concerns of the SERC membership is that the business of hazardous materials response is conducted at the local level by local teams and that the state does everything possible to help support these local entities. That support could be in the form of revenue, legislation providing immunity, or perhaps a disaster fund for covering costs when we can't find a spiller. Money seems to be the big issue right now. We've talked to a lot of local communities that are saying, 'we don't have any money, and we need some help.'"

Chief Johnson was asked about the threat of terrorism and weapons of mass destruction, and what has been the reaction in western Washington. "The City of Seattle is one of 27 cities that has been given a grant by the federal government to provide training in weapons of mass destruction. They are sharing that information with some of the rest of us in the hazardous materials response business. Interesting enough, several years back, a bombing incident occurred at the local Boeing plant in Auburn. I know the F.B.I. believes the Unabomber, Ted Kaczynski, was involved, and F.B.I. agents were out there in July or August 1997. I had some interest in the Unabomber because one of his last victims was a roommate of mine in college, a professional forester who ran the California State Forestry Association.

"We have some experience with explosives ourselves. Every year during the fireworks season we seize illegal fireworks, basically tennis balls packed with black powder and fuses. We have worked with the Federal Alcohol, Tobacco and Firearms agency on several occasions and uncovered factories within our business district that were building illegal fireworks. We learn as much as we can about explosives here in order to protect ourselves. When you go to a Haz Mat incident, unlike the aggressive suppression of a fire, you must take your time and be very careful. We work closely with our police department here in Auburn in trying to recognize the elements involved and try to use the City of Seattle responders' knowledge. While our team members may not be bomb disposal experts, they have a heightened awareness so that they can take appropriate actions at the scene."

The City of Auburn fire department has 73 personnel including the chief, assistant chief, and the fire prevention staff which consists of a fire marshal, assistant fire marshal, and two inspectors. They also have a public education officer, a training captain, and three suppression shifts. Each shift consists of a battalion chief, four captains, and several suppression firefighters. They have three fire stations, 31, 32, and 33. Station 31 is the headquarters station, about 10,000 square feet, which houses the administrative office, living quarters for the suppression crew, and a training office. The station's maintenance facility has a modular unit on the side that houses the paramedic facility. The minimum staffing here includes an engine company and an aid unit. Two captains are assigned to Station 31 because of what are called "credit shifts." To shorten the work week to 47 hours, each shift gets their sixth shift off.

Station 32 is a 7500-square-foot station located at 19th and R street SE. The minimum staffing includes a three-person engine company and an aid unit. Station 33 is located at the federal General Services Administration complex and was built to be a federal fire station, but closed down about 20 years ago. The station had been remodeled into offices, but Auburn Fire Department leases the property and remodeled the offices into living quarters for a fire station as the original apparatus bays were still there. This station covers a high value area, including the nearby Boeing plant, and its minimum staffing includes a three-person engine company, a backup engine company, and an aid unit. Auburn has a busy fire department. All engine companies and aid units are staffed by emergency medical technicians who do basic life support.

Friday, November 3, 1995 was just another day in Auburn until 9:41 a.m. when the City of Auburn fire department was dispatched to a "yellow plume of smoke" at the Boeing aircraft manufacturing plant located within the city limits. A toxic cloud would lead to the largest hazardous materials related mass casualty incident (MCI) in the state's history. Normally, about 7500 Boeing employees would be at this plant, one of seven in the Puget Sound area, but the company was undergoing a machinists' strike at the time. About 2300 employees were working at the site, while machinists in picket lines walked along the south and southeast sides of the 515 acre site. The machinists were right in the path of an orange toxic cloud about 75 to 300 feet high and up to a half-mile wide.

Supervisory personnel were handling some duties at the plant during the strike including the transfer of nitric-hydroflouric acid from a process pipeline into a 500 gallon portable tank not outfitted with a correct liner for the chemical involved. The 300 gallons of chemical mixture combined 67% nitric acid, 9% hydrofluoric acid, and 24% de-ionized water. The National Fire Protection Association has a standard system for the identification of fire hazards of materials known as the NFPA 704 system. The purpose of the NFPA 704 standard is to protect the lives of individuals who may encounter fires in an industrial plant or storage location where the fire hazards of materials may not be readily apparent. Health, flammability, and reactivity are rated by numbers 0 through 4, with 0 being the least dangerous and 4 being the most dangerous. NFPA 704 rates the health hazard of nitric acid as a 3 — "materials on which a short exposure could cause serious temporary or residual injury even though prompt medical treatment were given." NFPA 704 rates the health hazard of hydrofluoric acid as a 4 — "materials which on very short exposure could cause death or major residual injury even though prompt medical treatment were given."

Officials theorized that the chemical reaction that caused the toxic cloud was the emission of up to 750 pounds of nitrogen oxides from the tank for about one hour. Charles J. Baker, the author of "The Firefighters Handbook of Hazardous Materials," rates nitrogen oxide at a toxicity level to the lungs of 4, causing death or permanent injury. He also rates toxicity to skin as a 4.

Assistant Chief Russell J. Vandver of the City of Auburn fire department was a battalion chief at the time of the Boeing incident on November 3, 1995 and was the incident commander on the scene. Before he was promoted to assistant chief, he had been chief in charge of the special operations team for 12 years. "The special operations team has approximately 20 members and handles hazardous materials incidents and specialized land and water rescue including high and low angle rescue as well as surface and underwater rescue," begins Assistant Chief Vandver. "We have basically two branches, the land and water rescue branch and the hazardous materials branch. All members of the team are certified hazardous materials technicians, eight are certified in dive rescue, and several are trained in confined space entry. Eventually, all will be trained in this skill. The team originally started out as two separate divisions, but everybody has been cross-trained so that the entire team will respond to an incident if needed. We have hazardous materials technicians on-duty at all times. If we have an incident that is likely to involve Level A protective clothing or a prolonged operation, then we call the entire team back to duty. Members carry pagers so they can respond from home if necessary.

"On the morning of November 3, 1995, we were dispatched to a hazardous materials leak at the Boeing Company. For the initial dispatch, Engines 31 and 33 and an aid unit were sent to the scene. Engine 33 responds out of Station 33 which is right on the other side of the fence from the Boeing fire station. I was the on-duty battalion chief that day. The initial reports indicated some type of acid leak, and an orange or yellow cloud was forming at the plant. The captain on the scene got on the radio and reported to me that he had a large orange plume rising from a building. I couldn't see it until I got on C street and headed south. The plume was rising in the sky and heading toward southeast Auburn. I immediately called all off-duty personnel to report to the scene, and called the Kent Haz Mat, one of the Haz Mat teams with which we have a mutual aid agreement. Also, on the way down to the scene I called the police department to begin evacuation of the southeast area of town. Probably 30 seconds later, I rescinded that order because the police department has no protective equipment and officers should not go into the plume. I changed the order to 'deny access to.' Two main roads run through the path of the cloud, C street southwest and A street southeast. I called for roadblocks at 17th and A and 15th and C to stop traffic from going into that area.

"At the plant, I found the Boeing command post in chaos at first. I generated a second alarm, called for more help, and met with one of the Boeing facilities managers who had been with his Haz Mat team for some time. We agreed we would set up a Haz Mat operations area away from the command post so we were not stepping on each other. I was told immediately that the chemical was nitric-hydrofluoric acid. That was interesting because just the night before, I had a call from my brother who is a battalion chief in charge of hazardous materials with another city. He learned that a company in his city was using hydrofluoric acid. He was fairly new at it and asked if this was some bad stuff. I said, 'some of the worst stuff you could ever get into.' So, in my mind, I felt we had a severe incident. I appointed our Haz Mat team leader to take charge of hazardous materials operations and began making assignments for safety officer and for other positions.

"We had some discussion regarding two very large schools that were in the path of the plume, Mount Baker Junior High School and Riverside High School where my son was located. We notified the school district and told them to prepare for evacuation, but not to do anything yet as we might end up sheltering-in-place. After brief discussion, we decided we did not have enough time to get buses down there and get the people offloaded. We then directed the schools to shelter-in-place and shut down their ventilation systems, and we denied access to the rest of the area. Several individual residences and some shopping complexes are located in that area. We tried to warn people as much as possible. Fortunately, by the Boeing plant is a railroad yard, basically unoccupied, so the cloud would have to travel a fairly good distance before it hit populated areas again. The cloud had gone up quite high and was beginning to spread out. We hoped that the cloud wouldn't do too much damage and that it would dissipate.

"Boeing was on strike and had supervisors doing the workers' jobs. A supervisor took a 'pig,' a portable tank, over to an acid tank and was drawing off an amount to adjust the pH in the tank. Instead of using a lined tank, he used an unlined tank. As soon as he started drawing that acid solution, it hit mild steel, cooked off, and started blowing out the acid and vapors which formed the plume as everybody ran. Because of the size of the release and plume, our Haz Mat operations and decontamination areas were set up quite far back from the actual incident. The first entry team in Level A suits took a Boeing vehicle, either electric or propane powered, and drove it toward the other end of the building. They went in to do a recon, make sure there were not any victims down and see how much product had spilled. Then, they came out and briefed the Haz Mat officer. Over a period of time, five entries were made to solve the hazardous materials problem."

The hazardous materials control operations are run as a partnership between the City of Auburn fire department and the Boeing Company hazardous materials team. The first entry team was staffed by Neil Pederson of the Auburn fire department and Joe Admyers of Boeing. Their recon found no victims in the area, but a few gallons of the chemical mixture were found in a sump as well as some residue on the floor. Outside the building, they found a small amount of product in the tank plus some residue on the tank. Entry team two, composed of Gerry Maitland of the Auburn fire department and Joe Admyers of Boeing, drove a forklift into the building and disconnected a 50-foot hose that was still in place between a process pipeline and the 500-gallon tank. They plugged the hose, used the forklift to move the portable tank into a containment area, and left the building for the decon area. The third entry team, Perry Boogard of the Auburn fire department and Jeff Kinne of the Boeing Company, later entered the building to monitor the air inside. They found readings of up to 5 ppm of hydrofluoric acid in the sump area. After leaving the building, they went through the decon area. The assignment of entry team four, Kevin O'Brien of the Auburn fire department and Bob Smith of Boeing, was to wash down the area to further dilute the acid. The fifth and final entry team, Charlie Sustin and Jenette Ramos of the Boeing Company, entered the building just after 3:00 p.m. to sample the air again. Readings were below 1 ppm in all areas.

"We began our response mainly by concerning ourselves with the hazardous materials incident and setting up for that, then we started getting reports of injuries," continues Vandver. "We initially got a report that one Boeing person was sick. We called for a medic unit, then within a matter of minutes more victims arrived. At that point, I called for a MCI for 15 people. King County had been working on this program for about two years and had just implemented the MCI system to handle large numbers of sick or injured people. All you had to do was call the dispatch center and say, 'give me a MCI for so many people.' I increased the MCI call to 30 people. Chief Bob Johnson was at the scene, and I said to him at one point that we were going to go over 100 persons. He looked at me like I was out of my mind. My thinking was that with an off-site release and high volume news reports, we were going to get a lot of patients. We ended up with 125 persons being decontaminated, triaged, treated, and transported.

"We had so many news helicopters flying overhead that we were afraid they would crash into each other. I had one of the police sergeants call the Federal Aviation Administration and tell them to shut down the air space. We had four Haz Mat incidents at Boeing in the last few weeks. Tuesday night we had one, and I headed out as soon as it was dispatched. By the time I got there, the first news helicopter was already flying over the plant. The news stations listen to our scanners. Valley Com does police and fire dispatch for the City of Auburn and many other departments, basically for the entire south King County area."

"I actually wrote Boeing's hazardous materials response protocols about ten years ago when we were having some problems," relates Vandver. "If they must evacuate an area, or plan to activate their on-site Haz Mat team, they call 911 and have the City of Auburn fire department dispatched. They will have routine spills

TABLE 9.3

Mass Casualty Incident Plan

Level (No. of Patients)	Fire Units	Medic Units	MSO[a] Units	MSA[b] Units	Transport Units
Level 1 1 to 10	4 Fire Units: Aid/Engines/ Ladders	2	1	—	5 Ambulances
Level 2 11 to 19	8 Fire Units: Aid/Engines/ Ladders	3	2	1	10 Ambulances
Level 3 20 to 29	12 Fire Units 1 Task Force	4	2	1	15 Ambulances
Level 4 30 to 39	16 Fire Units 1 Task Force	5	2	1	20 Ambulances 1 Bus
Level 4 40 to 49	16 Fire Units 1 Task Force 1 Engine Strike Team	7	3	1	25 Ambulances 2 Buses
Level 4 50+	16 Fire Units 1 Task Force 2 Engine Strike Teams	7	4	1	25 Ambulances 3 Buses
Level 4 100+	16 Fire Units 3 Task Forces 3 Engine Strike Teams	9	5	1 King 1 Seattle	25 Ambulances 4 Buses

[a] MSO = Medical Service Officer

[b] MSA = Medical Service Administrator

Source: City of Auburn, Washington Fire Department

at the plant. If they have any type of spill at all, their protocols require that they call the fire department duty battalion chief directly. They might say, 'Here is a situation that we can handle. If you want to come out and check on what we are doing, fine. If not, we're going to go ahead and clean it up.' Boeing is a good company, they are well trained, and they are really good with their safety protocols and the way they deal with chemicals. I think they have a good track record."

This was a mass casualty incident which has its own protocols and standard operating guidelines whether it results from hazardous materials or some other type of natural or man-made disaster. The King County medical system employs paramedics who respond to a MCI whatever the cause may be. "Tom Gudmested, a MSO (medical service officer), established a medical division" says Vandver. "We have a layered system where fire department personnel are all trained in basic life support and are dispatched initially, or may have a medic unit dispatched with them, or may call a medic unit when they get to the scene. Medic personnel are not firefighters; they are pure paramedics. They belong to the firefighters' union, but they are a separate entity and are not attached to any specific fire department. For example, King County Medic 6 is based in one of Auburn's fire stations, but their dedicated response area includes other cities and a couple of fire districts as well. If one medic unit is out of service, Valley Com will dispatch the next closest medic unit so we could end up with Medic 5, 8, or 11. I can remember incidents where we had three medic units in town all at the same time.

"The paramedics under the medical service officer implemented the MCI protocols," says the assistant chief. "Patients were brought to a controlled entry point at the Boeing Fire Station where paramedics took them into the fire station, and ran them through the shower. The contaminated run-off water goes to the company's own treatment plant where hazardous wastes are treated. After that, the patients were triaged. They put different colored tarps on the ground and separated the patients: the red tarp for those in the most serious condition who would be treated and transported first, the yellow tarp for those who needed delayed treatment, and the green tarp for persons who needed minor treatment. Dr. Michael K. Copass, Director of the Trauma Center at Harborview Medical Center, is in charge of the medic program for all of King County. He was in touch with the MSO and helped to run the mass casualty incident. We faxed Dr. Copass the material safety data sheets on the chemicals involved, and several patients were treated with i.v. calcium drips. Symptoms of the exposed patients were mainly respiratory problems — irritation of the nose, or throat, watery eyes, some difficulty breathing, and a little bit of burning in the lungs."

Medical personnel attempt do the greatest good for the greatest number of patients during a mass casualty incident based on the incident command system. They use a system called "START" (Simple Triage and Rapid Treatment) that can be implemented by first-arriving EMTs or first responders rather than waiting for paramedics to arrive. Simple triage can begin by asking victims to walk to a certain spot which separates the "walking wounded" from the crowd, and makes it easier to identify the critical victims quickly. Triagers have about 20 seconds to triage each victim and move on to the next. The only treatment they should give is to open an airway or control major bleeding. They flag with different colored ribbons the victims who are deceased (black), are really sick or exhibiting trauma so as to require advanced life support (red), require immediate care but their condition is not life threatening (yellow), and have minor injuries or possible minor injuries that need to be noted (green). An appropriate colored ribbon is attached to the upper arm of a victim leaving enough tail on the ribbon so a treatment tag can be attached later.

A triage leader is appointed to designate and manage a funnel point, usually designated by a white flag at the entrance to the treatment area. Here, victims are numbered and treatment tags are attached. The triage leader coordinates movement of victims to the treatment area, if possible in order of priority as indicated by the colored ribbons tied to their arms.

A treatment leader is assigned to locate and set up the treatment area with regard to scene safety, ambulance access for loading, and location of the funnel point. The treatment leader assures that all patients are tagged and numbered, directs and supervises treatment areas, makes sure treatment for the degree of injury is provided, and ensures that the most critical victims move quickly from the treatment area to the transport area.

A transportation leader has been said to have the most difficult job in a mass casualty incident. He or she must be able to delegate duties to assure priority transport of critical victims, track ambulances assigned to staging, initiate medical communications, control hospital access, document the hospital destination of each patient, and provide patients' information (patient number, severity of condition, major injury type, status of i.v. or intubation, etc.).

A medical group supervisor will obtain a briefing from the incident commander, determine what resources are available to the medical group, designate triage, treatment, and transportation leaders, establish communications with all medical group leaders, determine the additional resources necessary, and provide updates to the incident commander.

In the Boeing mass casualty incident, the Boeing company's Emergency Medical Services Division responded with an ambulance, three nurses, and three physicians. Boeing medical staff assisted the medical group area and helped monitor rehabilitation of Haz Mat entry personnel. Gary Gordon, a toxicologist with Boeing emergency response, provided technical support to the Haz Mat Division. The Seattle fire department dispatched two medical services officers to Harborview Medical Center to provide assistance as fire department liaison officers at the trauma center. The fire department also provided the Haz Mat Unit to set up and operate a decon area at the Harborview emergency room entrance as an additional precaution against cross-contamination. All personnel who worked this incident were tracked via the passport accountability system originally developed by the Seattle fire department to maintain a vigilance over personnel deployment.

"Once things calmed down a bit, I asked the paramedics to employ a medical team to A street over which the cloud had passed," relates Vandver. "They went from business to business and found several people who had been exposed to the cloud and were showing signs and symptoms. These victims were brought back to the triage area. One problem we encountered involved identification. A mass casualty incident is designed to treat a lot of people in a short amount of time. The paramedics don't write down names. The patients get numbers on their hands and foreheads, and tags with their vital signs. Some people were upset because they called and wanted to know about a loved one or friend. A common question is what hospital did he or she go to. We really could not provide them with that information. We had 20 or 21 patients at Boeing a week ago at an acid spill, and we talked about the possibility of appointing someone to go with the transportation officer and gather the patients' names and note what hospital they went to. That would be nice to do, but it's a low priority when all you want to do is get them treated and get them transported to ten different hospitals. The City of Auburn fire department, and the King County medics, under the incident command system cooperate to run a mass casualty incident. As incident commander, I ran the MCI while King County medics handled decontamination, triage, treatment, and staged private ambulances to transport patients to area hospitals.

"Our assistant chief opened our emergency operations center and contacted the mayor, financial officer, the chief of police, a representative from the school district, someone from public works, and all the different people who are necessary to handle a large disaster in town. Basically, I took care of the incident scene and the assistant chief took care of the rest of the town by bringing in task forces from other areas to provide coverage for other incidents that might happen in town."

TABLE 9.4

City of Auburn Fire Department Special Operations Response Unit Equipment Inventory Haz Mat 31, Apparatus #24

Compartment 1
1 55 gallon overpack drum
1 25 gallon overpack drum
1 8 foot oil boom
1 hand truck
2 walkers (decon)
3 clipboards
Compartment 2
2 Surviv-Air® one-hour SCBAs
2 face masks w/radio
Compartment 3
8 one-hour spare SCBA bottles
1 2200 psi cylinder
Compartment 4
1 library shelf with books, Coast Guard CHRIS
  manuals, sewer maps, chemistry charts, local
  agreements, pre-fire plans, city and area maps, suit
  charts, Farm Chemical Handbook, inventory, phone
  lists, Boeing materials, binoculars.
Compartment 5 (counter area)
1 laptop computer with mouse
1 printer
1 stapler
1 tape dispenser
1 protective suit repair kit
1 propane bottle
1 box trash bags
1 box anti-fog cloths
Compartment 6
2 safety shields
10 orange arm bands
1 protective suit repair kit
1 box Ziplock bags
1 box trash bags
1 propane bottle
Compartment 7
1 file cabinet with team members' medical charts, meth
  labs, building plans, exposure charts, suit testing,
  SARA Title III information
1 instant camera
3 film packages
1 box perma film
6 VCR tapes
1 box filter masks
1 glass cleaner
1 tape dispenser
1 pair binoculars
1 bag filters MSA gas
1 padlock #9111
pens, pencils
Compartment 8
1 bag batteries
1 battery tester unit
3 bars soap
1 magnifying glass
1 box Ziplock baggies
8 aluminum rescue blankets
12 helmet shields
3 suit zipper wax

Compartment 20
2 large rolls Visgueen
Compartment 21
6 SL-20 flashlights with chargers
Compartment 22
4 hard hats
Compartment 23
4 Trellchem® Level A suits
4 Chemron Level A suits
2 spare face shields
Compartment 24
1 suit test kit
1 wool blanket
1 large decon pool
Compartment 25
1 mini-RAE box
1 gallon spray can
1 Gas Tech box
1 mercury spill kit
Compartment 26
3 garden hoses
Compartment 27
3 five-gallon buckets
1 portable pump
Compartment 28
7 disposable Level A suits
Compartment 29
sodium bicarbonate, TSP
soap
Compartment 30
4 disposable Level A suits
Compartment 31
1 decon kit with four brushes
1 gallon distilled water
1 quart vinegar
6 decon shower wands
Compartment 32
1 suit test kit
Compartment 33
2 patch kits
1 large patch
1 2200 psi air bottle
6 pairs rubber boots
Compartment 34
1 25-yard, 3/8 sisal rope
14 decon pools
6 full-face respirators
6 safety goggles
3 rolls of scene tape
Compartment 35
1 sample box
1 box radio equipment
8 radio headsets
Compartment 36
1 initial kit (numerous plugs and clamps)
Compartment 37
1 set command vests (IC, Ops, Safety, Medical,
  Decon

TABLE 9.4 (continued)

Compartment 9
2 boxes boot covers
1 five gallons METAL-X
Compartment 10
1 bag filter masks (11)
1 box pipettes
1 roll beaker cover
1 Trelleborg repair kit
6 rolls duct tape
Compartment 11
2 red tarps
12 traffic cones
Compartment 12
2 pairs rubber boots
Compartment 13
1 tool box
1 pipe wrench
1 hydrant gate valve
1 cheater bar
1 garden hose
Compartment 14
2 pairs rubber boots
Compartment 15
assorted reference books
1 pair binoculars
2 yellow legal pads
Compartment 16
1 power strip
2 radiation test kits
1 MSA gas meter
1 Gas Tech meter with tubes
1 mini-RAE meter
1 orange test kit, Draeger tubes, pH, oxidizer strips
Compartment 17
14 encapsulated Level B suits (Tyvek®/Saranex®)
Compartment 18
18 Level B Tyvek® suits/hoods
Compartment 19
2 aluminum flash suits
12 bags rice hulls
2 bags oil sweep
4 chairs
1 bag oil absorbent pads

Compartment 38
12 pairs neoprene gloves
Compartment 39
4 pairs nitrile gloves
4 pairs poly chlorinated gloves
Compartment 40
6 pairs Viton® butyl gloves
4 pairs butyl rubber gloves
Compartment 41
7 pairs nitrile gloves
Compartment 42
6 pairs neoprene gloves
Compartment 43
1 chock block
cribbing
Compartment 44
1 #20 dry chemical extinguisher
1 #20 CO_2 extinguisher
1 #10 halon extinguisher
Compartment 45
2 1-hour Surviv-Air® SCBAs
3 face masks
Compartment 46
1 propane hot water unit
1 weather tower
4 clipboards
1 SCBA checklist clip board
1 drum wrench
1 spare SCBA face piece
Compartment 47 (on top of response vehicle)
1 12-foot folding ladder
1 awning
1 incident command board
1 special ops. command board
Compartment 48 (front)
1 2200 watt Honda generator

A critique was held subsequent to the mass casualty incident. "Everyone who responded to the incident was invited to attend the critique; all the fire departments, Boeing representatives, the Valley Com dispatch center, the Federal Aviation Administration, all the police agencies in the area, anyone who was involved and had a need to know what was going on," remembers Vandver. "Not all of those invited chose to send representatives, but a pretty large group did show up. We talked about what went wrong and what went right. Currently, Boeing has very good Haz Mat teams in all their seven plants in the Puget Sound area. Boeing has its own fire brigade, the Boeing Fire Department. For a long time, their Haz Mat teams were under the facilities department, but now they are under the fire department. All Boeing firefighters are trained in hazardous materials. Others who staff the Boeing Haz Mat teams, such as plumbers and electricians, work in the production facilities and are hazardous materials technicians. Whenever we go to Boeing, we always do joint entries. An Auburn Haz Mat technician and a Boeing Haz Mat technician will make any entry together. They work as a team and depend on the Boeing employee to know the building and the production equipment."

Assistant Chief Vandver was asked if any procedures were changed as a result of the Boeing mass casualty incident. "When you have an incident of that magnitude, some things will go wrong. That's the nature of the beast. Overall, it proved to us that the structure of our team worked very well. The Washington Department of Labor and Industry audited us. This is a fire department's biggest fear — to have an L&I audit, they find something wrong, and you get fined. Actually, we did not get any citations or fines. Instead they wrote us a letter complimenting us on our structure. We had copies of that letter hanging all over our fire stations; we were very proud of it. The Department of Labor and Industry was very complimentary about our protocols for safety, what we did with our Level A entry personnel, our medical screening, and the Nomex we wear underneath our Level A clothing. Their letter basically reinforced that we were doing things correctly, that we had a good working relationship with the Boeing Company, and that we were able to handle the situation. It certainly reinforced the fact that the MCI system, while brand new, worked well. That was the biggest mass casualty incident combined with a hazardous materials response to date in the State of Washington to the best of my knowledge.

"We appointed a safety officer at this MCI in line with our protocols. This person has the authority to shut down the incident, to stop or change something, and to tell the incident commander, 'I don't think what you are doing is safe, and I am going to stop your operation.' Afterwards we realized that because of the magnitude of the incident and the area covered, we should have appointed a safety team. The safety officer would be in charge of safety but would have helpers who could observe different areas and report back with recommendations.

This mass casualty incident coupled with a hazardous materials release and toxic cloud required a response by approximately 250 fire and emergency services personnel and over 100 pieces of apparatus including 50 ambulances. Protocols and standard operating guidelines used followed the approved incident management system with use of the King County Fire Resource Plan, the regional portion of the Washington State Fire Resource Plan, and the Mass Casualty Incident Plan from the King County Fire Chiefs Association.

Contact: Fire Chief Robert Johnson, or Assistant Chief Russel J. Vandver, City of Auburn Fire Department, 1101 "D" Street NE, Auburn, WA 98002; 253-931-3060; 253-931-3055 (Fax).

10 Recent U.S. Incidents Involving Chemical Agents, Biological Materials, or Terrorist Actions

INTRODUCTION

Chemical and biological agents are most effective when employed against untrained or unprotected targets and victims. Civilian sites and innocent citizens offer the best targets for terrorist activities. Wherever you live or work, your best immediate protection against a terrorist act is a group of well-trained, fully equipped, and competently led first responders.

On October 4, 1992, an Israeli El Al Boeing 747-200F cargo plane crashed into a large apartment building in Amsterdam killing 43 people as well as four crew members aboard the aircraft. On October 30, 1998, the Dutch newspaper, *NRC Handelsblad*, said the plane contained about 50 gallons of dimethyl methylphosphonate (DMMP) and two other chemicals of the four required to make deadly sarin poison gas, the same chemical agent used in the subway incident in Tokyo on March 20, 1995. The DMMP, according to the newspaper, was enough to make over 500 pounds of sarin, and judging from the shipping papers was traveling from Solk-atronic Chemicals in Morrisville, PA to the Israeli Institute of Biological Research located in Ness Ziona near the city of Tel Aviv. The DMMP can also be used as a flame retardant in building materials, but the World Health Organization states it has induced cancer in laboratory mice. The Boeing also carried 800 pounds of depleted uranium. This type of uranium emits strictly low-level radiation, but, according to the U.S. Federal Aviation Administration, it should be handled only by personnel wearing proper protective clothing as it can cause cancer. Laboratory tests done in 1998 on 15 persons who were near the crash site yielded four samples with high uranium content. Israel has never admitted to making chemical weapons; any nuclear capability they have is open to question. Israel signed the Chemical Weapons Convention Treaty in 1996, but it has not been ratified by the Israeli Cabinet. The Dutch started a parliamentary inquiry in late October of 1998. Ongoing investigations imply that the El Al crash was a mistake or an attempted cover-up. Several items are reported to have been lost, stolen, or destroyed. The voice and flight-data recorder was not recovered. Police video tapes were erased before investigators had a chance to see them. Vital information related to the hazardous cargo remained confidential for years, and even today information is still unavailable for 20 tons of the 114 tons of cargo. Many survivors told police, doctors, and investigators that hours after the crash, when Dutch police had cleared the area of all workers and members of the media, persons in "moon suits" jumped from a helicopter into the debris searching for items of cargo rather than victims, and carried off such items in unmarked trucks. By mid-1998, 1200 residents of the Bijmermeer district of Amsterdam and safety workers assigned to the incident scene were complaining of physical and psychological ailments.

TERRORIST ACTIONS IN THE UNITED STATES: FACT OR FANTASY?

Across the midwest during 1994 and 1995, members of the Aryan Republican Army robbed 22 banks in 7 states from Nebraska to Ohio.

On December 6, 1994, Claude Daniel Marks and Donna Jean Wilmott, both top ten fugitives of the F.B.I. and supporters of the Armed Forces for National Liberation (FALN) and of the Prairie Fire Organizing Committee, a front for the extremist Weather Underground, surrendered to U.S. authorities. They had spent almost ten years living under aliases in Pittsburgh before they reappeared. On May 9, 1995, they both pled guilty to charges of conspiracy to violate laws prohibiting prison escape and other related activities. In 1985, Marks and Wilmott purchased from an undercover F.B.I. agent in Baton Rouge over 36 pounds of plastic explosives that they intended to use to help FALN leader Oscar Lopez escape from the U.S. penitentiary in Leavenworth, KS. The FALN is a clandestine, Puerto Rican terrorist group that since 1947 has been linked to over 130 bombings resulting in $3.5 million in damages, 5 deaths, and 84 injuries.

Ramzi Ahmed Yousef was caught in Islamabad, Pakistan on February 7, 1995 and then charged in New York the next day for his alleged involvement in the February 26, 1993 World Trade Center bombing. Later, he was also indicted for conspiring to bomb Philippine Airlines Flight 434 and to bomb several other U.S. air carriers in the Far East.

On February 28, 1995, after four members of a domestic extremist group manufactured the biological agent ricin with the intent to kill law enforcement officers, a Minneapolis jury convicted the members of a violating the Biological Weapons Anti-Terrorism Act of 1989.

On March 9, 1995, one of the F.B.I.'s top ten fugitives, Melvin Edward Mays, a member of the Chicago El Rukns street gang, was arrested and charged with over 40 federal counts related to conspiracy to conduct terrorist activities on behalf of the government of Libya. Mays made the mistake of purchasing an inert light anti-tank weapon from an undercover F.B.I. agent. Other members of the gang were convicted as well, the first time in U.S. history that American citizens had been found guilty of planning terrorist acts on behalf of a foreign government in return for money.

On April 12, 1995, Michael "Mixie" Martin, who was said to be a supporter of the Provisional Irish Republican Army (PIRA), pled guilty to conspiracy to obtain munitions and weapons. He and two other PIRA supporters had conspired to purchase 2900 detonators in Tucson in 1989 and a stinger missile in Florida in 1990. Martin was sentenced to jail for 16 months and was deported in 1996. The other two men were sentenced to 19 months in prison for placing explosives in a motor vehicle, possession of stolen property, and aid of a foreign government and were deported in 1996.

On April 19, 1995, a truck bomb made of ammonium nitrate and fuel oil destroyed the Alfred P. Murrah Federal Building in Oklahoma City, killing 168 people and injuring hundreds, the deadliest terrorist act so far committed in the United States.

On July 3, 1995, Rodney Coronado, who belonged to the Animal Liberation Front (ALF), was convicted of arson for a fire he started on February 2, 1992 at the Mink Research Facility at Michigan State University. He was sentenced to 57 months in jail, 3 years probation, and restitution of over $2,000,000.

On October 9, 1995, the Sunset Limited passenger train derailed near Hyder, AZ killing one person and seriously injuring 12 others. Someone had tampered with the tracks, causing the train to derail. Investigators found four typed letters that mentioned the Bureau of Alcohol, Tobacco, and Firearms, the F.B.I., Ruby Ridge, and Waco and were signed by the "Sons of the Gestapo." It remains unclear whether this incident was criminal sabotage or terrorism, but the F.B.I. is investigating this train wreck as a criminal matter and created a toll-free telephone number to appeal to the public for assistance.

In Vernon, OK during November of 1995, the leader of an Oklahoma militia was arrested and charged with planning a bombing spree.

Joseph Bailie and Ellis Hurst were convicted of attempting to blow up the Internal Revenue Service building in Reno, NV on December 18, 1995. A drum filled with 100 pounds of ammonium nitrate and fuel oil was found at the IRS building.

In Spokane, WA from April to July 1996, three bible-quoting men committed bank robberies and burned the offices of various businesses.

During July 1996, Federal agents arrested members of the so-called "Viper Militia" in Phoenix, AZ and seized over 300 pounds of ammonium nitrate, 70 automatic rifles, and 200 blasting caps.

In the fall of 1996, an unexplained incident occurred at a Dallas hospital when muffins and doughnuts were treated with shigella which causes dysentery. Laboratory staff were then sent e-mail messages inviting them to a free breakfast. A dozen of the 45 laboratory staff fell ill with severe intestinal symptoms.

On September 24 and 25, 1996, roughly 600 contract workers at a Georgia Gulf plant that manufactures PVC (polyvinyl chloride) were modernizing a vinyl chloride monomer unit when they and six company employees were sprayed with a cool mist. Within hours, the workers' skin began to blister in a delayed reaction. Some weeks later, analysts at OSHA's technical center in Salt Lake City found that the chemical mist contained mainly nitrogen mustard agent as well as a smaller amount of sulfur mustard agent. The same results were confirmed by the U.S. Army's Aberdeen Proving Ground in Maryland. Both sulfur mustard (H and HD), and nitrogen mustard (HN-1) agents can contaminate through inhalation, eye contact, skin contact, and ingestion. The U.S. Army Center for Health Promotion and Preventive Medicine explains that "HD is H that has been purified by washing and vacuum distillation to reduce sulfur impurities. HD is a vesicant (blister agent) and alkylating agent producing cytotoxic action on the hematopoietic (blood forming) tissues ... The rate of detoxification of HD in the body is very slow, and repeated exposure produces a toxic effect. The physiological action of HD may be classified as local and systemic. The local action results in conjunctivitis or inflammation of the eyes, erythema which may be followed by blistering or ulceration and inflammation of the nose, throat, trachea, bronchi, and lung tissue. Injuries produced by HD heal much more slowly and are more susceptible to infection than burns of similar intensity produced by physical means or by most other chemicals..."

For protective equipment required to handle nitrogen mustard (HN-1), the Army says, "Mandatory — Wear butyl toxicological agent protective gloves. Wear chemical goggles as a minimum; use goggles and face shield for splash hazard. Wear full protective clothing of M3 butyl rubber suit with hood, M2A1 boots, M3 gloves, treated underwear, and M9 mask and coverall (if desired)." No one had donned such personal protective equipment prior to this incident.

Georgia Gulf created mustard agents through an industrial process that handled ethylene dichloride (EDC), a liquid that attacks the liver, lungs, and kidneys and may cause cancer. Georgia Gulf paid a $103,000 fine to OSHA for its handling of hazardous chemicals. But two years later, the company is still making tris, better known as nitrogen mustard gas. Leading up to the September 24th incident, nitrogen mustard had collected as a solid in a device called a fin fan used to cool EDC and as a liquid in reactor 201. A company known as Hydro-Chem cleaned the clogged fin fan tubes with highly pressurized water allowing the residue to fall on contract workers who have now sued Georgia Gulf.

The company had deactivated the "Dopp kettle" used to collect hazardous waste from the reactor, according to the workers' lawyer. The kettle was taken out of service in the early 1980s because it did not work properly. The contract workers charged that Georgia Gulf failed to follow normal procedures by allowing them to be hydroblasted while they were working. A spokesman for the company said the company had no way of anticipating the risk because they did not know they were making tris. In a letter to the state Department of Environmental Quality, a Georgia Gulf engineer said, "there are a number of hazardous chemicals that are present in the reactor liquid," and that "tris is by no means the most hazardous."

The Chemical Weapons Convention outlaws the formal manufacture of mustard gas, but does not ban manufacture as an unavoidable by-product of industrial production as long as the mustard gas does not represent more than 3% of the total product.

In Atlanta during July of 1996 and February of 1997, pipe bombs exploded at Centennial Olympic Park, an abortion clinic, and a homosexual bar killing one and injuring more than 100.

On November 9, 1996, TWA's terminal at Kennedy Airport outside New York City was evacuated for an hour after shoes in a passenger's bag were found to have traces of nitrate used in both bombs and fertilizer. An X-ray screen device showed potential for a bomb, and a specially trained dog confirmed the machine's findings.

On January 26, 1997 at Tooele, UT, low levels of the nerve agent GB (sarin) were detected in an area of the Deseret Chemical Depot. Reportedly, no workers were in the area and no agent was released outside the facility. The plant had been processing nerve-agent-filled ton containers since January 17. In that period, 30 ton containers were drained and treated in the metal parts furnace, and 39,000 pounds of agent were destroyed in a liquid incinerator.

During March of 1997 in Kalamazoo, MI, Federal police arrested a local militia activist for allegedly giving 11 pipe bombs to a government informant and plotting to bomb government offices , armories, and a television station.

At Yuba City, CA in April 1997, police investigating an explosion that shattered area windows found 550 pounds of petrogel, a gelatin dynamite, kept by alleged militia members. Police arrested a "freeman" sympathizer after explosives stored outside his home exploded injuring the suspect and his two-year-old daughter. Several days later, two of the man's friends were found with 500 pounds of explosives and detonating caps in a motor home.

On April 8, 1997, air monitoring at Aberdeen Proving Grounds in Maryland found trace amounts of a nerve agent (tabun) after seven chemical rounds were detonated. All told, the facility destroyed 14 chemical rounds and 112 nonchemical rounds. The chemical rounds also included phosgene, a choking agent, and mustard, a blistering agent. The state was alerted and approved destruction of the final 7 chemical rounds two days later. Only 3 of 48 sensitive monitors near the detonation site picked up low amounts of tabun. Hand-held monitors detected no traces of the nerve agent, and monitors placed in homes five miles away showed no presence of the agent.

On April 17, 1997, a leaking manila envelope was sent to B'nai B'rith, a national Jewish service organization in Washington, D.C. A threatening letter enclosed warned that the envelope contained deadly anthrax bacillus. The incident proved to be a hoax.

The F.B.I. announced on April 22, 1997 that four people were arrested for plotting to blow up the Mitchell Energy and Development Corporation's natural gas plant. The suspects wanted to kill police and to cover the robbery of an armored truck. The robbery, from which the group expected to net $2 million, was planned to obtain money to buy weapons for future terrorist activities. Dallas police worked with an informer who had infiltrated the group. The accused planned to attach three bombs to storage tanks, one easily visible and the other two concealed, and then call in a bomb report. They hoped that police would spot the visible bomb, but would be killed later when the two hidden bombs went off.

In Denver on May 1, 1997, Ronald D. Cole, Wallace S. Kennett, and Kevin I. Terry were arrested after federal agents seized rocket fuel, land mines, AK-47s, and munitions from their rental home. Kennett and Cole said they are members of the Branch Davidians, a cult in which several members died when federal agents burned their headquarters in Waco, TX on April 19, 1993. Authorities declined to say whether arrests were related to the Oklahoma City bombing trial then underway in Denver's federal courthouse. Prosecutors say Timothy McVeigh blew up the federal building in retaliation for the deaths of more than 80 Branch Davidians, and that in the months before the bombing in Oklahoma City, McVeigh attempted to purchase rocket fuel. Cole distributed literature in Denver when McVeigh's trail opened March 31, 1997.

About 14,000 people who received significant doses of radioactive material from the Hanford Nuclear Reservation, near Richland in southern Washington state, between 1945 and 1951 should be found and offered regular medical evaluations according to the Agency for Toxic Substances and Disease Registry. This government agency says the U.S. Department of Energy should sponsor the program to look for thyroid cancer and other radiation-related conditions affecting the gland. The monitoring program is estimated to cost $4 million, plus an estimated $9.6 million to operate in its first year. These costs would not include funds for medical care.

Mir Aimal Kansi, suspected as the lone gunman who killed two CIA agents in front of the Central Intelligence Agency building in Langley, VA, was captured in Pakistan on June 7, 1997 and brought back to the United States. Actually, he was "bought" out of Pakistan. The CIA paid $3.5 million to the fugitive's own bodyguards for his return.

The New Jersey Department of Environmental Protection announced in June 1997 that virtually all private wells in South Jersey are delivering radioactive water and should be tested to determine the severity the problem. The state printed thousands of booklets called, "The Homeowner's Guide to Radioactive Drinking Water" and made them available through county health departments. The booklet tells readers that "exposure to radium over a long period of time is believed to increase one's lifetime risk of cancer," including bone and sinus cancer. The contamination is thought to come from ancient rocks in the Cohansey Aquifer and has been affecting the water for thousands of years.

During June of 1997, police arrested two brothers wanted for a shooting incident with Ohio police officers near Cincinnati. A search of their vehicle produced weapons, bulletproof vests, F.B.I. baseball caps, and U.S. marshal badges. One of the brothers was charged with possession of weapons that had been stolen allegedly from an Arkansas gun dealer found murdered along with his wife and daughter in June of 1996.

In July of 1997, the U.S. Army planned to transport 241,328 "binary" chemical weapons (two chemicals that must be mixed together to form a deadly tool; weapons in storage contain only one of these chemicals) from storage space at the Umatilla Weapons Depot in Oregon to an ammunition plant in Hawthorne, NV for destruction. The project will take several years to complete. Every week six trucks will each carry 400, 155 mm, chemical projectiles designed to be fired by cannon. Each truck will be tracked by satellite carried on-board.

At Fort Hood, TX in July of 1997, an antigovernment group was planning an attack on the fort acting under the impression that Army bases are training United Nations troops to stage a coup. Seven people were arrested and machine guns and pipe bombs were seized.

An emergency alarm system was shut down for at least five days on July 22, 1997. The Oregon Emergency Management Agency oversees the operation of 42 sirens and 9 reader boards as part of the Chemical Stockpile Preparedness Program's outdoor warning system at the Umatilla Chemical Depot near Hermiston. The emergency system linked to the Army nerve gas storage area was shut down in two counties when air-cooling systems overheated at two or three remote transmission sites and caused "irregularities" in the system's operation.

On August 5, 1997 in Washington, D.C., two Pentagon Defense Protective Service security guards wrestled to the ground an apparently deranged man, Steve Maestas of Covina, CA. Maestas had pulled a loaded gun and tried to enter the Pentagon with a knapsack full of 12 ten-round clips of 9 mm ammunition. This was his third attempt to enter the building during the morning rush hour.

On August 10, 1997 in Westminster, CA, a man was killed by a bomb in a parcel he found outside his residence. William Bays died from massive chest and head injuries. Two friends who were standing nearby suffered ear injuries but were otherwise unharmed. The package was not mailed or delivered by a package service, but it reportedly had a name on it. A hail of metal fragments tore through a cluttered garage where the package was opened, leaving stored items peppered with small holes. No motive or suspects are known.

On August 13, 1997 in Wheeling, WV, Floyd "Ray" Looker pleaded guilty to selling copies of blueprints of the Federal Bureau of Investigation's fingerprint complex to what he believed was a terrorist group that planned to blow up the center. He sold the copies for $50,000 to an undercover F.B.I. agent. James "Rich" Rogers, a firefighter accused of making the photocopies of the blueprints, is scheduled to stand trial.

On August 19, 1997 in Colebrook, NH, Carl Drega, 67, apparently became incensed by local government officials in northern New Hampshire. He went on a rampage with an AR15 rifle and killed two state troopers, a District Court Judge, and a newspaper editor. Drega then burned his house. He fled to Brunswick, VT, wounding a New Hampshire Fish and Game officer on the way, and established an ambush site. A police dog warned police that something was wrong, but three officers were wounded before finally killing Drega. When officials visited his isolated property in the town of Columbia, NH they found 86 empty pipe bombs, 400 pounds of ammonium nitrate, 61 gallons of diesel fuel to set it off, gunpowder, three Kevlar helmets, one semiautomatic rifle, a maze of tunnels, and birdhouses wired with electronic listening devices. Drega had a long-running feud with local officials over property rights and other issues, and after the incident officials worried about his activities as an employee at local nuclear power plants (Vermont Yankee, Pilgrim, and Indian Point). On September 5, 1997 the Nuclear Regulatory Commission completed an inquiry "to determine if the access programs, as implemented, identified information that should have precluded Drega from being granted unescorted access," and reported the three nuclear plants had followed federal security regulations.

On August 29, 1997, indictments were returned in New York charging Gazi Ibrahim Abu Mezer, 23, and Lafi Khalil, 22, with conspiring to explode a pipe bomb in the New York City subway. Their confederate and roommate informed police of their location and plans to detonate bombs in the busy Atlantic Avenue subway station and on a commuter bus. Prosecutors stated that a global investigation failed to link the plot to any known terrorist group. Mezer and Khalil were arrested in a Brooklyn apartment on July 31, 1997 where police found components of one or more pipe bombs. Both suspects were shot during their arrest when one of them reportedly tried to reach a nail-studded device resembling a pipe bomb. Abu Mezer had been detained by federal authorities earlier in the year, but was given 60 days of freedom — until August 23, 1997 — when he was arrested with Lafi Khalil for planning to bomb the New York City subway. In the previous months, Mezer had been caught trying to illegally cross the border from Canada to the United States three times. However, Canada refused to accept him back the third time, and Mezer subsequently requested political asylum in the United States.

On October 1, 1997, a leaking container of hazardous pesticide — one of ten 50-pound bags illegally placed on an American Airlines passenger aircraft — produced fumes and sickened passengers at Miami International Airport just before takeoff to Ecuador. The passengers were evacuated and put on another flight. The ten packages were covered with plastic wrapping that hid the "Hazardous Materials" labels, they were not packaged securely, and they contained an amount of Dowicide A pesticide more than 200 times the maximum permitted on a passenger flight (a passenger is allowed to bring 2.2 pounds of Dowicide on a flight if it is properly packaged and labeled). A courier paid $800 in extra baggage fees for 22 bags and boxes, ten of which contained Dowicide. The courier told the F.B.I. what he earned for his work: a $100 discount on his airline ticket.

On October 1, 1997 in Pearl, MS, 16-year-old Luke Woodham was the first to start a series of killings in U.S. high schools and middle schools that has continued to the present day. He killed his mother at home; then he went to his high school and killed three students and wounded seven.

Six marines from Camp Lejuene, one captain, and seven civilians were arrested on October 16, 1997 after a nationwide investigation of the misappropriation of military weapons including rocket launchers, machine guns, mines, mortars, and grenades. Five marines held in custody are suspected of independently hiding extra weapons from training exercises and seeking buyers. Federal agents acting as middlemen bought the stolen explosives and other weapons and then sold them to gun enthusiasts searching for greater firepower.

Republic of Texas leader Richard McLaren and top lieutenant Robert Otto were convicted October 31, 1997 of abducting a couple. The incident led to a week-long armed standoff with authorities. McLaren and Otto were found guilty of organized criminal activity in the April 27, 1997 abduction that lasted until May 3, 1997 when as many as 300 state troopers and Texas Rangers caused them to lay down their guns.

In November of 1997, skinheads in Denver, CO shot at cops and bystanders, killing an African immigrant and a police officer.

During November of 1997, *E. coli* bacteria was discovered in hoses used to fill water tanks on Amtrak trains at a maintenance facility in Miami, FL after routine tests found bacteria in train drinking water. The hoses were replaced and water tanks on 250 passenger rail cars were flushed and disinfected.

The Army's chemical weapons incinerator at the Deseret Chemical Facility in Tooele County was cited on November 17, 1997 for 25 state hazardous-waste regulations identified during its first year of operation. None of the violations was serious enough to warrant closing the facility noted the Utah Division of Solid and Hazardous Waste, and most were discovered by workers and reported to the state.

Some of the most serious violations carry fines of up to $10,000 per incident. The most serious incident probably occurred on January 26, 1997 when doors left open for maintenance work allowed chemical agent to leak into an observation corridor.

On December 1, 1997, Michael Carneal, age 14, killed three of his fellow students at a morning prayer meeting in West Paducah, KY. Another student was able to overcome Carneal when he stopped to reload.

On December 12, 1997 in Little Rock, AR, two white supremacists were charged with murder, racketeering, and conspiracy for planning to overthrow the federal government and replace it with an Aryan People's Republic.

Also on December 12, 1997 in Philadelphia, PA, a man was charged with leaving pipe bombs at various businesses and painting swastikas on politicians' offices.

On the weekend of December 13, 1997, 600 pounds of ammonium nitrate fertilizer mixed with fuel oil were stolen from Jerico Services Facility in Weeping Water, NE. Twelve 50-pound bags trade-named "Pellite" were missing from a semitrailer, although no dynamite or blasting caps which could be used as detonators were missing from the site. The Federal Bureau of Alcohol, Tobacco and Firearms has posted a $5000 reward for information, and can be called at 1-888-ATF-BOMB.

In December 1997, a ship carrying three containers of methyl bromide, a poisonous gas, was met at Port Elizabeth, NJ by state and federal authorities after crew members of the *Teval* reported an odor during a crossing from Spain. The ship flying the flag of Malta was halted at Ambrose Light while a commercial response contractor hired by Lykes Lines went aboard and conducted a series of air quality tests outside the containers. All tests were negative, and the ship was allowed to dock at Port Elizabeth.

Recently, the media has touted irradiation as a solution to food contamination. In the mid to late 1980s, the U.S. Department of Energy sent hundreds of radiation-containing capsules around the country to attempt to find an industrial use for radioactive strontium and cesium waste they had stockpiled after 50 years of making nuclear weapons. The department did not want to call the radioactive materials "wastes" or to develop a disposal plan. The central idea to this effort was to use such radiation for sterilizing food or medical instruments. In 1988, at an irradiation plant in Decatur, GA, one capsule developed a pinhole leak and 0.02% of its contents escaped. The D.O.E. spent more than four years and $47 million to clean up the leak and make the facility safe once again. All the capsules sent around the country were brought back to the Department of Energy's Hanford Site at Richland, WA to be tested. Since 1988, 16 of approximately 2000 capsules have failed the test and have been segregated. Maintenance for all the capsules costs about $10 million a year. The D.O.E. was to decide by the end of 1997 whether to declare the waste as "waste." On February 12, 1999, the Department of Agriculture Secretary announced at a meeting of the National Beef Cattlemen's Association in Charlotte, NC that his department has approved the controversial process for using nuclear energy to treat potentially contaminated meat.

On December 18, 1997 at the Malton Canada Post office in Toronto, a letter carrier noticed a hole in a parcel that was labeled as containing at least 17 different infectious bacteria including influenza, gonorrhea, and hepatitis. A spokesman for the Toronto local chapter of the Canadian Union of Postal Workers stated that management did not take any precautionary measures, and that it was an employee who called 911 and the fire department. Only after police and firefighters arrived at the postal station were workers told to leave the building. The package was traveling from Minnesota to PML Microbiologicals in Mississauga, Ontario. Labels on the package noted that if the package was damaged, the Center for Disease Control in Atlanta should be called. The Canada Post bans perishable biological substances "except when sent between officially recognized laboratories." The hole was in an outside package, and the inner package had not leaked. A vice-president with the union local said, "I never thought they would allow something like this through the mail."

In January of 1998 at Riverside, CA, a former Air Force Ordnance expert who worked as a safety and quality control officer for Allied Technology Group was charged with second-degree murder for allegedly allowing a live military shell to be taken from the cleanup of live-fire areas at the Army National Training Center at Fort Irwin and to be delivered to Dick's Auto Wrecking in Fontana as scrap metal. One 105-millimeter shell exploded and killed a worker who was attempting to dismantle it with a blowtorch. The safety officer falsely certified that he had inspected demilitarized scrap and concluded it no longer contained explosive material. County and federal investigators found 54 additional pieces of live ammunition at the Fontana wrecking yard, including 30 that were considered potentially lethal if they exploded. The defendant was held in jail on $250,000 bail.

The Salt Lake Tribune reported in its January 1, 1998 issue that a newly uncovered document shows the U.S Army had conclusive proof a deadly nerve agent (Agent VX) was in grass and snow eaten by 6000 sheep that died in Skull Valley in 1968. Agent VX is a colorless to amber liquid with no noticeable odor, a vapor density of 9.2, and a median lethal dosage of 100 (mg-min/m³). It has very high eye and skin toxicity, its rate of action is very rapid, and it produces casualties when inhaled or absorbed. The Army had proof for many years that the nerve agent was found where the 6000 sheep died in western Utah on March 14, 1968 — apparently after a low-flying aircraft sprayed nerve agent in a target area at the Dugway Proving Ground about 27 miles west of Skull Valley. "Agent VX was found to be present in snow and grass samples that were received approximately three weeks after the sheep incident," said a newly-located report prepared in 1970 by the Army's Edgewood Arsenal and obtained by the *Salt Lake Tribune*. The 1970 report concluded, "...it is possible that the quantity of VX originally present was sufficient to account for the death of the sheep." The military still refuses to accept responsibility for the accident. However, federal testing of recently discovered sheep burial pits at Skull Valley is scheduled to begin within the next few months, 30 years after the deaths.

A Middlebury College freshman appeared in U.S. District Court in Burlington, VT on January 7, 1998 to answer to charges of handling explosives after his duffel bag was found smoking at an airport in St. Louis. A St. Louis County bomb squad checked the bag and found homemade explosives that were sensitive to friction and impact. The student, Timothy Boarini, of Iowa City, was allowed to continue his trip without the bag, and filed a claim for missing luggage. Early in the morning of his court appearance, he was picked up by the F.B.I. at his dormitory room. If convicted, the student could be jailed for up to ten years and fined $250,000.

On February 9, 1998 in Downers Grove, IL, a high school student swallowed potassium cyanide, seriously injuring himself and sending more than 40 students and staff to hospitals after they inhaled fumes from his vomit. The potentially lethal vapors seeped into two classroom laboratories when the chemical mixed with the victim's stomach acids. No explanation was given as to why the boy ingested potassium cyanide just as classes were changing.

On February 9, 1998, Peter Howard, 44, pleaded guilty to attempted arson and use of an explosive device and was sentenced to 15 years in prison for trying to blow up an abortion clinic with a truck filled with tanks of propane and gasoline. Wearing a crash helmet and earplugs, he had driven a pickup truck into the Family Planning Associates in Bakersfield, CA and tried to light a fire, but was stopped by a security guard before he was able to set off an explosion.

A federal alert was issued after 1800 pounds of ammonium nitrate and fuel oil mixture disappeared February 16, 1997 from a locked bunker at C&K Coal Co. in Sligo, PA, about 60 miles northeast of Pittsburgh. Investigators have no suspects, but are attempting to learn the identities of two men who were thought to be involved and were seen near a green pickup truck over the weekend.

Early in February 1998, the National Academy of Sciences reported at the D.O.E.'s center for anthrax research, using a refined type of DNA analysis, found a piece of evidence behind the leak of anthrax at a biological laboratory in Siberia. The leak killed at least 42 people in the city of Sverdlovsk in April 1979. Soviet officials stated the tragedy was caused by tainted meat. Scientists estimate billions of invisible spores were swept aloft from the lab on a strong southerly wind. The13 body tissue samples that now sit in a freezer at the Los Alamos National Lab are evidence that at least four strains wafted over Sverdlovsk. A researcher in the lab believes the presence of multiple strains indicates the Soviets were making bio agents in Sverdlovsk in 1979 even after signing the Biological Weapons Convention Treaty.

A sailor was sentenced on February 12, 1998 in Norfolk, VA for stealing plastic explosives from the aircraft carrier *John C. Stennis*. He received nine months confinement, a bad conduct discharge, and reduction to the lowest enlisted grade. Another sailor was found guilty of the same charges and sentenced to serve ten months at the Norfolk Naval Air Station brig.

Larry Wayne Harris and William Leavitt were charged February 19, 1998 in Las Vegas under a federal law that prohibits the production and possession of biological agents for use as weapons. Harris had previously been on probation in 1995 after pleading guilty to illegally obtaining bubonic plague through the mail. An F.B.I. affidavit stated that one of the suspects told an informant he had "military grade anthrax" in his Mercedes. The informant noted he had seen eight to ten bags marked "biological" in the trunk of the car. Actually, the suspects had a legal anthrax vaccine that is readily available for inoculation of farm animals. The F.B.I. had 70 agents armed with few facts investigating this case. Both suspects were released.

Early in March of 1998 in Illinois, federal agents raided the homes of Dennis McGiffen of Wood River, Wallace Weicherding of Salem, and Ralph Bock of Brighton and reportedly found a machine gun, a pipe bomb, and hand grenades. The suspects were held without bail until their trial. The suspects, members of the

"New Order," reportedly planned to rob banks and armored cars to acquire funds, to bomb public buildings, to kill a prominent Southern Poverty Law Center lawyer and an unidentified federal judge, and to contaminate a large water supply with cyanide.

Will A. Wiltz, a ski technician for former U.S. Olympic gold medalist Tommy Moe, and for Skis Dynastar of Colchester, VT, could owe as much as $350,000 in civil fines as a result of Federal Aviation Administration action. On March 3, 1998, Wiltz took Flight 921 from Dulles International Airport in Washington, D.C. to Denver, CO. As a baggage handler loaded a metal box containing a butane torch, two canisters of butane gas, and a can of liquid acetone onto a conveyor belt, the box exploded and blew off its lid which struck the baggage agent. Wiltz, suspected of being a terrorist, was questioned for nine hours, the plane was evacuated, and the baggage agent was hospitalized. The F.A.A. cited Haz Mat regulations that forbid the transportation of such items on passenger-carrying aircraft. A spokesman for Skis Dynastar was not aware that carrying such items was against F.A.A. rules. Screening checked baggage is the task of the airlines.

On March 8, 1998, Daniel Rudolph videotaped himself as he intentionally severed his hand with a circular saw, saying "This is for the F.B.I. and the media." The F.B.I. had questioned the victim about his brother, Eric Rudolph, who is a suspect in the bombing of the New Women All Women Clinic, on January 29, 1998 — the first bombing of a U.S. abortion clinic that killed one person and injured another. Authorities believe Eric Rudolph may be hiding in the mountains near his hometown of Murphy, NC.

On March 10, 1998 in Phoenix, employees at Recovery Technologies, a collection agency, received a letter supposedly containing the biological agent, anthrax. Police officers and firefighters closed Thomas Road, from 32nd to 38th streets, and evacuated the Sunshine Square shopping area and several restaurants nearby. Nine employees at the agency and one police officer were decontaminated, given quick medical exams at the scene, sent to a hospital where potential victims were placed in an isolation room, and then interviewed by F.B.I. agents. The letter was sent to a local laboratory and tested.

A federal grand jury in Newark, NJ stated on March 24, 1998 that Daniel J. Malloy, a wealthy military weapons supplier, and Joseph Balakrisha Menom, a co-conspirator in Singapore, conspired to ship to Iran 20 batteries needed to power Iran's supply of Phoenix air-to-air missiles. Under the Federal Arms Export Act, it is illegal to provide weapons or spare parts for weapons to countries that support terrorism.

On March 24, 1998 in Jonesboro, AR, Mitchell Johnson, age 13, and Andrew Golden, age 11, set off a fire alarm at their school. When the children went outside to the playground, the boys opened fire on them killing four students and a teacher.

In Cleveland during March of 1998, a four-year-old boy was caught for a second time bringing a loaded handgun to a day-care center to show his classmates. He had possession of a 9 mm automatic gun with one bullet in the chamber and another 13 bullets in the magazine.

At 3:20 a.m. Monday, March 30, 1998, an improperly drained nerve-gas bomb caused an automatic shutdown of a $650 million chemical weapons incinerator at the Army's Tooele Chemical Agent Disposal Facility in Utah. A drain probe failed to empty a 500-pound MC-1 bomb of its 220 pounds of liquid GB (sarin), leaving about 60 pounds of sarin in the bomb. A metal parts furnace is not designed to burn that much agent. Because nerve agent burns hotter than other materials, the bomb caused the metal parts furnace to heat to 2200 degrees — well above its normal temperature of 1600 degrees. The overheating promoted an immediate automatic shutdown of the metal parts furnace. To date, the facility has destroyed 386 of the 4077 MC-1 bombs abandoned at the facility.

On May 8, 1998, ATF (U.S. Bureau of Alcohol, Tobacco and Firearms) agents stated that four men in central Florida planned to use pipe bombs to create confusion while they robbed banks. A federal court complaint indicated that 14 bombs were to be used on major highways in the City of Orlando, including some along the access road to Walt Disney World. Suspects included Todd Vanbiber, Brian Pickett, Christopher Norris, and Deena Wanzie. Some bombs had already been made using timers and batteries.

In Springfield, OR on May 21, 1998, Kip Kinkel, age 15, killed his parents at home and then went to school to kill two students and wound 22 others. When apprehended, he said, "Shoot me!"

In Billings, MT during the last week in May of 1998, a second trial began for 12 "Freemen" who held the F.B.I. at bay in an armed stand-off for 81 days. They were charged with committing wire, bank, and mail fraud; armed robbery of a television news crew; and threatening to kill a federal judge. Six "Freemen" were tried in March 1998, and five were convicted.

Richard Shotts was killed by an explosion in his garage in Danville, IL on May 28, 1998 before he could explain why he had previously placed three bombs in two churches and a garage, killing one person and injuring 34. Investigators from the ATF interviewed more than 1500 people, analyzed 1000 pieces of evidence, found more than 60 similarities among the three bombs, and were on their way to apprehend Shott when the explosion occurred. No one knows Shott's motive for the bombings, or whether the explosion that killed him was an accident or a suicide.

Three men armed with automatic weapons and dressed in camouflage clothing killed a Cortez, CO police officer by firing through the officer's cruiser window when he stopped them on suspicion that they were driving a stolen water truck on May 30, 1998. The gunmen stole a second truck, were chased by other police officers, shot seven police cars, and wounded two Montezuma County sheriff's deputies. The fugitives, described as "survivalists," abandoned the second truck and fled on foot into rugged country. National Guard helicopters with heat-seeking infrared sensors, F.B.I. SWAT teams, and National Guard troops joined search efforts in a sparsely populated area about 50 miles northwest of Cortez. One of the men probably committed suicide, and the other two were never located in the following 12 months.

A Palestinian, Mohammed Rashid, 51 years of age, was secretly taken from Egypt on June 3, 1998 and flown to Washington, D.C. to stand before a federal court 15 years after he was first sought by U.S. authorities. Rumors circulated that another country that did not wish to be named assisted the United States in Rashid's capture. He faces a nine-count indictment for murder, sabotage, and other crimes connected to the 1982 mid-air bombing of a Honolulu-bound Pan Am flight that killed one passenger and wounded 15 others. During the 1980s, Rashid was described as one of the leaders of the "15th of May" organization, an Iraq-based, Palestinian terrorist group that was linked to approximately 20 attacks.

A section of the Norfolk Naval Air Station on July 1, 1998 was ordered off-limits to outside personnel after a Federal Express driver entered the base to pick up a metal footlocker for shipment. When the driver put the shipment on his hand cart to carry it to his truck, liquid seeped from the top of the locker which was marked "biohazardous" material. Shipping papers identified the contents as samples of suspected biological agents, possibly including anthrax, from the 1991 Persian Gulf War with Iraq. Air station officials initially were unable to explain what was actually in the footlocker and how the locker came to be in the base warehouse. The container had been flown to Norfolk Naval Air Station three days earlier aboard a chartered DC-8 from Bahrain, an island in the Persian Gulf used by the United States for cargo shipments related to United Nations sanctions against Iraq. Naval firefighters and Haz Mat personnel immediately closed Gate 22 to prevent all traffic from approaching the terminal which is used for daily flights overseas.

Two Haz Mat inspectors from the Virginia Department of Emergency Service were called to the scene. Three employees were decontaminated and taken to the Portsmouth Naval Medical Center and held for observation. The "leak" was eventually proven to be uncontaminated water. The containers inside held samples from a site where Iraq claims to have destroyed warheads that previously contained chemical and biological agents. The United Nations Special Commission (UNSCOM) had taken samples from the site in early May, and DNA testing had shown they were positive for the past presence of fragments of the anthrax genone which poses no health hazard.

Sometime in late June 1998, other samples were taken from the evidence site and sent from Bahrain. UNSCOM issued the following statement: "UNSCOM has no expectation that the second set of samples contain live agents because of the method of its chemical neutralization and destruction, as well as the results of previous analysis." No defense official would say why a substance as lethal as anthrax was shipped into the United States. The particular container was owned by the United Nations and shipped by the Air Force to a Navy base for shipment to an Army laboratory at Aberdeen Proving Grounds for analysis. The Navy was unaware they were transporting the remnants of biological weapons into the United States, so they responded as they would to any emergency biological incident.

In Brownsville, TX on July 1, 1998, federal marshals arrested Johnny Wise, 72, Jack Abbott Grebe Jr., 43, and Oliver Dean Emigh, 63, on charges of conspiracy to use weapons of mass destruction. Federal prosecutors accused them of threatening to use biological weapons against the Internal Revenue Service director, the Federal Bureau of Investigation director, Attorney General Janet Reno, and President Bill Clinton. Prosecutors presented no witnesses to support their contention that the three men posed a danger, nor did they say whether the men possessed any weapons. Grebe's lawyer asked the court whether the federal criminal justice system had reached the point where a person could be jailed by the government for saying something, but that the government was not required to tell the defendant's lawyer what he allegedly said. Two weeks later, an affidavit stated Wise and Grebe told an F.B.I. informant that they planned to modify a Bic lighter so it would expel air instead of propane. They planned to glue a hypodermic to the opening of the lighter and insert a cactus needle coated with a biological agent such as anthrax, botulism, or the AIDS virus.

In West Blocton, AL, a man dressed in military fatigues and a breathing apparatus fired shots at an armored delivery vehicle parked outside a bank in early July of 1998, then retreated into woods where four pipe bombs had been installed. Police officers spotted one bomb and abandoned the chase then called in state and federal explosive ordnance units. There were no arrests.

On July 8, 1998, a foul-smelling acid was spilled at four abortion clinics in Houston, TX. Evacuation was ordered at three of the clinics, and ten people were treated at the scene for inhalation problems. The acid was reportedly the same as used in recent attacks in New Orleans, LA and in central Florida.

On July 11, 1998 on the south side of Chicago, some children found a suspicious box containing several vials of an unidentified liquid attached to an aerosol container and a threatening note. As officials examined it the device exploded, sending ten firefighters and seven police officers to the hospital. They were all released after tests showed the liquid was not life-threatening and no one was reporting symptoms.

On July 13, 1998, Joshua England pleaded guilty to shooting three teenagers in a crowd outside a nightclub in Pelion, SC because they were black.

On July 24, 1998, Russell E. Weston set off a metal detector as he passed through the east front entrance of the nation's Capitol building, known as the "Document Door." In the ensuing gun battle, he killed two police officers and wounded a tourist. Weston was wounded in the chest, shoulder, thigh, and buttocks inside the offices of House Majority Whip Tom DeLay.

In Morgantown, WV, 25 tons of ammonium nitrate disappeared from Bruceton Farm Supplies located in Preston County on July 29, 1998. The F.B.I. believed the disappearance of the fertilizer resulted from simple theft rather than terrorist activity and offered a reward of $10,000 for information.

On August 2, 1998, a stolen pickup truck crashed through the doors of the Tippecanoe County courthouse in Lafayette, IN during what local officials called a terrorist activity. The fire was put out within an hour, 25 people were evacuated from the area, the courthouse was extensively damaged, and F.B.I. and ATF agents are looking for the driver of the 1979 Ford pickup. The truck was carrying drums of accelerant, and wire hanging from the vehicle may have been a fuse.

On August 23, 1998, Kathryn Schoonover was arrested at the Marina Del Rey, CA post office after she stuffed 100 envelopes with the chemical agent, cyanide, to appear as free samples of a nutritional supplement. Inside the envelopes, sodium cyanide had been placed in clear plastic envelopes with an authentic advertising brochure attached which touted health and diet products. The letters were addressed to people who worked in medicine or law enforcement. A bystander reported seeing Schoonover at a post office counter wearing protective gloves as she took a substance in a container marked "Poison" and placed it into envelopes. The gray, window envelopes were business-size, with printed labels and no return address. Police had hazardous materials experts test the substance which proved to be sodium cyanide.

Nearby streets and a shopping mall near the area were evacuated for five hours. Schoonover was eventually charged with attempted murder. Cyanide blocks the use of oxygen in body cells and thus causes asphyxiation in each cell. The cells of the brain and the heart are most susceptible.

On August 27, 1998 in Orlinda, TN, a truck tanker hauling 36,000 pounds of sodium cyanide for DuPont Co. in Memphis overturned on I-65. Responders were able to secure the scene and prevent any leakage. The tank was not breached and no product was lost at the wreck site. DuPont sent a Haz Mat team. A driver for Miller Transporters of Jackson, MS was treated for injuries at a local hospital.

After years of resistance, the federal government, as of September 1998, is considering providing potassium iodide pills which will block the radiation that causes thyroid cancer to people living near nuclear power plants. The government, which already has stockpiles of the drug for emergency workers, has long argued that potassium iodide might cause dangerous side effects and give people a false sense of security during a nuclear incident. Critics have held that government has kept the drug out of local areas where nuclear power plants are located because its distribution would promote fears of a nuclear incident.

In late September of 1998, the Army approved a plan by Aberdeen Proving Ground to destroy stockpiled World War II mustard agent by removing the deadly blistering agent from storage containers, adding boiling water to neutralize the mustard, and adding bacteria to help the mustard biodegrade. The remaining sludge would be shipped to a landfill. The neutralized water would be dumped into the Bush River. Initial plans called for incinerating the mustard, but citizen opposition canceled that option. This new plan is viewed by some as a less offensive alternative. It still has to be approved by Maryland and the EPA. If approved, it will then probably be used at seven other sites in the United States where mustard gas is stored.

Kelvin Smith, a U.S. wildlife officer, pleaded guilty on September 30, 1998 to lying to F.B.I. agents about his involvement in the military training of Moslem extremists convicted of plotting bombings in New York City. During four weekends in January and February of 1993, Smith used his Pennsylvania farm as a guerrilla training location for eight Moslem terrorists, followers of Sheikh Omar Abdel-Rahman, training them in assault weapons, cut-and-run shooting, night-time assaults, hand-to-hand combat, and martial arts. He was also charged with dumping assault rifles in the Delaware River. He even used his U.S. Fish and Wildlife Service van to transport his trainees. Five of the eight trainees were eventually convicted of plotting to bomb the New York headquarters of the United Nations, the Lincoln and Holland Tunnels, and the George Washington Bridge. The charges against Smith carry a maximum penalty of 20 years in prison and a $1 million fine, but sentencing guidelines are likely to keep Smith in jail for less than two years. Currently, Smith is on unpaid administrative leave as a special agent for the U.S. Fish and Wildlife Service.

In mid-October 1998, the F.B.I. opened its National Domestic Preparedness Office which "will assume overall responsibility for coordinating the government's efforts to prepare America's communities for terrorist incidents involving weapons of mass destruction," according to Attorney General Janet Reno. The new office opened in response to local responders' and officials' requests for federal training, equipment, and funds for response to terrorist actions. In the recent past, at least 40 federal agencies or offices were funded or instructed to use their own funds to assist state and local response officials in responding to terrorist actions involving nuclear, biological, and chemical weapons. The Board of Directors of the International Association of Fire Chiefs passed a resolution in September 1998 that when catastrophic events occur in local communities, the local fire department will respond and that the federal government should recognize, accept, and act on that reality. The board will create initiatives identifying the federal responsibility to fund local fire service preparedness and response to natural or technological catastrophic events. Funding would include equipment, training, and staffing needs. Federal resources would enable the local fire service agencies, as first responders, to act in a safe and timely manner to minimize injuries, loss of life, and damage to property.

In Eugene, OR on October 15, 1998, Jeffery Pickering made no statements before a federal magistrate when charged with threatening president Bill Clinton. Earlier in June, Clinton flew to Eugene Airport after two people were killed at the Springfield, OR school shooting. A day preceding the president's arrival in Oregon, someone called officials and mentioned "two by the airport." Two bombs were found in a pipe behind a museum on airport property. Pickering was held without bail for a grand jury the following week. The suspect's brother reported him to the F.B.I.

Seven separate fires on October 19, 1998 caused approximately $12 million in damage to buildings and chairlifts on Vail mountain in Colorado. It was perhaps one of the most expensive night's work in the war between environmentalists and large corporations. The Earth Liberation Front, said to have been involved in arson incidents across the Northwest, claimed responsibility for the fires.

On October 28, 1998 in Little Rock, AR, a man was convicted of violating the Federal Access to Clinic Entrances Act for parking Ryder rental trucks outside two abortion clinics.

On October 30, 1998, five abortion clinics in three states received letters with powder inside the envelope that was stated to be deadly anthrax. Thirty-one persons at a Planned Parenthood clinic in Indianapolis were stripped and decontaminated, including a postal delivery person and two police officers. The incubation time for anthrax is one to six days. Those contaminated would take antibiotics for four weeks and perhaps take anthrax vaccine.

The United States has experienced several anthrax hoaxes since late October 1997. In Coppell, TX on December 4, 1998, the North Texas Joint Terrorism Force responded after a suspicious package with a note attached stating it contained anthrax was found in a sorting bin at a regional postal distribution center. Because of this and similar incidents, the U.S. Postal Service has issued a nationwide alert. A number of envelopes with warnings, such as "You have been exposed to anthrax," were sent from Fort Worth, TX which is near Coppell. In Bloomington, IN an anthrax hoax note was found at a high school. In Indianapolis an anthrax note was found at a Bank One branch, while at St. Matthew Catholic School a letter was received stating the envelope contained anthrax. In October, an abortion clinic in Indianapolis received an anthrax letter as well. The fire department at Seymour, south of Indianapolis, reported a threat of anthrax exposure at a Wal-Mart. A secretary at Redeemer Lutheran Church in Rochester, NY opened an envelope said to contain anthrax postmarked at Fort Worth. A similar letter postmarked at Fort Worth was sent to Queen of Martyrs Catholic Church in Cheektowaga in Erie County near Buffalo, NY. At Cheektowaga, three priests, two secretaries, a cook, a housekeeper, a town police officer, and a U.S. postal inspector were decontaminated and treated with antibiotics. Also treated were up to six members of the Erie County Hazardous Materials Team who worked with the possible victims in the church rectory. In another case, an identical letter with the threat, "You have been exposed to anthrax," was received at the headquarters of the Pro-Life Action League in Chicago. Also, the offices of *Ocean Drive Magazine* in South Beach, Miami were evacuated when an anthrax note was sent through the mail. It is a crime to use a weapon of mass destruction, but is also a crime to *threaten* to use a WMD.

On November 18, 1998 in Kalamazoo, a member of a Michigan militia was convicted of plotting to blow up an Internal Revenue Service office and a television station and of threatening to kill federal officials.

In East St. Louis, on December 5, 1998, one member of a white supremacist group was sentenced to five years and ten months in jail for stockpiling weapons to start a race war. Wallace S. Weicherding was found guilty in August of conspiracy to possess and manufacture illegal firearms and destructive devices.

Eleven pipe bombs were found in the San Diego area of southern California during a 12-hour period on December 6, 1998. All the bombs were thought to be made by the same person who spread them throughout the area. The dispersion seemed haphazard; a couple of the bombs were left on highways.

In Troy, MI near Detroit, hundreds of workers at Electronic Data Systems Corporation were sent home on December 17, 1998 after a bomb notice was received by telephone. The caller said the bomb was a protest against the U.S. air strikes on Iraq. No bomb was found, and no one was arrested.

Southern California experienced 18 anthrax threats over a period of three months. On December 14, 1998, a Perris School District secretary opened a letter addressed to the superintendent that claimed the envelope contained anthrax bacteria. Eighteen people and two responding firefighters were decontaminated and quarantined. Also during December 1998, a Riverside elementary school teacher received a thank-you card containing a moist towelette with a warning that it contained anthrax. On December 17, 1998 in Westwood Village, 21 office workers in a 21-story office building were decontaminated and quarantined. Officials stated they immediately evacuated the tower and turned off the ventilation system. The 21 people were taken to a secure room in the basement, tested for traces of anthrax, and decontaminated by employees from the county health department. Later in the day, when it appeared the incident was a hoax, they were provided with new clothing and transported by van to the UCLA Medical Center under F.B.I. escort, questioned by F.B.I. agents, decontaminated a second time, given inoculations, and held for the rest of the day under observation.

On December 18, 1998, Harvey Craig Spelkin was accused of calling the U.S. Bankruptcy Court in Woodland Hills, CA and stating that anthrax had been placed into the air conditioning system of the court where he was involved in a bankruptcy case that same day. He was charged under the Federal Biological Weapons Anti-Terrorism Act of 1989 in which threatening an anthrax attack is punishable by up to ten years in federal prison. If the hoax caused someone to die of a heart attack, the perpetrator could face capital murder charges. Spelkin was released on a $50,000 bond.

Two Van Nuys courthouses in the Los Angeles area were evacuated on December 21, 1998 when a person called 911 and stated, "Anthrax has been released in the Van Nuys courthouse," causing officials to evacuate and quarantine up to 2000 people from both the new and the old Superior court buildings.

On December 23, 1998, the Times Warner Cable Company in Chatsworth, CA received a telephone threat of a "biological agent." Authorities evacuated and quarantined about 70 employees for several hours and gathered air samples. On Christmas Eve 1998, people at a Mervyn's store in Palm Desert, CA were rinsed with a bleach solution after receiving an anthrax threat over the telephone. On December 26, 1998, the Glass House nightclub in Pomona, CA received a threat by telephone from a male who stated that a significant quantity of anthrax would be released into the air. Approximately 800 persons were quarantined for four hours. L.A. County firefighters, and the F.B.I.'s domestic terrorism task force searched the air conditioning system, filters, and vents. All tests were negative. On January 4, 1999, Gilbert-East High School received an anthrax threat by telephone around noon. A young-sounding male, perhaps a student, called the school several times reporting that anthrax spores had been released within the ventilation system. Response authorities held a threat assessment and found "the threat lacked credibility."

False anthrax threats are costing federal and local governments millions of dollars. The U.S. Attorney's office in Rochester, NY received two hoaxes within two weeks. The first incident occurred on December 31, 1998 when a secretary opened a letter announcing, "You have now been infected by the anthrax virus, smile," with a smiley face drawn below the text. She was treated at Rochester General Hospital's decon unit and placed on antibiotics. The second incident, on January 14, 1999, required treatment for two persons after they opened letters claiming to be contaminated with anthrax. These two victims were also treated at Rochester General's decon unit and placed on antibiotics.

In San Francisco on December 31, 1998, local police and federal ATF agents arrested a veteran employee of Pacific Gas and Electric Company after a bomb and hundreds of pounds of bomb-making materials, including ammonium nitrate and calcium nitrate, were found at two utility company warehouses. The person arrested did not appear to have a political agenda, and may simply have enjoyed making small explosives and fireworks which he sold to coworkers. He was held on $1 million bail on three counts of having an explosive device.

In the Borough of Queens on January 4, 1999, someone drove a car through a steel, protective gate into the offices of an Arabic newsletter then set the car afire. The fire caused minor damage, and no injuries or arrests occurred.

On January 10, 1999, Delta Flight 9912, about to take off from Bradley International Field in Windsor Locks, CT enroute to West Palm Beach, was evacuated when a passenger reading an in-flight magazine found a message supposedly written by a Middle Eastern terrorist group. State police with bomb-sniffing dogs searched the plane for an hour but found nothing.

On January 28, 1999, Walter W. Johnson of Capital Heights, MD was apprehended when he tried to enter the Capitol building in Washington, D.C. He had his ticket to the impeachment trial of the president as well as an 18-inch knife, a handbook related to terrorism, a price list of weapons, a *Soldier of Fortune* magazine and a bag of explosives that showed up on an X-ray machine. A judge held Johnson without bond on charges of carrying a dangerous weapon, and he was held for psychiatric evaluation.

In Bellevue, NE on January 28, 1999, a 20-year-old security guard at the First National Bank branch apparently faked a bomb threat that led officials to close businesses and streets. Police searched for a bomb for more than an hour, but soon realized the security guard had not received a call on the telephone he said he had. The guard faces felony charges of making a terrorist threat.

On January 30, 1998, Eric Robert Rudolph, a suspect in the explosion at the 1996 Atlanta Summer Olympic Games and the bombings of several abortion clinics, disappeared in western North Carolina and has been sought by 200 to 400 heavily-armed federal and local officials. Armed with night detection gear and tracking dogs, the officials have been unable to find Rudolph in the Nantahala National Forest. This has been one of the most intense manhunts in F.B.I. history.

On February 12, 1999 in Atlanta, 21 persons were hospitalized when an unknown perpetrator sprayed an unknown agent in a commuter train car. The victims had chest pains, respiratory problems, and eye irritation and were treated with oxygen and decontaminated with water and bleach solution before being removed to a hospital. When released from the hospital, victims had to wear hospital scrub clothing since the F.B.I. had confiscated all clothing until the spray could be indentified.

In Asheville, NC on March 13, 1999, a bomb partially exploded outside an abortion clinic but caused limited damage and no injuries. The bomb was described as "large" by the police chief who said, "It appears it could have caused a lot of damage" if it had completely exploded. A month earlier the clinic had received a package said to contain anthrax.

On March 20, 1999, Serbian Orthodox churches in Sacramento, Milwaukee, Chicago, and Indianapolis received a message calling on all Serbian-Americans to return violence for violence when the United States bombed Yugoslavia. The F.B.I. distributed copies of the FAX to military bases, nuclear weapons labs, and other sensitive installations by an e-mail warning system.

On March 29, 1998 in Essex, VT, an eight-year-old student at Founders Memorial School wrote a threat saying a bomb would go off in the school between March 29 and March 30, 1999. The note was found on March 29th and the school was evacuated for an hour while police searched the building, but no bomb was found. The boy is too young to be charged with a crime.

On April 1, 1999 in State College, PA, three homemade bombs created from plastic soda bottles filled with water and dry ice exploded at a high school, injuring a science teacher who was treated and released. One of the "bombs" exploded in a plastic trash can, one under a bench and another inside the boys' rest room. A fourth device was found inside another trash can.

A spokeperson for the Animal Liberation Front on April 6, 1999 said the group claimed credit for ransacking 12 University of Minnesota laboratories and releasing 27 pigeons, 48 mice, 36 rats, and 5 salamanders. The ALF caused an estimated $1 million in damage by wrecking computers and lab equipment to protest the use of animals in scientific research.

On April 20, 1999 at Columbine High School in Littleton, CO, Eric Harris, age 18, and Dylan Klebold, age 17, killed 12 of their fellow students and one teacher before entry was made by police SWAT teams. The duo then committed suicide. Three days after the murders, 14 people remained hospitalized, 8 of them in critical or serious condition.

On April 27, 1999 in Fairport, NY, a 12-year-old boy, who had plotted for months to blow up his school, had his home raided by police who found gunpowder, propane, bomb-making books, and notebooks listing plans to destroy the Johanna Perris Middle School. The boy was immediately suspended by the school, but may escape criminal charges. Officials are seeking to have a family court judge order a psychiatric evaluation. The boy told officials he was angered by other students who had teased him about being small. The boy was not identified because of his age.

In Bakersfield, CA on April 29, 1999, a 13-year-old child with a .40-caliber semiautomatic handgun, a "hit list" of 30 classmates and teachers and a note that said, "They deserved to die," was detained at Sierra Middle School. The boy's father was arrested for allegedly leaving the handgun where his child had access to it. The boy was not identified because of his age.

11 Research Sources and Resources

MANUALS USED BY HAZARDOUS MATERIALS RESPONSE PERSONNEL

A manual, guidebook, or handbook may be used by first responders during the initial stages of a hazardous materials incident. First responders or hazardous materials response teams (HMRTs) may obtain chemical information from computer software, an emergency center, poison control centers, physicians, toxicologists, chemists, or from Material Safety Data Sheets. However, chemical response manuals such at the following are often carried on the response vehicle. Many HMRTs require in their standard operating guidelines that at least three research sources agree on the actions to take when responding to a specific chemical or agent.

Fire Protection Guide on Hazardous Materials
National Fire Protection Association, 1 Batterymarch Park, P.O. Box 9101, Quincy, MA 02269-9101; 800-344-3555

Includes information on fire hazards properties (Section 325) and hazardous chemical data (Section 49).

NFPA Hazardous Materials Response Handbook
National Fire Protection Association, 1 Batterymarch Park, P.O. Box 9101, Quincy, MA 02269-9101; 800-344-3555

Contains text and commentary on the following NFPA /ANSI standards: 471 — Recommended Practice for Responding to Hazardous Materials Incidents, 472 — Standards for Professional Competence of Responders to Hazardous Materials Incidents, and 473 — Competencies for EMS Personnel Responding to Hazardous Materials Incidents. This handbook also includes sections on how to start a Haz Mat team, the national response team's Haz Mat emergency planning guide, the Sacramento protocol for Haz Mat response, chemical compatibility of protective clothing, and decontamination procedures.

North American Emergency Response Guidebook
Developed by U.S. Department of Transportation, Transport Canada, and the Secretariat of Transport and Communications of Mexico. In the United States, response personnel should be given the new edition free-of-charge, but it will also be available from commercial suppliers. It contains information on shipping documents, identification and safety precautions, the hazard classification system, protective actions, protective clothing, fire and spill control, and isolation/protective action distances. It also contains 172 guides for different classes of chemicals that provide information on potential hazards, public safety, and emergency response.

Emergency Handling of Hazardous Materials in Surface
    Transportation

Bureau of Explosives Publications, P.O. Box 1020, Sewickley, PA 15143; 412-741-1096; 412-741-0609 (Fax)

    Designed for first responders, this manual was updated in 1998. It provides commodity-specific descriptions and response information for all the U.S. Department of Transportation-listed hazardous materials and for many specifically named chemicals transported under a generic DOT description. Materials regulated only by Canada and the International Maritime Organization (IMO) are also included. Over 3600 individual regulated chemicals are covered. Features of this manual include: basic properties of the listed chemicals; recommended methods of dealing with the hazardous materials in the early stages of an emergency; a listing of emergency environmental mitigation procedures; first aid information; suggested chemical compatible protective equipment for some of the commodities.

Emergency Action Guides

Bureau of Explosives Publications, P.O. Box 1020, Sewickley, PA 15143; 412-741-1096; 412-741-0609 (Fax)

    The purpose of the Hazardous Materials Emergency Action Guides is to provide detailed information about the hazardous materials commodities most frequently carried by rail transport. Each of the 134 guides provides 6 pages of basic data for about 98% of the total volume of hazardous materials carried by rail.

NOISH Pocket Guide To Chemical Hazards (NIOSH Publication
    No. 78-210)

U.S. Government Printing Office, Washington, D.C. 20402; 202-512-1803

    This manual is a source of general industrial hygiene and medical surveillance information for 397 individual chemicals or chemical types found in the work environment.

CHRIS (Chemical Hazards Response Information System)

Developed by the U.S. Coast Guard and available through:
U.S. Government Printing Office, Washington, D.C. 20402; 202-512-1803

    Although the CHRIS system is composed of four separate manuals, only Manual 2.2 — Hazardous Chemical Data — is of interest here. Manual 2.2 lists the specific chemical, physical, and biological data for 1000 chemicals.

Quick Selection Guide to Chemical Protective Clothing

Kluwer Academic Publishing, 101 Philip Drive, Norwell, MA 02061; 781-871-6600

Managing Hazardous Materials Incidents (Volume I, Emergency
    Medical Services),

Managing Hazardous Materials Incidents (Volume II, Hospital
    Emergency Departments),

Managing Hazardous Materials Incidents (Volume III, Medical
Management Guidelines for Acute Chemical
Exposure)

Information Center, Agency for Toxic Substances and Disease Registry, 1600 Clifton Road, NE (E57) Atlanta, GA 30333; 888-422-8737 or 404-639-6360

Volume I includes information on emergency medical services response, with sections on hazard recognition, principles of toxicology, personal protection and safety principles, assessment, decon, treatment, transport, planning, and more. Volume II contains specific information for emergency department response to hazardous materials incidents. Volume III is a guide for health care professionals and outlines medical management for acute chemical exposures.

THE INTERNET

In the past, training and response information about nuclear/biological/chemical (NBC) materials was strictly the purview of the military. For years, such information was restricted as it contained technical or operational information for official government agency use only. The Tokyo subway incident, the bombing of the World Trade Center in New York City, the destruction of the federal office building in Oklahoma City, and other terrorist incidents within the United States caused the government to change its methods of handling NBC information making it more available to response personnel.

Hazardous materials response teams, trainers, and consultants should be aware that previously "secret" NBC information dealing with use, epidemiology, sampling, identification, defense, detection, protective equipment, decontamination, treatment, and mass casualty management is now available from government agencies, industry, commercial firms, and private interests on the Internet. Listed below are some of the websites that provide military field manuals, information on chemical and biological substances, and assistance to first responders and Haz Mat teams.

ACDA Homepage

http://www.acda.gov/

Home page of the Arms Control and Disarmament Agency. Provides information on nuclear, biological, and chemical weapons and threats they pose.

American Chemical Society

http://www.acs.org

The producer of the world's largest and most comprehensive databases of chemical information.

American Conference of Governmental Industrial Hygienists (ACGIH)

http://www.acgih.org

Publications, events, leadership, links, classified, etc.

American Industrial Hygiene Association (AIHA)

http://www.aiha.org

Frequently asked questions (FAQ) about industrial hygiene and AIHA, consumer information, calendar, laboratory and scientific information, site index, etc.

American National Standards Institute (ANSI)

http://www.ansi.org

Provides national and international standards-related activities regarding ANSI, standards information, conformity assessment, events, news, reference library, searches, databases, etc.

American Society for Testing and Materials (ASTM)

http://www.ia-usa.org/K0043.htm

A voluntary group in which members devise consensus standards for materials characterization and use, ASTM Standards, etc. ASTM provides a forum for producers, users, ultimate consumers, and others to write standards for materials, products, systems, and services. The society publishes standard test measures, specifications, practices, and guides. This site includes sub-sites for what's new, national, geographic, product, stage, FAQ, glossary, participation, and search.

Biological and Toxic Weapons Verification Program (Federation of American Scientists)

http://www.fas.org/bwc

Briefing papers, negotiations, associated issues, project papers, biological weapons, conventions, links, etc.

Canadian Centre for Occupational Health and Safety (CCOHS)

http://www.ccohs.ca

Products and services, Canadian Centre INFOweb, occupational safety and health answers, education and training, about CCOSH, Internet directory, 100,000 Material Safety Data Sheets, etc.

Centers for Disease Control and Prevention (CDC)

http://www.cdc.gov

About CDC, data and statistics, funding, health topics A-Z, in the news, other sites/links, publications/software/products, training and education, travelers' health, etc.

Chemical and Biological Defense Information Analysis Center

http://www.cbiacarmy.mil/index.html

The CBIAC operated by Battelle Memorial Institute is a Department of Defense (DOD) information analysis center. Established in 1986, the CBIAC serves as the DOD focal point for information related to chemical warfare and chemical and biological defense (CW/CBD) technology. The main interests of CBIAC are chemical and physical properties of CW/CBD materials, chemical identification, combat effectiveness, counter proliferation, counter terrorism, decontamination, domestic preparedness, environmental fate and effects, force protection, medical effects and treatment, toxicology, warning and identification, nuclear/biological/chemical survivability, demilitarization, manufacturing processes for NBC defense systems, etc. The center collects, reviews, analyzes, synthesizes, and otherwise treats information pertaining to chemical and biological warfare. Links to many other related sites.

Chemical and Biological Weapons Nonproliferation Project
(The Henry L. Stimson Center)

http://www.stimson.org/cwc/bwagent.htm

This page is a clearinghouse for information on nonproliferation. It comprises a home page, about us, what's new, search, publications, projects, etc.

Counterproliferation/Chemical and Biological Defense

http://www.acq.osd.mil/cp

Homepage of the Deputy Assistant to the Secretary of Defense for Counterproliferation/Chemical and Biological Defense. Includes summary of activities, the Pentagon's Chemical and Biological Defense Program, and downloadable reports.

Chemical Manufacturers Association (CMA)

http://www.cmahq.com

CMA home page, Responsible Care®, about CMA, what's new, publications, compliance center, workshops/seminars, news and information, issue advocacy, CHEMTREC®, CHEMSTAR®, ChemEcology, and health research.

Defense Advanced Research Projects Agency

http://www.darpa.mil/

This home page of DARPA describes basic and applied research and development projects being performed for the Defense Department. Provides a link to Biological Warfare Defense Program.

Defense Special Weapons Agency

http://www.dna.mil

Provides information on the agency's mission, director, programs, and the Defense Nuclear Weapons School.

Dugway Proving Ground

http://www.dugway.army.mil/

Homepage of the U.S. Dugway Proving Ground, location of many field tests of chem/bio defense equipment. Also contains historical, chemical, and biological warfare information.

Emergency Net News

http://www.emergency.com/ennday.htm

Provides emergency news from around the world.

Hanford Nuclear Site

http://www.handford.gov/

The U.S. Department of Energy's plutonium production complex covering 560 square miles in Washington state is the world's largest environmental cleanup project (includes Hanford homepage page, site information, programs, opportunities, public involvement, resource center, what's new, etc.).

Harvard Sussex Program on CBW Armament

http://fas-www.harvard.edu/~hsp/

Promotes the global elimination of chemical and biological weapons.

Health of Chemical-Biological Defense in the U.S. Military (A White paper by the NBC Industry Group)

http://www.nbcindustrygroup.com/white.htm

Idaho National Engineering and Environmental Laboratories

http://www.inel.gov/

About INEEL, engineering and science, environment, national programs, and opportunities.

The International Tanker Owners' Pollution Federation (ITOPF)

http://www.itopf.com

The ITOPF homepage deals with information on response strategies, historical data on oil spills, fate and effects of oil spills, planning for oil spills, compensation schemes, cleanup techniques, and other factors.

Medical Radiological Defense

http://www.afrri.usuhs.mil/

Provides information on medical radiobiological research and education activities of the Armed Forces Radiobiological Research Institute.

Medical Research and Materiel Command

http://140.139.42.108/home.html

Provides information on medical, chemical, and biological defense research programs and more.

National Emergency Management Association (NEMA)

http://www.nemaweb.org

NEMA is a professional association of state and pacific Caribbean insular state emergency management directors seeking to provide leadership and expertise in comprehensive emergency management, to serve as a vital information and assistance resource for state and territorial directors and their governors, and to forge strategic partnerships to advance continuous improvements in emergency management. NEMA's homepage has sub-sites for committees, conference, membership, state contacts, file library, forum, and feedback.

National Fire Protection Association (NFPA)

http://www.NFPA.org

Information on fire investigations, Oklahoma federal building bombing, hazardous materials and chemical protective clothing standards, training materials and equipment, etc.

National Institutes of Health (NIH)

http://www.nih.gov

Contains a welcome message, news and events, health information, funding opportunities, scientific resources, links, publications, and information for employees.

National Institute for Occupational Safety and Health (NIOSH)

http://www.cdc.gov/niosh

Publications, databases, topic index, health hazard evaluations, training, state activities, extramural programs, what's new, conferences, press releases, Federal Register notices, highlights, about NIOSH, employment and fellowships, and links.

National Library of Medicine

http://www.nlm.nih.gov

The world's largest medical library provides health information through MED-LINE/MEDLINEplus, library services with catalog, databases, publications, training/grants, research programs, announcements, exhibits, hot topics, and general information.

National Response Center

http://www.nrc.uscg.mil/

The National Response Center is the sole federal point of contact for reporting oil and chemical spills by telephoning 1-800-424-8802. On its homepage, you can view NRC information, how to report a spill, legislative requirements, Chemical/Biological Hotline, using NRC data, statistics, organization, links, monthly briefing, management, webmaster, and NRT/EPA/U.S. Coast Guard Internet homepages.

National Safety Council (NSC)

http://www.nsc.org

The NSC distributes CAMEO software and provides technical support for this system which integrates a chemical database, emergency response information, an air dispersion model and local maps with a data management capability. CAMEO is the predominant chemical response software for firefighters and other first responders.

Hazardous Materials Response and Assessment Division, National Ocean Service, National Oceanic and Atmospheric Administration

http://response.restoration.noaa.gov/

This website can provide hazardous materials responders with valuable data and information. NOAA scientists assigned to the Hazardous Materials Response and

Assessment Division (NOAA HAZMAT) respond to dozens of oil spills and other hazardous materials releases each year, help emergency planners prepare for potential accidents, and create software and other products to assist people in responding to hazardous materials incidents. They provide two collections of materials of interest to Haz Mat responders, Aids for Oil Spill Responders and Aids for Chemical Accident Responders. Especially helpful is the Chemical Reactivity Worksheet— a free program you can download and use to find out about the reactivity of chemicals. It contains a database of over 4000 common hazardous chemicals and includes a way for you to virtually "mix" chemicals to find out what dangers could arise from accidental mixing. NOAA provides information on how to acquire any aids available.

North Carolina Emergency Management Division

http://www.dem.dec.state.nc.us
State Emergency Response Commission, regional Haz Mat teams, etc.

Nuclear, Biological, Chemical Industry Group

http://www.erols.com/nbcgroup/
Homepage of the NBC Industry Group, an association of organizations supporting NBC defense, domestic preparedness, and the Chemical Weapons Convention.

Organization for the Prohibition of Chemical Weapons

http://www.opcw.nl/chemhaz/nerve.htm
Information, documents, the Chemical Weapons Convention, fact-finding files, links, a complete report on nerve agents, etc.

Outbreak — Chemical and Biological Agents Internet

http://www.outbreak.org/cgi-unreg/dynaserve.exe/cb/index.html
Chemical and biological agents (dengue, Ebola, hantavirus, plague, smallpox, staphylococcus, yellow fever, etc.), FAQ, active and historical outbreaks, resource center, reading list, etc.

Program Manager for Chemical Demilitarization

http://www-pmcd.apgea.army.mil/
Provides information on the Chemical Stockpile Disposal Program, the Non-Stockpile Chemical Material Program, the Alternative Technology Program, the Chemical Stockpile Emergency Preparedness Program, and the Cooperative Threat Reduction Office.

The PTS-OPCW-PrepCom Homepage

http://www.opcw.nl/

The homepage for the Provisional Technical Secretariat, the Organization for the Prohibition of Chemical Weapons Convention. Provides detailed information about the treaty and more.

Safety Equipment Institute (SEI)

http://www.seinet.gov

The SEI has adopted new testing procedures for chemical and biological terrorists-incident protective clothing used by emergency responders. The new criteria include meeting the requirements for inward leakage using a chemical surrogate that mimics penetration of biological agents into protective ensembles. Ensembles must also meet minimum protective levels against cyanogen chloride, lewisite, sarin, sulfur mustard, and V-agent. This site features the following sub-sites: about SEI, board of directors, certified products list, suppliers, standards and testing agencies, news releases, SEI staff, and related links.

Agency for Toxic Substances and Disease Registry — Haz Mat (ATSDR)

http://atsdr.cdc.gov/mmg.html

"Medical Management Guideline for Acute Chemical Exposures" was developed by ATSDR to aid emergency department physicians and other emergency healthcare personnel to manage acute exposures resulting from chemical incidents, to decontaminate patients, to protect themselves and others from contamination, to communicate with other involved personnel, to transport patients to a medical facility, and to provide competent medical evaluation and treatment to exposed persons. Selected chemicals include benzene, formaldehyde, phosgene, ammonia, hydrogen peroxide, hydrogen sulfide, xylene, arsine, methyl bromide, and others.

U.S. Army Chemical School

http:www.mcclellan.army.mil/

Homepage for Fort McClellan, AL. Provides information on the U.S. Army Chemical School at Fort McClellan, one of the most advanced and sophisticated training centers for chemical and biological defense.

The U.S. Army Medical Department and School

http://www.armymedicine.army.mil/armymed/

Provides extensive information about the Army's medical department. Includes information on doctrine development and the use of medical products for victims of weapons of mass destruction.

U.S. Army Medical Research and Materiel Command
(USAMRMC)

http://www.matmo.org/gobook/gobook.html

This website provides information on both chemical and biological defense products. The chemical pages provide information on pretreatment, antidotes, and skin decontaminants of the following approved products: aerosolized atropine (MANAA), convulsant antidote for nerve agent (CANA), nerve agent antidote kit (Mark I), nerve agent pretreatment pyridostigmine (NAPP), and skin decontamination kit (M291). Information includes countermeasures, status, availability, manufacturer, point of contact, product description, effectiveness, dose and administration, side effects, shipping/handling requirements, other available countermeasures, and contingency protocol. The information can be printed or downloaded from this website.

The biological pages provide information on the following vaccines and biological substances: anthrax, plague, smallpox, and botulinum toxoid vaccine-pentavalent. Information includes countermeasures, status, expected route of exposure, availability, manufacturer, point of contact, product description, effectiveness, dose and administration, side effects, shipping/handling requirements, other countermeasures, and contingency protocol. The information can be printed or downloaded.

U.S. Army Medical Research Institute of Chemical Defense

http://chemdef.apgea.army.mil/

Provides research information to advance the medical prevention and treatment of chemical warfare casualties. It also offers medical chemical defense literature and conducts the training program, "Medical Management of Chemical Casualties," which can be obtained from this site. This agency's mission is to protect the warfighter so that a sustainable force can be projected anywhere at anytime. The Institute develops pretreatments and antidotes and provides instruction to protect and treat casualties.

U.S. Army Medical Research Institute of Infectious Diseases
(USAMRIID)

http://www.usamriid.army.mil

This organization's homepage is the location of much of the science and technology research efforts for medical biological defense. The Institute, located at Fort Detrick in the foothills of western Maryland's Catoctin Mountains near the city of Frederick, conducts research to develop strategies, products, information, procedures, and training programs for medical defense against biological warfare threats. It is the lead medical research laboratory for the U.S. biological defense research program.

U.S. Army Soldier and Biological Chemical Command (SBCCOM)

http://www.sbccom.army.mil/

Contains information about SBCCOM, command information, research and development, strategy, press releases, publications, installation guides, products, programs and technology demonstrations, facilities, business and partnering opportunities, assembled chemical weapons assessment, chemical stockpile program, DOD combat feeding program, domestic preparedness, joint service materials group, soldier enhancement program, community education, consumer research, hazardous materials packaging and transport, environmental clean-up, modeling and simulation, testing and analysis, chemical/biological helpline, chemical/biological hotline, Materials Safety Data Sheets for chemical agents, etc.

U.S. Army Weapons Systems

http://www.dtic.mil/

Provides links to information on nuclear, biological, and chemical detection.

U.S. Central Intelligence Agency (U.S. CIA)

http://www.odci.gov/cia

About the CIA, what's new, publications, FAQ, speeches/testimony/press releases/statements, document release center, related links, etc.

U.S. Department of Defense - Defense Link (U.S. DOD)

http://www.dtic.mil/defenselink/

The primary homepage of the Department of Defense (news articles, daily activities summary, contracts, briefing slides, news archive, speeches, links to DOD organizations, detailed search help, other news sources, news photos, etc.).

U.S. Department of Energy (U.S. DOE)

http://www.doe.gov/

News and information, people and pages, science education, content map, and search activities.

U.S. Department of Energy (U.S. DOE)

http://www.rw.doe.gov

Reports, testimony, current events, Yucca Mountain, websites, etc.

U.S. Department of Energy (U.S. DOE) Technical Information
Service

http://tis.eh.doe.gov

A source for environment, safety, and health information (management tools, comments, welcome tour, ES&H searches, DOE Tech Standard 1098-99 "Radiological Control," etc.).

U.S. Environmental Protection Agency (U.S. EPA)

http://www.epa.gov/

Searches, concerned citizens, researchers and scientists, small business/industry, state/local/tribal, about EPA, projects and programs, other resources, news and events, laws and regulations, databases and software, money matters, and publications.

U.S. Environmental Protection Agency (U.S. EPA) Chemical
Emergency Preparedness and Prevention

http://www.epa.gov/swercep

EPA's CEPPO provides leadership, advocacy, and assistance to prevent and prepare for chemical emergencies, to respond to environmental crises, and to inform the public about chemical hazards in their community. It deals with accident investigation histories, links to other websites, prevention, preparation, response, counterterrorism, laws and regulations, databases and software, search, emergency planning, and Y2K.

U.S. Environmental Protection Agency (U.S. EPA) Emergency
Response Notification System

http://www.epa.gov/ERNS

Information maintained at the National Response Center on oil and hazardous substance discharge and releases. Its sites include a table of contents, top ten spills, spills in the news, ERN quick facts, search ENS, fact sheets, etc.

U.S. Environmental Protection Agency Hazardous Substances
Research Centers (HSRC)

http://www.hsrc.org

The Hazardous Substance Research Centers conduct an active program of basic and applied research, technology transfer, and training. Activities are conducted regionally by five multi-university centers that focus on different aspects of hazardous substance management. The HSRC homepage has sections for national, centers, contact information, publications, research, and web links.

U.S. Federal Bureau of Investigation (U.S. FBI)

http://www.fbi.gov/

This Internet site for the FBI includes the top ten most wanted fugitives, crime alerts, seeking information, kidnapping/missing persons, parental kidnappings, unknown suspects, press room, major investigations, library, uniform crime reports, programs and organizational initiatives, and a kids' and youth education page.

U.S. Federal Emergency Management Agency (U.S. FEMA)

http://www.fema.gov

Mitigation, map service center, "know your risks," nationwide urban search and rescue (USAR) task forces, training, disaster recovery, etc.

U.S. Food and Drug Administration (U.S. FDA)

http://www.fda.gov

The goal of the FDA is to foster the national capabilities needed to respond to potential chemical and biological threats from bioterrorism, including development of new vaccines and drugs, safeguards for the food supply, and research for diagnostic tools and treatment of disease outbreaks. At the present time, the FDA is requesting a total of $13.4 million for bioterrorism. The FDA site contains information such as dockets, press releases, enforcement, reports, "FDA Consumer Magazine," FDA backgrounder, and the Center for Biological Evaluation and Research.

U.S. Department of Health and Human Services (U.S. DHHS)

http://www.os.dhhs.gov/

There are 13 different agencies under the control of the U.S. Department of Health and Human Services including the Agency for Toxic Substances and Disease Registry (ATSDR), the Centers for Disease Control and Prevention (CDC), the Food and Drug Administration (FDA), and the National Institutes of Health (NIH) profiled in this section. U.S. DHHS has the following sections on its homepage: about HHS, healthfinder and human services information, news and public affairs, research/policy/administration, what's new, search, gateways, and HHS agencies on the Internet.

U.S. Department of Justice (U.S. DOJ)

http://www.usdoj.gov/

The U.S. Department of Justice homepage contains the following items: what's new, organizations and information, press room, Freedom of Information Act, publications and documents, business with DOJ, community support and grants, fugitives and missing persons, and justice for kids and youth.

U.S. National Response Team (U.S. NRT)

http://www.nrt.org

Containing preparedness and response links, the National Response Center is the sole federal point for reporting the release of oil or hazardous chemicals, including chemical and biological agents, into U.S. waterways and environment.

U.S. Occupational Safety and Health Administration (U.S. OSHA)

http://www.osha.gov/

OSHA is an agency under the U.S. Department of Transportation with safety and health regulatory and enforcement over most industry, business, and states within the United States. OSHA's homepage deals with programs, state plans, news releases, speeches, testimony, publications, compliance links, Federal Register, standards, interpretations, directives, manuals, statistics, technical links, training, software, and other subjects.

U.S. Department of Transportation (U.S. DOT)

http://www.dot.gov/

The U.S. DOT is responsible for the transportation of hazardous materials through some of its 13 agencies such as the Federal Aviation Administration, the Federal Highway Administration, the Federal Railroad Administration, the Maritime Administration, the Research and Special Programs Administration, and U.S. Coast Guard. The U.S. DOT homepage deals with information about DOT, news and happenings, dockets/rules/references, doing business with DOT, programs and initiatives, technology matters, Y2K, etc.

TELEPHONE HOTLINES

Chemical Emergency Preparedness Hotline: (CERCLA; SARA Title III)	1-800-535-0202
Chemical Transportation Emergency Center: (CHEMTREC 24 hour, emergency only)	1-800-424-9300
Domestic Preparedness CB Helpline:	1-800-368-6498
Emergency Planning & Community R.T.K. Act (EPCRA):	1-800-424-9346
Environmental Justice Hotline:	1-800-962-6215
Hazardous Waste Ombudsman:	1-800-262-7937

National Institute of Occupational Safety & Health: 1-800-356-4674
(NIOSH Hotline)

National Pesticide Telecommunication Network: 1-800-858-7378

National Response Center CB Hotline: 1-800-424-8802
(24-hour hotline for use in reporting oil spills, chemical
spills, pipeline spills, transportation incidents,
and terrorist actions or releases)

Occupational Health and Safety Administration Hotline: 1-800-321-6742
(OSHA 24-hour hotline)

U.S. Department of Transportation Hotline: 1-202-366-4488
(Questions on regulation)

U.S. Environmental Protection Agency RCRA Hotline: 1-800-424-9346
(Superfund, Hazardous Waste)

U.S. Environmental Protection Agency Small Business
Hotline: 1-800-368-5888

Toxic Substances Control Act Assistance Information
Service: 1-202-554-1404
(TSCA)

Appendix I

Material Safety Data Sheet for Mustard Gas (HD)

SECTION 1   CHEMICAL PRODUCT AND COMPANY IDENTIFICATION

MDL INFORMATION SYSTEMS, INC.   EMERGENCY TELEPHONE NUMBER:
1281 Murfreesboro Road, Suite 300   1-800-424-9300 (NORTH AMERICA)
Nashville, TN 37217-2423   1-703-527-3887 (INTERNATIONAL)
1-615-366-2000

SUBSTANCE: MUSTARD GAS

TRADE NAMES/SYNONYMS:
BIS(2-CHLOROETHYL) SULFIDE; 2,2'-DICHLORODIETHYL SULFIDE; DISTILLED MUSTARD;
AGENT HD; SULFUR MUSTARD; YPERITE; KAMSTOFF 'LOST'; S-LOST; DICHLORODIETHYL
SULFIDE; DICHLOROETHYL SULFIDE; 1,1'-THIOBIS(2-CHLOROETHANE);
BIS(BETA-CHLOROETHYL) SULFIDE; IPRIT; BETA, BETA'-DICHLORODIETHYL SULFIDE;
SULFUR MUSTARD GAS; S MUSTARD; C4H8CL2S; OHS15220; RTECS WQ0900000

CHEMICAL FAMILY: organic sulfur compounds

CREATION DATE: Jul 15 1986
REVISION DATE: Jun 01 1999

SECTION 2   COMPOSITION, INFORMATION ON INGREDIENTS

COMPONENT: MUSTARD GAS
CAS NUMBER: 505-60-2
EC NUMBER: Not assigned.
PERCENTAGE: 100.0

SECTION 3   HAZARDS IDENTIFICATION

NFPA RATINGS (SCALE 0-4): HEALTH=4 FIRE=1 REACTIVITY=0

EC CLASSIFICATION (CALCULATED):
  T+ Very Toxic
  T Toxic
  C Corrosive

  R 26/27-34-49

EMERGENCY OVERVIEW:
COLOR: yellow
PHYSICAL FORM: liquid, crystals
ODOR: pleasant odor
MAJOR HEALTH HAZARDS: potentially fatal if inhaled or on contact with the
  skin, respiratory tract burns, skin burns, eye burns, mucous membrane burns,
  cancer hazard (in humans)

POTENTIAL HEALTH EFFECTS:
INHALATION:
  SHORT TERM EXPOSURE: burns, vomiting, loss of voice, difficulty breathing,
  death

LONG TERM EXPOSURE: cancer
SKIN CONTACT:
  SHORT TERM EXPOSURE: burns, blisters, nausea, vomiting, death
  LONG TERM EXPOSURE: same as effects reported in short term exposure
EYE CONTACT:
  SHORT TERM EXPOSURE: burns, tearing
  LONG TERM EXPOSURE: same as effects reported in short term exposure
INGESTION:
  SHORT TERM EXPOSURE: burns, nausea, vomiting
  LONG TERM EXPOSURE: same as effects reported in short term exposure

CARCINOGEN STATUS:
OSHA: N
NTP: Y
IARC: Y

SECTION 4   FIRST AID MEASURES

INHALATION: When safe to enter area, remove from exposure. Use a bag valve
  mask or similar device to perform artificial respiration (rescue breathing)
  if needed. Keep warm and at rest. Get medical attention immediately.

SKIN CONTACT: Remove contaminated clothing, jewelry, and shoes immediately.
  Wash with soap or mild detergent and large amounts of water until no
  evidence of chemical remains (at least 15-20 minutes). Get medical attention
  immediately.

EYE CONTACT: Wash eyes immediately with large amounts of water or normal
  saline, occasionally lifting upper and lower lids, until no evidence of
  chemical remains. Get medical attention immediately.

INGESTION: Get medical attention immediately.

NOTE TO PHYSICIAN: For inhalation, consider oxygen. For skin contact, consider
  sodium bicarbonate solution. For eye contact, consider sodium bicarbonate
  solution or pontocaine. For ingestion, consider gastric lavage. Avoid
  stimulants.

SECTION 5   FIRE FIGHTING MEASURES

FIRE AND EXPLOSION HAZARDS: Slight fire hazard.

EXTINGUISHING MEDIA: regular dry chemical, carbon dioxide, water, regular foam

Large fires: Use regular foam or flood with fine water spray.

FIRE FIGHTING: Move container from fire area if it can be done without risk.
  Do not scatter spilled material with high-pressure water streams. Dike for
  later disposal. Use extinguishing agents appropriate for surrounding fire.
  Avoid inhalation of material or combustion by-products. Stay upwind and keep
  out of low areas.

FLASH POINT: 219 F (104 C)
FLAMMABILITY CLASS (OSHA): IIIB

SECTION 6   ACCIDENTAL RELEASE MEASURES

WATER RELEASE:
Subject to California Safe Drinking Water and Toxic Enforcement Act of 1986
(Proposition 65). Keep out of water supplies and sewers.

OCCUPATIONAL RELEASE:
Do not touch spilled material. Stop leak if possible without personal risk.
Small spills: Absorb with sand or other non-combustible material. Collect
spilled material in appropriate container for disposal. Small dry spills: Move
containers away from spill to a safe area. Large spills: Dike for later
disposal. Keep unnecessary people away, isolate hazard area and deny entry.
Notify Local Emergency Planning Committee and State Emergency Response
Commission for release greater than or equal to RQ (U.S. SARA Section 304). If
release occurs in the U.S. and is reportable under CERCLA Section 103, notify
the National Response Center at (800)424-8802 (USA) or (202)426-2675 (USA).

SECTION 7   HANDLING AND STORAGE

Store and handle in accordance with all current regulations and standards.
Notify State Emergency Response Commission for storage or use at amounts
greater than or equal to the TPQ (U.S. EPA SARA Section 302). SARA Section 303
requires facilities storing a material with a TPQ to participate in local
emergency response planning (U.S. EPA 40 CFR 355.30).

SECTION 8   EXPOSURE CONTROLS, PERSONAL PROTECTION

EXPOSURE LIMITS:
MUSTARD GAS:
  No occupational exposure limits established.

VENTILATION: Provide local exhaust or process enclosure ventilation system.
  Ensure compliance with applicable exposure limits.

EYE PROTECTION: Wear splash resistant safety goggles with a faceshield.
  Provide an emergency eye wash fountain and quick drench shower in the
  immediate work area.

CLOTHING: Wear appropriate chemical resistant clothing.

GLOVES: Wear appropriate chemical resistant gloves.

RESPIRATOR: Under conditions of frequent use or heavy exposure, respiratory
  protection may be needed. Respiratory protection is ranked in order from
  minimum to maximum. Consider warning properties before use.
  Any supplied-air respirator with a full facepiece that is operated in a
    pressure-demand or other positive-pressure mode.
  Any self-contained breathing apparatus that has a full facepiece and is
    operated in a pressure-demand or other positive-pressure mode.
For Unknown Concentrations or Immediately Dangerous to Life or Health -

Any supplied-air respirator with full facepiece and operated in a
pressure-demand or other positive-pressure mode in combination with a
separate escape supply.
Any self-contained breathing apparatus with a full facepiece.

SECTION 9   PHYSICAL AND CHEMICAL PROPERTIES

PHYSICAL STATE: liquid
COLOR: yellow
TEXTURE: oily
PHYSICAL FORM: liquid, crystals
ODOR: pleasant odor
MOLECULAR WEIGHT: 159.08
MOLECULAR FORMULA: C4-H8-CL2-S
BOILING POINT: 423 F (217 C)
FREEZING POINT: 57 F (14 C)
VAPOR PRESSURE: 0.09 mmHg @ 30 C
VAPOR DENSITY: Not available
SPECIFIC GRAVITY (water=1): 1.2741
WATER SOLUBILITY: very slightly soluble
PH: Not available
VOLATILITY: Not available
ODOR THRESHOLD: Not available
EVAPORATION RATE: Not available
COEFFICIENT OF WATER/OIL DISTRIBUTION: Not available
SOLVENT SOLUBILITY:
  Soluble: fat solvents, organic solvents

SECTION 10   STABILITY AND REACTIVITY

REACTIVITY: Stable at normal temperatures and pressure.

CONDITIONS TO AVOID: Avoid heat, flames, sparks and other sources of ignition.
  Dangerous gases may accumulate in confined spaces. May ignite or explode on
  contact with combustible materials.

INCOMPATIBILITIES: oxidizing materials

MUSTARD GAS:
  STRONG OXIDIZERS: Fire and explosion hazard.

HAZARDOUS DECOMPOSITION:
  Thermal decomposition products: oxides of carbon, sulfur, chlorine, acid
    halides

POLYMERIZATION: Will not polymerize.

SECTION 11   TOXICOLOGICAL INFORMATION

MUSTARD GAS:
IRRITATION DATA:

2000 mg/m3 skin-man severe; 65 ug skin-human; 100 mg/m3 eyes-man moderate;
200 mg/m3 eyes-rabbit mild; 200 mg/m3 eyes-rabbit mild
TOXICITY DATA:
  23 ppm/10 minute(s) inhalation-human LCLo; 64 mg/kg skin-human LDLo; 100
  mg/m3/10 minute(s) inhalation-rat LC50; 5 mg/kg skin-rat LD50; 1500 ug/kg
  subcutaneous-rat LD50; 700 ug/kg intravenous-rat LD50; 120 mg/m3/10
  minute(s) inhalation-mouse LC50; 92 mg/kg skin-mouse LD50; 20 mg/kg
  subcutaneous-mouse LD50; 8600 ug/kg intravenous-mouse LD50; 70 mg/m3/10
  minute(s) inhalation-dog LC50; 20 mg/kg skin-dog LD50; 5 mg/kg
  subcutaneous-dog LDLo; 200 ug/kg intravenous-dog LD50; 80 mg/m3/10 minute(s)
  inhalation-monkey LC50; 70 mg/m3/10 minute(s) inhalation-cat LC50; 280
  mg/m3/10 minute(s) inhalation-rabbit LC50; 40 mg/kg skin-rabbit LD50; 20
  mg/kg subcutaneous-rabbit LD50; 1100 ug/kg intravenous-rabbit LD50; 200
  mg/m3/10 minute(s) inhalation-guinea pig LC50; 20 mg/kg skin-guinea pig
  LD50; 20 mg/kg subcutaneous-guinea pig LD50; 190 mg/m3/10 minute(s)
  inhalation-domestic animal LC50; 50 mg/kg skin-domestic animal LD50; 40
  mg/kg subcutaneous-domestic animal LD50; 19500 ug/kg/13 week(s) intermittent
  oral-rat TDLo; 100 ug/m3/6 hour(s)-52 week(s) intermittent inhalation-dog
  TCLo
CARCINOGEN STATUS: NTP: Known Human Carcinogen; IARC: Human Sufficient
  Evidence, Animal Limited Evidence, Group 1; EC: Category 1; TRGS 905: K 1
  Several studies have shown an increased incidence of respiratory tract
  cancer in men exposed to mustard gas. Inhalation or intravenous
  administration to mice resulted in an increased incidence of lung tumors.
LOCAL EFFECTS:
  Corrosive: inhalation, skin, eye, ingestion
ACUTE TOXICITY LEVEL:
  Highly Toxic: inhalation, dermal absorption
TUMORIGENIC DATA:
  100 ug/m3 inhalation-rat TCLo/1 year(s) intermittent; 1250 mg/m3
  inhalation-mouse TCLo/15 minute(s) continuous; 6 mg/kg subcutaneous-mouse
  TDLo/6 week(s) intermittent; 600 mg/kg intravenous-mouse TDLo/6 day(s)
  intermittent
MUTAGENIC DATA:
  mutation in microorganisms - Salmonella typhimurium 10 ug/plate (+S9);
  mutation in microorganisms - Escherichia coli 8 mg/L (+S9); DNA repair -
  Escherichia coli 5 ug/well; specific locus test - Drosophila melanogaster
  parenteral 300 umol/L; cytogenetic analysis - Drosophila melanogaster
  inhalation 10 pph 15 minute(s)-intermittent; sex chromosome loss and non
  disjunction - Drosophila melanogaster inhalation 10 pph 15
  minute(s)-intermittent; dominant lethal test - Drosophila melanogaster
  inhalation 10 pph 15 minute(s)-intermittent; heritable translocation test -
  Drosophila melanogaster inhalation 10 pph 15 minute(s)-intermittent;
  mutation in microorganisms - Neurospora crassa 200 umol/L (-S9) 30
  minute(s); DNA damage - Saccharomyes cerevisae 500 umol/L; DNA damage -
  human other cell types 1 umol/L; DNA damage - human HeLa cell 2 mg/L;
  unscheduled DNA synthesis - human HeLa cell 200 mg/L; DNA inhibition - human
  other cell types 500 nmol/L; DNA inhibition - human HeLa cell 75 mg/L; other
  mutation test systems - human HeLa cell 75 mg/L; cytogenetic analysis -
  human leukocyte 4 mg/L; sister chromatid exchange - human leukocyte 8 mg/L;
  cytogenetic analysis - rat lymphocyte 20 ng/L; dominant lethal test - rat
  inhalation 100 ug/m3 12 week(s)-continuous; micronucleus test - mouse
  intraperitoneal 8 mg/kg; micronucleus test - mouse oral 10 mg/kg; DNA adduct
  - mouse Ascites tumor 2 mg/kg; DNA damage - mouse lymphocyte 1 mg/L;
  cytogenetic analysis - mouse leukocyte 20 ug/L; mutation in mammalian
  somatic cells - mouse leukocyte 20 ug/L; host-mediated assay - mouse
  leukocyte 100 mg/kg; host-mediated assay - mouse Escherichia coli 8 mg/kg;

unscheduled DNA synthesis - mammal lymphocyte 750 nmol/L; DNA damage -
chicken leukocyte 30 mmol/L
REPRODUCTIVE EFFECTS DATA:
    20 mg/kg oral-rat TDLo 6-15 day(s) pregnant female continuous; 25 mg/kg
    oral-rat TDLo 10 week(s) male; 68800 ug/kg oral-rat TDLo multigenerations;
    11200 ug/kg oral-rabbit TDLo 6-18 day(s) pregnant female continuous

HEALTH EFFECTS:
INHALATION:
    ACUTE EXPOSURE:
        MUSTARD GAS: The local effects are usually delayed for several hours and
        may include rhinitis, laryngitis, cough, dyspnea and severe respiratory
        impairment. Necrosis of the respiratory tract and pulmonary lesions are
        possible. If sufficient amounts are absorbed, systemic effects may include
        malaise, vomiting and fever. With amounts approaching the lethal dose,
        injury to the bone marrow, lymph nodes and spleen may occur, represented
        by leukopenia. Secondary infections are common.

    CHRONIC EXPOSURE:
        MUSTARD GAS: Humans chronically exposed to mustard gas have been shown to
        suffer from an increase in chronic bronchitis and cancers of the
        respiratory tract. Mice chronically exposed were shown to have a
        significant increase of lung tumors. The rate of detoxification is very
        low, and small, repeated exposures may be cumulative in their effects.

SKIN CONTACT:
    ACUTE EXPOSURE:
        MUSTARD GAS: The local effects are usually delayed for several hours and
        may include erythema possibly followed by ulceration and blistering.
        Absorption through intact skin may cause systemic effects such as malaise,
        nausea and vomiting. Leukopenia with involution of the lymph nodes may
        occur and secondary infection is common. Data indicate that very small
        amounts may be absorbed through the skin, resulting in death.

    CHRONIC EXPOSURE:
        MUSTARD GAS: The rate of detoxification is low, and small, repeated
        amounts may be cumulative in their effects.

EYE CONTACT:
    ACUTE EXPOSURE:
        MUSTARD GAS: Exposure to vapor has induced uncomfortable and temporarily
        disabling reactions. In mild cases the vapor burns are usually superficial
        and the injured epithelium is replaced by normal cells. More severe
        injuries produced by contact with the liquid has caused the corneas to
        cloud and in some cases they become opaque. A porcelain white area in the
        episcleral tissues may develop and is subject to repeated ulceration and
        gradual deterioration of the cornea over many years. These recurrences of
        ulceration are accompanied by photophobia, pain and lacrimation.

    CHRONIC EXPOSURE:
        MUSTARD GAS: Depending on concentration and duration of exposure, repeated
        or prolonged contact may result in symptoms as those in acute exposure.

INGESTION:
    ACUTE EXPOSURE:
        MUSTARD GAS: May cause nausea and vomiting and delayed ulceration and
        perforation of the gastrointestinal tract.

CHRONIC EXPOSURE:
  MUSTARD GAS: A 42 week, two generation study in which rats were gavaged
  with o.03, 0.1 or 0.4 mg/kg for 13 weeks resulted in reduced growth in
  both generations at the 0.4 mg/kg level and a dose-related lesion of the
  squamous epithelial mucosa of the forestomach; benign neoplasms of the
  forestomach were found in about 10% of the intermediate and high dose
  groups.

SECTION 12   ECOLOGICAL INFORMATION

Not available

SECTION 13   DISPOSAL CONSIDERATIONS

Dispose in accordance with all applicable regulations.

SECTION 14   TRANSPORT INFORMATION

U.S. DOT 49 CFR 172.101 SHIPPING NAME-UN NUMBER:
Forbidden materials

U.S. DOT 49 CFR 172.101 HAZARD CLASS OR DIVISION:
Forbidden

LAND TRANSPORT ADR/RID: No classification assigned.

AIR TRANSPORT IATA/ICAO: No classification assigned.

MARITIME TRANSPORT IMDG: No classification assigned.

SECTION 15   REGULATORY INFORMATION

U.S. REGULATIONS:
  TSCA INVENTORY STATUS: Y

  TSCA 12(b) EXPORT NOTIFICATION: Not listed.
  CERCLA SECTION 103 (40CFR302.4): N
  SARA SECTION 302 (40CFR355.30): Y
    MUSTARD GAS: 500 LBS TPQ
  SARA SECTION 304 (40CFR355.40): Y
    MUSTARD GAS: 500 LBS RQ
  SARA SECTION 313 (40CFR372.65): Y
    MUSTARD GAS
  SARA HAZARD CATEGORIES, SARA SECTIONS 311/312 (40CFR370.21):
    ACUTE: Y
    CHRONIC: Y
    FIRE: N
    REACTIVE: N

```
  SUDDEN RELEASE: N
  OSHA PROCESS SAFETY (29CFR1910.119): N
STATE REGULATIONS:
  California Proposition 65: Y
   Known to the state of California to cause the following:
      MUSTARD GAS
         Cancer (Feb 27, 1987)
EUROPEAN REGULATIONS:
  EC NUMBER: Not assigned.

  EC RISK AND SAFETY PHRASES:
    R 26/27        Very toxic by inhalation and in contact with skin.
    R 34           Causes burns.
    R 49           May cause cancer by inhalation.

    S 1            Keep locked-up.
    S 2            Keep out of reach of children.
    S 4            Keep away from living quarters.
    S 13           Keep away from food, drink and animal feeding stuffs.
    S 20           When using do not eat or drink.
    S 24           Avoid contact with skin.
    S 25           Avoid contact with eyes.
    S 26           In case of contact with eyes, rinse immediately with plenty
                   of water and seek medical advice.
    S 35           This material and its container must be disposed of in a
                   safe way.
    S 36           Wear suitable protective clothing.
    S 39           Wear eye/face protection.
    S 45           In case of accident or if you feel unwell, seek medical
                   advice immediately (show the label where possible).
    S 46           If swallowed, seek medical advice immediately and show this
                   container or label.
-----------------------------------------------------------------------------
SECTION 16    OTHER INFORMATION
-----------------------------------------------------------------------------

COPYRIGHT 1984-1999 MDL INFORMATION SYSTEMS, INC.  ALL RIGHTS RESERVED.
```

Source: Used with permission from MDL Information Systems, Inc., Nashville, TN.

Appendix II

Selected Laws Related to Terrorism

Selected Laws Related to Terrorism

Trade and Foreign Assistance Legislation

Foreign Assistance Act of 1961, as Amended	Prohibited the provision of U. S. assistance to foreign countries whose governments support terrorism (22 U.S.C. 2371, as amended).
Arms Export Control Act, as Amended (Formerly the Foreign Military Sales Act of 1968)	Prohibited various transactions with foreign countries that support acts of terrorism, such as exports of any munition items or the provision of credits, guarantees, or other financial assistance to those countries (22 U.S.C. 2780, as amended).
International Financial Institutions Act (1977)	Directed that the U.S. government, while participating in enumerated international financial institutions, shall seek to channel assistance to countries other than those whose governments provide refuge to individuals that commit acts of international terrorism by hijacking aircraft (Title VII, P.L. 95-118).
1978 Amendments to the Bretton Woods Agreements Act	Required the U.S. Executive Director to the International Monetary Fund to oppose the extension of any financial or technical assistance to any country that supports terrorist activities (P.L. 95-435).
Export Administration Act of 1979	Listed compatibility with U.S. efforts to counter international terrorism as a factor in determining whether certain controls should be imposed for a particular export license on foreign policy grounds (P.L. 96-72, sec.6).
International Security and Development Cooperation Act of 1985	Authorized the President to ban the import into the United States of any good or service from any country that supports terrorism or terrorist organizations (Part A of Title V, P.L. 99-83).
Iraq Sanctions Act of 1990	Classified Iraq as a terrorism-supporting foreign country and imposed U.S. export controls and foreign assistance sanctions (P.L. 101-513, sec. 586).
Iran-Iraq Arms Non-Proliferation Act of 1992	Suspended foreign assistance military and dual-use sales to any foreign country whose government knowingly and materially contributes to Iran's

	or Iraq's efforts to acquire advanced conventional weapons (Title XVI, P.L. 102-484).
1996 Amendment to Export-Import Bank Act	Restricted the President from granting special debt relief regarding any Export-Import Bank loan or guarantee to any country whose government has repeatedly supported acts of international terrorism (P.L. 103-87, sec. 570).
Middle East Peace Facilitation Act of 1994	Allowed the President to suspend for 6-month periods, until July 1995, any previously passed restrictions on U. S. assistance to the Palestinian Liberation Organization (Part E of title V, P.L.103-236).
Spoils of War Act of 1994	Prohibited the transfer of spoils of war in the possession of the United States to any country that the Secretary of State has determined to be a nation whose government has repeatedly supported acts of international terrorism (Part B of title V, P.L. 103-236).
Foreign Operations, Export Financing, and Related Programs Appropriations Act for Fiscal Year 1995	Prohibited the direct funding of any assistance or reparations to certain terrorist countries such as Cuba, Iraq, Libya, Iran (Title V, P.L. 103-306).
1996 Amendments to the Foreign Assistance Act of 1961 and the Arms Export Control Act	Removed certain restrictions on the manner in which antiterrorism training assistance could be provided (Chapter 3 of title I, P.L. 104-164).
1996 Amendments to the Trade Act of 1974	Required the President to withhold General System of Preferences designation as a beneficiary developing country entitled to duty free treatment, if the country is on the Export Administration Act's terrorist list, or if the country has assisted any individual or group that has committed an act of international terrorism (P.L. 104-295, sec. 35).

Appendix II
Selected Laws Related to Terrorism

| Iran and Libya Sanctions Act of 1996 | Required the President to impose sanctions against companies that make investments of more than $40 million in developing Iran's or Libya's oil resources (P.L. 104-172, sec. 5). |

State Department and Related Foreign Relations Legislation

Act for the Protection of Foreign Officials and Official Guests of the United States (1972)	Established as a federal crime the murder or manslaughter of foreign officials and official foreign guests (Title I, P.L. 92-539).
Act for the Prevention and Punishment of Crimes Against Internationally Protected Persons (1976)	Provided federal jurisdiction over assaults upon, threats against, murders of, or kidnapping of U.S. diplomats overseas (P.L. 94-467).
Act for the Prevention and Punishment of the Crime of Hostage-Taking (1984)	Imposed punishment for taking a hostage, no matter where, if either the terrorist or the hostage is a U.S. citizen, or if the purpose is to influence the U.S. government (Part A of ch. XX, P.L. 98-473).
1984 Act to Combat International Terrorism	Offered cash awards to anyone who furnishes information leading to the arrest or conviction of a terrorist in any country, if the terrorist's target was a U.S. person or U.S. property (Title I, P.L. 98-533).
Omnibus Diplomatic Security and Antiterrorism Act of 1986	Provided extraterritorial criminal jurisdiction for acts of international terrorism against U.S. nationals (Title XII, P.L. 99-399).
Antiterrorism Act of 1987	Prohibited U.S. citizens from receiving anything of value except information material from the Palestine Liberation Organization, which has been identified as a terrorist organization (Title X, P.L. 100-204).
PLO Commitments Compliance Act of 1989	Reaffirmed a U.S. policy that any dialogue with the Palestinian Liberation Organization be contingent upon certain commitments, including the

302

Emergency Response to Chemical and Biological Agents

Appendix II
Selected Laws Related to Terrorism

organization's abstention from and renunciation of all acts of terrorism (Title VIII, P.L. 101-246).

Immigration Act of 1990	Required the exclusion or deportation from the U. S. any alien who the U. S. government knows or has reason to believe has engaged in terrorist activities (P.L. 101-649, sec. 601 and 602).
Federal Courts Administration Act of 1992	Provided civil remedies for U. S. nationals or their survivors for personal or property injury due to an international terrorism act; granted U. S. district courts jurisdiction to hear cases (Title X, P.L. 102-572).
Antiterrorism and Effective Death Penalty Act of 1996	Established procedures for removing alien terrorists from the United States; prohibited fundraising by terrorists; prohibited financial transactions with terrorists (Title IV, P.L.104-132).

Aviation Security

Federal Aviation Act of 1958	Authorized the Federal Aviation Administration (FAA) Administrator to prescribe such rules and regulations as necessary to provide adequately for national security and safety in air transportation; prohibited the air transportation of explosives and other dangerous articles in violation of a FAA rule or regulation (P.L. 85-726, sec. 601 and 902).
Anti-Hijacking Act of 1974	Established a general prohibition against aircraft piracy outside U.S. special aircraft jurisdiction; allowed the President to suspend air transportation between the United States and any foreign state that supports terrorism (Title I, P.L. 93-366).
Air Transportation Security Act of 1974	Authorized screening of passengers and their baggage for weapons (Title II, P.L. 93-366).
Aircraft Sabotage Act of 1984	Prohibited anyone from setting fire to, damaging, or destroying any U.S. aircraft (Part B of ch. XX, P.L. 98-473).

Foreign Airport Security Act of 1985	Required FAA to assess foreign airport security procedures and the security procedures used by foreign air carriers serving the United States. (Part B of title V, P.L. 99-83).
Aviation Security Improvements Act of 1990	Implemented many recommendations of the President's Commission on Aviation Security and Terrorism to improve aviation security and consular affairs assistance (Titles I and II of P.L. 101-604).
Federal Aviation Reauthorization Act of 1996	Mandated the performance of an employment investigation, including a criminal history record check, of airport security personnel (Title III, P.L. 104-264).

Other Legislation

International Security and Development Cooperation Act of 1981	Required the President to submit a report to Congress describing all legislation and all administrative remedies that can be employed to prevent the participation of U.S. citizens in activities supporting international terrorism (P.L. 97-113, sec. 719).
Convention on the Physical Protection of Nuclear Material Implementation Act of 1982	Prohibited a person from engaging in the unauthorized or improper use of nuclear materials (P.L. 97-351, sec. 2).
National Defense Authorization Act for Fiscal Year 1987	Required Department of Defense (DOD) officials to ensure that all credible, time-sensitive intelligence received concerning potential terrorist threats be promptly reported to DOD headquarters (P.L. 99-661, sec. 1353).
Undetectable Firearms Act of 1988	Prohibited the import, manufacture, sale, and shipment for civilian use of handguns that are made of largely nonmetallic substances (P.L. 100-649, sec. 3).
Biological Weapons Antiterrorism Act of 1989	Prohibited a person from knowingly producing or possessing any biological agent or toxin for use as a weapon or knowingly assisting a foreign state or organization to do so (P.L. 101-298, sec. 3).

National Defense Authorization Act for Fiscal Year 1994	Required certain defense contractors to report to DOD each commercial transaction with a terrorist country; expressed the sense of Congress that FEMA should strengthen interagency emergency planning for potential terrorists' use of chemical or biological agents or weapons (P.L. 103-160, sec. 843 and 1704).
Violent Crime Control and Law Enforcement Act of 1994	Made it a federal crime to intentionally destroy or damage a ship or its cargo or to perform an act of violence against a person on board a ship (P.L. 103-322, sec. 60019).
Antiterrorism and Effective Death Penalty Act of 1996	Expanded and strengthened criminal prohibitions and penalties pertaining to terrorism; established restrictions on the transfer and use of nuclear, biological and chemical weapons, as well as plastic explosives (Titles II, III, V, and VII, P.L. 104-132).
National Defense Authorization Act for Fiscal Year 1997	Established the Domestic Preparedness Program to strengthen U.S. capabilities to prevent and respond to terrorist activities involving WMD; authorized DOD to take the lead role and provide necessary training and other assistance to federal, state, and local officials (Title XIV of P.L. 104-201, commonly known as Nunn-Lugar-Domenici).
Omnibus Consolidated Appropriations Act (1997)	Provided substantial funding for multiple federal agencies to combat terrorism, in response to the President's request (see individual agency appropriations acts within P.L. 104-208).
Emergency Supplemental Appropriations for Additional Disaster Assistance, for Antiterrorism Initiatives, for Assistance in the Recovery From the Tragedy That Occurred at Oklahoma City, and Rescissions Act (1995)	In response to the tragedy of the Oklahoma City federal building bombing, provided substantial emergency funding for various federal agencies to combat terrorism (Title III, P.L. 104-19).

Source: From "Selected Laws Related to Terrorism," COMBATING TERRORISM — Federal Agencies' Efforts to Implement National Policy and Strategy, United States General Accounting Office, Report to Congressional Requesters (GAO/NSIAD), September 1997.

Appendix III

Sample Jurisdiction Emergency Operations Plan

FUNCTIONAL ANNEX: TERRORISM

MISSION

To develop a comprehensive, coordinated, and integrated response capability, involving all levels of government, to effectively assess the threat of and vulnerability to terrorism acts within the community, as well as prevent, mitigate against, respond to, and recover from an actual terrorist incident that may occur.

SITUATION

All communities are vulnerable to acts of terrorism

Intelligence gathering and tactical capabilities vary between jurisdictions as well as levels of government.

The fact that an emergency or disaster situation was a result of a terrorist act will not always be evident during the initial emergency response phase, and may not be determined until days, weeks, or months after the event has occurred.

Terrorist events will occur with little or no warning and involve one or more of a variety of tactics to include but not be limited to bombing, chemical, biological, and nuclear incidents, hostage taking, etc.

The local and state emergency response organization must develop the tactical capability to quickly recognize and respond to the range of potential tactics that could be employed locally as well as regionally.

The effects of a chemical, biological, or nuclear terrorist act will likely overwhelm local, regional, and state capabilities.

ORGANIZATION

The local emergency services organization is based on a broad, functionally oriented, multi-hazard approach to disasters that can be quickly and effectively integrated with all levels of government. In the initial stages of response to and recovery from a terrorist event, the existing local emergency services organization will provide the framework under which local resources will be deployed and coordinated.

Upon arrival of regional, state and federal resources, command and control of response and recovery operations will be structured under a unified command organization that will include but not be limited to the following: the local Director/Coordinator of Emergency Services, the State Coordinating Officer (SCO), the Federal Coordinating Officer (FCO), FBI-Special Agent-In Charge,

Virginia State Police, and a State On-Scene Coordinator. The State On-Scene Coordinator will initially be either the VDES Regional Coordinator or the Hazardous Materials Officer who arrives on the scene first. The designation of the State On-Scene Coordinator may change depending on the type of incident and as more senior officials arrive at the scene. The Unified Command organization will be modified to include representatives from other emergency support functions (e.g. fire, health, public works, communications) as well as private industry depending on the following factors: the terrorist tactic(s) employed, the challenges presented to the emergency management community in responding to and recovering from the tactic(s), the target group involved, and the community impacted (see Tab 1).

The Federal Bureau of Investigation (FBI), by Presidential Directive, is in charge of the response to a terrorist incident. The Federal Emergency Management Agency, in coordination with the Virginia Department of Emergency Services and Sample Jurisdiction's Emergency Services, will support the FBI in coordinating and fulfilling non-law enforcement response and recovery missions. The Virginia State Police in coordination with local law enforcement personnel, will be supporting the FBI in their functional responsibilities.

CONCEPT OF OPERATIONS

Hazards Analysis/Plan Development

The Coordinator of Emergency Services, in coordination with local, regional, and state law enforcement officials, will conduct a hazards analysis to identify groups that may pose a threat to the community, as well as facilities or activities that may be at risk or potential targets of terrorist acts. A capability assessment will be conducted to identify what resources will be needed to effectively respond to and recover from the potential situations identified. A listing of resources available within the jurisdiction, as well as in and outside of the region, from public and private sources, will be developed. Potential target facilities and activities should be evaluated in terms of what measures could be implemented to mitigate against potential acts of terrorism. Facility plans should be developed, reviewed, and tested in coordination with the appropriate local, state, and federal government agencies.

Incident Management System

In responding to any emergency or disaster situation within Sample Jurisdiction, the Incident Management System will be utilized to effectively organize and integrate multiple disciplines into one multi-functional organization. This command system is comprised of five functions which include the following: command, operations, planning, logistics, and finance/administration. An Incident Commander is responsible for ensuring that all functions identified above are effectively working in a coordinated manner to fulfill the established objectives and overall management strategy that were developed for the emergency at hand. An Operations Chief, who reports directly to the Incident Commander, is designated to conduct the necessary planning to ensure operational control during emergency operations.

Site Assessment/Security

To ensure public safety, as well as facilitate response and recovery initiatives, security and access control measures in and around the disaster site will be implemented immediately by first responders. The area will be quickly evaluated in terms of public health and safety considerations in order to identify the need to implement any protective actions, as well as the use of protective equipment by response personnel entering the area, in order to conduct life-saving activities. Once it is suspected or determined that the incident may have been a result of a terrorist act, the Sample Jurisdiction's Coordinator of Emergency Services will notify the Virginia Department of Emergency Services, who will in turn notify the appropriate state and federal agencies. Local law enforcement will immediately begin working with the Director of Emergency Services, the Fire Chief, and other emergency support functions on scene to ensure that the crime scene is preserved to the maximum extent possible.

Unified Command

Response to a suspected, threatened or actual terrorist event will typically involve multiple jurisdictions and levels of government. These situations will be managed under a Unified Command organization which is illustrated in Tab 1. Members of the Unified Command are jointly responsible for the development of objectives, priorities, and an overall strategy to effectively address the situation. The Unified Command Organization will be structured very similar to the Incident Management System already in place and functioning at the local level.

All agencies involved in the emergency response report to one Incident Command Post and follow one Incident Action Plan similar to a single command structure. The Operations Section Chief, who is designated by the Unified Command, will be responsible for the implementation of the plan. The designation of the Operations Section Chief is based on a variety of factors that may include, but not be limited to, such things as existing statutory authority, which agency has the greatest involvement, the amount of resources involved, or mutual knowledge of the individual's qualifications. A Joint Information Center will be established to support the Unified Command. It will be composed of Public Information Officers from essentially the same organizations that are represented in the Unified Command.

Tactical Support

Once federal authorities have been notified of a suspected, threatened, or actual terrorist incident, a federal interagency Domestic Emergency Support Team will be rapidly deployed to the scene. This team will be comprised of members who have the technical expertise to deal with a full range of terrorist tactics to include biological, chemical, and nuclear incidents. In the case of an incident involving nuclear materials, weapons, or devices, the Department of Energy's Nuclear Emergency Search Team (NEST) will be deployed to provide the necessary technical assistance in responding to and recovering from such events. Local, regional, and state specialized teams (e.g., hazmat, crime narcotics, gang, hostage etc.) who have skills, equipment, and expertise to support these operations, will assist these teams as directed.

Preserving the Crime Scene

Due to the very nature of terrorist acts involving a variety of tactics, law enforcement personnel will work in tandem with one or more emergency support functions to preserve the crime scene, while carrying out life saving actions, implementing the necessary protective actions, developing strategies to protect response personnel, and in defining and containing the hazard. Therefore, while responding to the incident and carrying out their functional responsibilities, first responders become potential witnesses, investigators, and sources of intelligence in support of the crime scene investigation. As such, they must be trained in looking at the disaster area as a potential crime scene that may provide evidence in determining the cause of the event and identifying the responsible party(ies). Responders must also be aware that the crime scene may harbor additional hazards to responders as they carry out their responsibilities. Emergency Support Functions will have to review and modify their response procedures to ensure that the crime scene can be preserved to the extent possible without compromising functional responsibilities or standards of service.

Accessibility Policies

Once the life-saving activities and investigation of the crime scene are completed and the area is considered safe, the area will be made accessible to damage assessment teams, restoration teams, property owners, insurance adjusters, media etc. However, access to the area may still be limited depending on the extent of damage sustained, general conditions of the area, and who is requesting access. Accessibility and reentry policies will be developed, in cooperation with the appropriate local, state, and federal officials, to define who will be given access to the damaged areas, any time restrictions regarding access, whether escorts will be necessary, and what protective equipment will be required, if any, to enter the area. Methods to facilitate identification and accountability of emergency workers, media, insurance adjusters, and property owners will also be developed for safety and security purposes, utilizing some system of colored badges, name tags, arm bands, etc. Security personnel will be responsible for enforcing these policies and procedures. Areas on site that pose a potential hazard or risk will be identified and cordoned off with the appropriate isolation and warning devices.

Training/Exercising

Trained and knowledgeable response personnel are essential in effectively assessing the scene and recognizing situations that may be of a suspicious nature or that could pose additional harm to responders as well as the general population. Sample jurisdiction will ensure that all response personnel have a basic course to enhance their awareness and to enhance recognition of such situations. Also, responders will be trained to fulfill their functional responsibilities in the context of a crime scene environment that may pose a variety of unique health, safety, and environmental challenges. Responders will have a thorough understanding of their responsibilities in responding to a terrorist act, as well as how their role and responsibilities interface with other state and federal components of the terrorist response and recovery team.

To ensure an effective response capability, Sample Jurisdiction's training for terrorist incidents will be integrated with state and federal training programs and based on state and federal guidance. Training will focus on tactical operations for explosive, chemical and biological agents, hostage taking, skyjacking, barricade situations, kidnapping, assaults and assassinations. Local specialized crime units such as gang, organized crime, narcotics, as well as hazmat teams will have skills that terrorism training can use and build upon.

EMERGENCY MANAGEMENT ACTIONS - TERRORISM

1. Normal Operations

   a. Establish the Unified Command Management System as the organizational framework that representatives of local, state, federal government will operate under while responding to and recovering from acts of terrorism.

   b. Identify critical systems/facilities within the community, assess their vulnerability to terrorist actions, and develop and implement the necessary mitigation and response strategies.

   c. Assess local and regional resource capabilities in context of potential terrorist tactics.

   d. Identify sources of special equipment and services to address shortfalls identified in capability assessment.

   e. Develop protective actions for response personnel, as well as the general population at risk, to follow in responding to a suspected or actual terrorist event involving a variety of tactics.

   f. Provide guidance for potential terrorist targets to follow in assessing their vulnerability to such events as well as in developing mitigation strategies and response capabilities.

   g. Coordinate and integrate planning efforts of critical public and private systems and facilities in order to ensure an effective response to, recovery from, and mitigation against terrorist attacks.

   h. Enhance and broaden local and regional response capabilities by developing a training program that integrates local, state, and federal resources.

   i. Ensure local and regional capability to effectively address mass casualty and mass fatality incidents involving both uncontaminated and contaminated victims.

   j. Develop the necessary decontamination, contamination containment, and monitoring procedures to ensure the safety of response personnel, the evacuated population, and the general population in situations involving chemical, biological, and radiological agents.

EMERGENCY MANAGEMENT ACTIONS - TERRORISM (continued)

k. Identify and address potential legal, environmental, and public safety health issues that may be generated by such events.

l. Prepare mutual aid agreements with surrounding jurisdictions to augment local resources.

m. Develop and coordinate the necessary prescripted announcements with the Public Information Office regarding the appropriate protective actions for the various terrorist tactics and situations that may confront the jurisdiction.

2. Increased Readiness

Although terrorist actions occur with little or no warning, there may be situations where notice of terrorist actions may be received by the jurisdiction, target facility, or individual(s) just prior to occurrence.

a. Alert appropriate local, state, and federal agencies that have the expertise, resources, and responsibility in mitigating against, responding to and recovering from such events.

b. Attempt to prevent the event from occurring by locating and eliminating the hazard, and identifying and apprehending responsible party (ies).

c. Notify the public of the threat, implement and advise the risk population of the necessary protective actions to take in context of anticipated event.

d. Stage resources out of harm's way and in areas that can be effectively mobilized.

3. Emergency Operations

a. Dispatch emergency response teams to disaster area.

b. Establish a command post and utilize the Unified Command Management System to effectively integrate and coordinate response resources and support from all levels of government.

c. Provide for the security of evacuated areas, critical facilities, resources, and the impacted area to protect the crime scene and facilitate response and recovery efforts.

EMERGENCY MANAGEMENT ACTIONS - TERRORISM (continued)

    d. Develop, implement, and enforce accessibility policies that will define who will be given access to the damaged and impacted areas, whether there are any time restrictions regarding access, whether escorts will be necessary, and what protective gear and identification will be required, if any, to enter these areas.

    e. Implement the necessary traffic control measures that will facilitate evacuation from the risk area and enhance and complement site security measures following the event.

    f. Activate mutual aid agreements as necessary.

    g. Establish Joint Information Center to coordinate the timely and appropriate release of information during the response and recovery phases.

    h. Coordinate and track resources (public, private) and document associated costs.

4. Recovery

    a. Upon completion of the crime investigation, restore the scene to condition prior to event.

    b. Continue to monitor area as necessary for any residual after-effects.

    c. Maintain protective actions as situation dictates.

    d. Continue to coordinate and track resources and document costs.

    e. Continue to keep public informed of recovery developments.

    f. Develop and implement long-term environmental decontamination plan, as necessary, in coordination with the appropriate local, state and federal government agencies.

Source: Courtesy of the Commonwealth of Virginia Department of Emergency Services.

Appendix IV

Certificate of Training — Medical Management of Biological Casualties

U.S. ARMY
MEDICAL RESEARCH INSTITUTE
OF
INFECTIOUS DISEASES

CERTIFICATE OF TRAINING

This is to certify that

John Cashman

has successfully completed all requirements including an examination of the

MEDICAL MANAGEMENT OF BIOLOGICAL CASUALTIES

Given by satellite broadcast
On 16, 18 and 19 September 1997

DAVID R. FRANZ, Colonel, VC
Commander

Source: Certificate of Training, "Medical Management of Biological Casualties,"
U.S. Army Medical Research Institute of Infectious Diseases, Fort Detrick, MD.

Glossary

μg/g: Microgram per gram, one part per million (ppm).

μg/l : Microgram per liter, one part per billion (ppb).

Absorbent materials: A material designed to pick and hold liquid hazardous materials to prevent a spread of contamination.

Acetylcholinesterase: An enzyme that hydrolyzes the neurotransmitter acetylcholine. The action of this enzyme is inhibited by nerve agents.

Acute: Health effect that occurs over a short term; brief and severe as opposed to chronic.

Acute exposure: A single encounter to toxic concentrations of a hazardous material or multiple encounters over a short period of time.

Adsorption: The attraction and accumulation of one substance on the surface of another.

Aerosol: A fine aerial suspension of liquid (fog or mist) or solid (dust, fume, or smoke); particles sufficiently small in size to be stable.

After action report: A post incident analysis report gathered by a responsible party or responding agency after termination of a hazardous materials incident, describing actions taken, materials involved, impacts, and similar information.

Air monitoring: To observe, record and/or detect pollutants in ambient air.

Anhydrous: Without water, dry. Describes a substance in which no water molecules are present.

Antibiotics: Substances produced by and obtained from living cells, such as bacteria or molds, capable of destroying or weakening bacteria. Examples of antibiotics would include penicillin and streptomycin.

Antidote: A remedy to relieve, prevent, or counteract the effects of a poison.

Antiserum: A serum containing an antibody or antibodies produced from animals or humans that have survived exposure to an antigen.

ANSI: American National Standards Institute, a private organization that is engaged in creating voluntary standards or characteristics and performance of materials, products, systems, and services.

Aquifer: A permeable geologic unit with the ability to store, transmit, and yield fresh water in usable quantities.

Area plan: A plan established for emergency response to a release or threatened release of a hazardous material.

Assessment: To determine the nature and degree of hazard of a hazardous material or a hazardous materials incident from a safe vantage point.

ASTM: American Society for Testing and Materials, a voluntary group in which members devise consensus standards for materials characterization and use.

Atropine: Sometimes used as an antidote for nerve agents. It inhibits the action of acetylcholine by binding to acetylcholine receptors.

BAL: British Anti-Lewisite. Dimercaprol, a treatment for toxic inhalations that displaces arsenic bound to enzymes.

Biochemical oxygen demand: A numerical estimate of contamination in water expressed in milligrams per liter of dissolved oxygen. A measure of the amount of oxygen consumed in biological processes that break down organic matter in water.

Biological agent: A microorganism that causes disease in people, plants, or animals or causes the deterioration of material.

BLEVE: Boiling Liquid Expanding Vapor Explosion.

Blister agents: Substances that cause blistering of the skin. Exposure is through liquid or vapor contact with any exposed tissue.

Blood agents: Substances that injure a person by interfering with cell respiration disrupting the exchange of oxygen and carbon dioxide between blood and tissues.

Boiling point: The temperature at which the vapor pressure of a liquid equals the atmospheric pressure so that the liquid becomes a vapor. Knowing the boiling point, you can estimate the persistence of a chemical agent under a given set of conditions because the vapor pressure and the evaporating tendency vary inversely with its boiling point. Chemical agents with high boiling points tend to be persistent, while agents with low boiling points are normally non-persistent.

Boom: A floating physical barrier serving as a continuous obstruction to the spread of a contaminant.

Bootie: A sock-like over-boot protector worn to minimize contamination.

Botulism: Poisoning by toxin derived from *Clostridium botulinum*.

Boyles Law: The volume of gas is inversely proportional to its pressure at constant temperature.

Breakthrough time: The elapsed time between initial contact of a hazardous chemical with the outside surface of a protective clothing material and the time at which the chemical can be detected at the inside surface of the material.

Buddy system: The organizing of employees into work groups so that each employee in the work group is designated to be observed by at least one other employee in the work group.

CAER: Community Awareness and Emergency Response: a program developed by the Chemical Manufacturers Association to provide guidance for chemical plant managers to assist them in cooperating with local communities to develop integrated hazardous materials response plans.

CAM: Chemical Agent Monitor used by the U.S. military; it detects chemical agent vapors and provides a readout of the relative concentration of the vapors present.

CAMEO: Computer Aided Management of Emergency Operations a computer database for storage and retrieval of pre-planning data for on-scene use at hazardous materials incidents.

CANA: Convulsant Antidote for Nerve Agent, also called diazepam.

CANUTEC: Canadian Transport Emergency Center: a 24 hour, government sponsored hot line for chemical emergencies.

Carbamates: Organic chemical compounds that can be neurotoxic by competitively inhibiting acetylcholinesterase binding to acetylcholine.

Carcinogen: A material that has been found to cause cancer in humans or in animals.

Cascade system: Several air cylinders attached in series to fill SCBA (self-contained breathing apparatus) bottles.

CAS registration number: Chemical Abstract Service. An assigned number used to identify a chemical. CAS numbers identify specific chemicals and are assigned sequentially; the number is a concise, unique means of material identification. A product of more than one component will have a specific number for each component.

Catastrophic incident: An event that significantly exceeds the resources of a jurisdiction.

CERCLA: Comprehensive Environmental Response, Compensation and Liability Act. Known as CERCLA, or the Superfund amendment, this federal law deals with hazardous substances releases to the environment and the cleanup of hazardous waste sites.

CFR: Code of Federal Regulations. A collection of federal regulations established by law.

CG Phosgene.

Charles Law: The volume of gas is directly proportional to its absolute temperature at constant pressure.

CK: Cyanogen chloride.

Chemical agent: A chemical substance that is intended for use in military operations to kill, seriously injure, or incapacitate people through its physiological effects (including blood, nerve, choking, blister, and incapacitating agents). Not included in this category are riot control agents, chemical herbicides, or smoke and flame materials.

Chemical agent detector paper: (ABC-M8 VGH) M8 detector paper comes in booklets of 25 sheets; it detects and identifies liquid V or G type nerve agents or H type blister agents.

Chemical agent GA: The chemical Ethyl N, N-dimethylphosphoramidocyanidate (CAS# 77-81-6) also known as tabun, is a nerve agent.

Chemical agent GB: The chemical Isopropyl methyl phosphonofluoridate (CAS# 107-44-8) also known as sarin, is a nerve agent.

Chemical agent GD: The chemical Pinacolyl methyl phosphonofluoridate (CAS# 96-64-0) also known as soman, is a nerve agent.

Chemical agent H: Levinstein mustard (CAS# 471-03-4) is a mixture of 70% bis (2-chloroethyl) sulfide and 30% sulfur impurities produced by the Levinstein process and is a blister agent.

Chemical agent HD: Distilled mustard, or bis (2-chloroethyl) sulfide, (CAS# 505-60-2) is mustard (H) that has been purified by washing and vacuum distillation to reduce sulfur impurities. Agent HD is a blister agent.

Chemical agent HT: Agent T is bis [2-(2-chloroethylthio) ethyl] ether (CAS# 63918-89-8) and is a sulfur, oxygen, and chlorine compound similar in structure to HD. It is 60% HD and 40% T with a variety of sulfur contaminants and impurities. It is a blister agent.

Chemical agent L: (lewisite) Agent L is a blister agent, Dichloro 2-chlorovinyldichloroarsine, (CAS# 541-25-3).

Chemical name: The scientific designation of a chemical or a name that will clearly identify the chemical for hazard evaluation purposes.

Chemical surety: Controls, procedures, and actions that contribute to the safety, security, and reliability of chemical agents and their associated weapon systems throughout their life cycle without degrading operational performance.

Chemical warfare: All aspects of military operations involving the use of lethal munitions/agents and the warning and protective measures associated with such offensive operations.

CHEMTREC: The Chemical Transportation Emergency Center located in Washington, D.C.; a public service provided by the private Chemical Manufacturers Association, provides emergency response information and assistance 24-hours a day for hazardous materials responders.

Choking agents: Substances that cause physical injury to the lungs by exposure through inhalation. Death results through lack of oxygen.

Cholinesterase: (ChE) An enzyme that catalyzes the hydrolysis of acetylcholine to choline (a vitamin) and acetic acid.

Chemical degradation: A chemical action involving the molecular breakdown of a material due to contact with a chemical. The action may cause personal protective equipment to swell, shrink, blister, discolor, become brittle, sticky, soft or to deteriorate. These changes permit chemicals to get through the suit more rapidly or increase the probability of permeation.

Chemical Hazards Response Information System/Hazard Assessment Computer System: (CHRIS/HACS) Developed by the Coast Guard, HACS is a computerized model of the CHRIS manuals. It is used by federal on-scene coordinators during a chemical spill or response.

Chemical penetration: The movement of material through a suit's closures, such as zippers, buttonholes, seams, flaps, or other design features. Abraded, torn, or ripped suits will also allow penetration.

Chemical protective clothing material: Any material or combination of materials used in an item of clothing for the purpose of isolating parts of the wearer's body from contact with a hazardous material.

Chemical protective suit: A single or multi-piece garment constructed of chemical protective clothing materials designed and configured to protect the wearer's torso, head, arms, legs, hands, and feet.

Chemical resistance: A material's ability to resist chemical attack. Resistance is dependent on the method of test and is measured by determining the changes in physical properties. Time, temperature, stress, and reagent, may all be factors that affect the chemical resistance of a material.

Chemical resistant materials: Materials that are specifically designed to inhibit or resist the passage of chemicals into and through the materials by the processes of penetration and permeation.

Chronic toxicity: Adverse health effects from repeated doses of a toxic chemical or other toxic substance over a relatively prolonged period of time, generally greater than one year.

Code of Federal Regulations: The federal government's official publication of federal regulations. Volumes are divided into 50 titles according to subject matter. Titles are divided into chapters which are divided into parts and sections.

Compatibility charts: Permeation and penetration data supplied by the manufacturers of protective clothing to indicate chemical resistance and breakthrough time of various garment materials as tested against a battery of chemicals.

Competence: Having skills, knowledge, and judgment necessary to perform certain objectives in a satisfactory manner.

Concentration: The amount of a chemical agent present in a unit volume of air, usually expressed in milligrams per cubic meter (mg/m^3).

Containment: All activities necessary to bring the incident to a point of stabilization and to establish a degree of safety for emergency personnel greater than existed upon arrival.

Contamination: A substance or process that poses a threat to life, health, or the environment. The deposit and/or absorption of NBC contamination on and by structures, areas, personnel and objects.

Contamination control line: The established line around a contamination reduction zone that separates it from the support zone.

Contingency plan: A pre-planned document defining specific responsibilities and tasks and presenting an organized and coordinated plan of action to limit potential pollution in case of fire, explosion, or discharge of hazardous materials.

Control: The procedures, techniques, and methods used in the mitigation of a hazardous materials incident, including containment, confinement, and extinguishment.

Control zones: The designation of areas at a hazardous materials incident based upon safety and the degree of hazard.

Coordination: To bring together in a uniformed and controlled manner the functions of all agencies on scene.

Corrosive: A substance that causes visible destruction or visible changes in human skin tissues at the site of contact.

Cost recovery: The procedure that allows for the agency having jurisdiction to pursue reimbursement for all costs associated with a hazardous materials incident.

Cryogenic: Gases, usually liquefied, that induce freezing temperatures of $-150°F$ and below such as liquid oxygen, liquid helium, liquid natural gas, and liquid hydrogen.

CX: Phosgene oxime dichloroforoxime, a blister agent.

Decontamination: The physical and/or chemical process of reducing and preventing the spread of contamination from persons and equipment used at a hazardous materials incident. The process of making any person, object, or area safe by absorbing, destroying, neutralizing, making harmless, or removing the hazardous material.

Decontamination corridor: A corridor that acts as a protective buffer; it bridges the hot zone and the cold zone and is located in the warm zone within which decontamination stations and personnel are located to apply decontamination procedures.

Degradation: The process of decomposition. A chemical action involving the molecular breakdown of protective clothing material due to contact with a chemical. Degradation is evidenced by visible signs such as charring, shrinking, or dissolving. Testing clothing for weight of thickness changes, or loss of tensile strength, will also reveal degradation.

Degree of hazard: A relative measure of how much harm a substance can cause.

Dermal exposure: Exposure to or absorption through the skin.

Detection: The determination of the presence of a chemical agent.

Detection in the field: Methods and apparatus for detecting and/or identifying chemical agents in the field.

Detoxification rate: The rate at which the body's own action will overcome or neutralize chemicals or toxins.

Dike: An embankment or ridge, natural or man made, used to control the movement of liquids, sludges, solids, or other materials. An overflow dike would be constructed to allow uncontaminated water to flow unobstructed over the dike keeping the contaminant behind the dike. An underflow dike allows the uncontaminated water to flow unobstructed under the dike keeping the contaminant behind the dike.

Direct reading instrument: A portable device that measures, and displays, in a short period of time, the concentration of a contaminant in the environment.

Dispersion: To spread, scatter, or diffuse through air, soil, surface or ground water.

Disposal drum: A specially constructed drum used to overpack damaged or leaking containers of hazardous materials for shipment.

Diversion: The intentional movement of a hazardous material in a controlled manner so as to relocate it in an area where it will pose less harm to the community and the environment.

DKIE: Decontamination Kit, Individual Equipment.

Dose: The amount of substance ingested, absorbed, and/or inhaled per exposure period.

Dose rate: How fast a dose is absorbed or taken into a body.

DOT identification numbers: Four-digit numbers proceeded by UN (United Nations) or NA (North American) that are used to identify the particular hazardous materials for regulation of transportation. HMRTs often refer to such numbers to identify specific chemicals listed in the Department of Transportation's "Emergency Response Guidebook." For example, the UN code number for chlorine trifluoride is UN1749.

Double gloving: A set of gloves worn in addition to the already in-place protection.

DP: Diphosgene, a choking agent.

DS2: Decontaminating Solution No. 2. A military decon solution for the battlefield (skin contact with DS2 must be avoided at all times, do not breath fumes, contact with liquid or vapors can be fatal).

ED: Ethyldichloroarsine, a blister agent.

Emergency operations plan: A document that identifies the available personnel, equipment, facilities, supplies, and other resources in the jurisdiction and states the coordinated actions to be taken by individuals and government services in the event of natural, man-made, or attack-related disasters.

Emergency response: Response to any occurrence which has or could result in a release of a hazardous substance.

Emergency response guidebook: A manual for first responders to use during the initial phase of a hazardous materials/dangerous goods incident developed under the supervision of the Office Hazardous Materials Issues and Training, Research and Special Programs Administration, U.S. Department of Transportation (D.O.T.).

Emergency response plan: A plan that establishes guidelines for handling hazardous materials incidents as required by 29 CFR 1910.120.

Entry point: A specified and controlled access into a hot zone at a hazardous materials incident.

Entry team leader: The entry leader is responsible for the overall entry operations of assigned personnel within the hot zone.

Etiologic agent: A viable microorganism or its toxin that causes, or may cause, human disease.

Evacuation: To quickly and calmly leave an area in order to avoid exposure to a potentially harmful situation.

Explosive: A material that releases pressure, gas, or heat suddenly when subjected to shock, heat, or high pressure.

Explosive ordnance disposal: The detection, identification, field evaluations, rendering safe, recovery, and final disposal of unexploded ordnance or munitions chemical agents.

Exposure: The subjection of a person to a toxic substance or harmful physical agent through any route of entry.

Exposure routes: The major routes of exposure include ingestion, inhalation, and absorption though the skin.

Filter: A High-Efficiency Particulate Air (HEPA) filter is at least 99.97% efficient in removing particles with a diameter of 0.3 microns.

First responder, Awareness level: Individuals who are likely to witness or discover a hazardous substance release and who have been trained to initiate an emergency response sequence by notifying proper authorities.

First responder, Operations level: Individuals who respond to releases or potential releases of hazardous substances as part of the initial response to the site for the purpose of protecting nearby persons, property, or the environment from the effects of the release. They are trained to respond in a defensive fashion without actually trying to stop the release. Their function is to control the release from a safe distance, keep it from spreading, and prevent exposures.

Flammable: A material that catches on fire easily or spontaneously under conditions of standard temperature and pressure.

Flammable (explosive) range: The range of gas or vapor concentration (percentage by volume in air) that will burn or explode if an ignition source is present. Limiting concentrations are commonly called the lower explosive limit and the upper explosive limit. Below the lower explosive limit, the mixture is too lean to burn; above the upper explosive limit, the mixture is to rich to burn.

Flaring: A process that is used with high vapor pressure liquids or liquefied compressed gases for the safe disposal of the product. Flaring is the controlled burning of material in order to reduce or control pressure and/or to dispose of a product.

Flash point: The lowest temperature at which a flammable liquid gives off sufficient vapor near its surface or within a vessel to form an ignitable mixture with air.

Formula: An expression of the constituents of a compound by symbols and figures.

Freezing point: Temperature at which a liquid changes to a solid or which crystals start to form as a liquid is slowly cooled. Also the melting point — the temperature at which a solid changes to a liquid as it is slowly heated.

Full protective clothing: Protective gear, to include SCBA, and designed to keep gases, vapor, liquids, and solids from any contact with the skin while preventing ingestion or inhalation.

Fully encapsulating suits: Chemical protected suits that are designed to offer full body protection, include SCBA, are gas tight, and meet the design criteria as outlined in NFPA Standard 1991.

Fumes: Tiny solid particles formed by the vaporization of a solid which then condense in air.

Fungus: A general term to denote a group of eukaryotic protist, including mushrooms, yeasts, rusts, molds, smuts, etc., which are characterized by the absence of a rigid cell wall composed of chitin, mannans, and sometimes cellulose.

GA: Ethyl N,N-dimethylphosphoramidocyanidate or tabun, a nerve agent. (CAS# 77-81-6)

Gas: A state of matter in which the material is compressible, has a low density and viscosity, can expand or contract greatly in response to changes in temperature and pressure, and readily and uniformly distributes itself throughout any container.

GB: Isopropyl methyl phosphonofluoridate or sarin, a nerve agent, more toxic than tabun or soman. (CAS# 107-44-8)

GB2: A binary nerve agent.

GC: Gas chromatography.

GD: Pinacolyl methyl phosphonofluoridate or soman, a nerve agent. (CAS# 96-64-0)

Grounding: A safety practice to conduct any electrical charge to the ground, preventing sparks that could ignite a flammable material.

G-series nerve agents: A series of nerve agents developed by the Germans: tabun (GA), sarin (GB), and soman (GD).

H: Levinstein mustard, a blister agent. (CAS# 471-03-4)

HD: Distilled mustard, a blister agent. (CAS# 505-60-2)

Half-life: The time in which the concentration of a chemical in the environment is reduced by half.

Hazard assessment: A process used to qualitatively or quantitatively assess risk factors to determine incident operations.

Hazard class: A series of nine descriptive terms that have been established by the UN Committee of Experts to categorize the hazardous nature of chemical, physical, and biological materials. These categories are flammable liquids, flammable solids, explosives, gases, oxidizers, radioactive materials, corrosives, poisonous and infectious substances, and dangerous substances.

Hazardous material: A substance which by its nature, containment, and reactivity has the ability to inflict harm during an accidental occurrence; characterized as being toxic, corrosive, flammable, reactive, an irritant, or a strong sensitizer and thereby posing a threat to health and the environment when improperly managed.

Hazardous materials incident: The uncontrolled release, or potential release, of a hazardous material from its container into the environment.

Hazardous materials response team (HMRT): An organized group of trained response personnel operating under an emergency response plan and appropriate standard operating procedures, who are expected to control actual or potential situations in which close approach to leaking or spilled hazardous substances may be required.

HCN: Hydrogen Cyanide.

Hot zone: The area adjacent to and surrounding a hazardous materials incident that extends far enough to prevent the effects of hazardous materials releases from endangering personnel outside the zone; also known as the restricted zone or the exclusion zone.

H-series agents: A series of persistent blister agents including distilled mustard (HD) and the nitrogen mustards (HN-1, HN-2, and HN-3).

HTH: Calcium hypochlorite.

Hydrolysis: Process of an agent reacting with water. It does not materially affect the agent cloud in tactical use, because the rate of the chemical action is too slow.

IDLH: Immediately Dangerous To Life or Health. An atmospheric concentration of any toxic, corrosive, or asphyxiant substance that poses an immediate threat to life or would cause irreversible or delayed adverse health effects or would interfere with an individual's ability to escape from a dangerous atmosphere.

Incident: An event involving the release or potential release of a hazardous material.

Incident action plan: A plan which is initially prepared at the first meeting of emergency personnel who have responded to an incident. The plan contains general control objectives reflecting overall incident strategy and specific action plans.

Incident commander: The incident commander will assume control of the incident scene beyond the first responder incident level, and must demonstrate competency in the Incident Command System. The incident commander is responsible for developing an effective organizational structure, allocating resources, making appropriate assignments, managing information, and continually attempting to mitigate the incident. The employer shall certify that the incident commander meets the requirements of 29 CFR 1910.120.

Incident Command System (ICS): An organized system of responsibilities, roles, and standard operating procedures used to manage and direct emergency operations.

Incident safety officer: The incident safety officer is a position mandated by Occupational Safety and Health laws. The position is attached to the incident commander, and should be filled by the person with the most knowledge about the various safety aspects at a hazardous materials scene. As provided under OSHA law, the incident safety officer has the power and authority to alter, suspend, or terminate the operation when, in his/her opinion, the conditions are unsafe.

Industrial agents: Chemicals manufactured for industrial purposes rather than to specifically kill or maim human beings. Hydrogen cyanide, cyanogen chloride, phosgene, and chloropicrin are industrial chemicals that can be military agents as well. Many herbicides and pesticides are industrial chemicals that also can be chemical agents.

Ingestion: Swallowing (such as eating and drinking). Chemicals can get into or onto food, drink, utensils, cigarettes, or hands where they can be ingested.

Inhalation: Breathing. Once inhaled, contaminants can be deposited in the lungs, taken into the blood, or both.

L: Dichloro-2-chlorovinyldichloroarsine or lewisite, a blister agent. (CAS# 541-25-3)

Labpack: Generally refers to any small containers of hazardous waste in an over-packed drum, but not restricted to laboratory wastes.

LC_{50}: The lethal concentration of a toxicant in air which is lethal to 50% of the exposed population.

LCt_{50}: Refers to the lethal concentration time fifty, the concentration time that will kill 50% of the exposed population.

LD_{50}: Lethal dose fifty. The lethal dose of a toxicant, consumed orally or absorbed through the skin, that is lethal to 50% of the exposed population, also known as median lethal dose.

Leak control compounds: Substances used for the plugging and patching of leaks in non-pressure and some low-pressure containers, pipes, and tanks.

Leak control devices: Tools and equipment used for patching and plugging of leaks in non-pressure and some low-pressure containers, pipes, and tanks.

Level of protection: In addition to positive pressure breathing apparatus, designations of types of personal protective equipment to be worn based on NFPA standards.

Level A: Vapor protective suit for hazardous chemical emergencies.

Level B: Liquid splash protective suit for hazardous chemical emergencies.

Level C: Limited use protective suit for hazardous chemical emergencies.

Level one incident: Hazardous materials incidents that can be contained, extinguished, and/or abated using equipment, supplies, and resources immediately available to first responders having jurisdiction, and whose qualifications are limited to and do not exceed the scope of the training explained in 29 CFR 1910.

Level two incident: Hazardous materials incidents that can only be identified, tested, sampled, contained, extinguished, and/or abated utilizing the resources of a

HMRT (hazardous materials response team), that require the use of specialized chemical protective clothing, and whose qualifications are explained in 29 CFR 1910.

Level three incident: A hazardous materials incident that is beyond the controlling capability of a HMRT (technician or specialist level), whose qualifications are explained in 29 CFR 1910, and/or must be additionally assisted by qualified specialty teams or individuals.

Liquid density: Liquid density of a chemical agent measures the weight of the agent compared to water, which has a density of 1.0. Specific gravity, the ratio of the mass of a unit volume of a substance to the mass of the same volume of a standard substance, usually water, at a standard temperature, is the result of this comparison. Liquid agent forms layers in water — the greater the density, the more it will sink, while less dense agents will float. Nerve agents are roughly the same density as water so they tend to mix throughout the depth of water.

Local Emergency Planning Committee (LEPC): A committee appointed by the state emergency response commission, as required by Title III of SARA, to formulate a comprehensive emergency plan.

Lower Explosive Limit (LEL): The lowest concentration of gas or vapor by percent of volume in air that will burn or explode if an ignition source is present at ambient temperatures. The LEL is constant up to 250°F.

M18A2: A kit used by military personnel consisting of portable tests capable of detecting selected choking agents, blood agents, nerve agents, and blister agents.

M256-series chemical agent detector kit: A kit used by military personnel to detect and identify field concentrations of nerve, blister, or blood agent vapors. The kit consists of 12 samplers/detectors and a packet of M8 detector paper.

M272 water testing kit: A kit used by military personnel to detect and identify dangerous levels of common chemical warfare agents in water sources.

M291 skin decontamination kit: This kit used by military personnel is used to decontaminate a soldier's hands, face, ears, and neck.

M295: Decontamination Packet, Individual Equipment (DPIE).

M34 soil sampling kit: This kit used by military personnel is to sample soil, surface matter, and water.

M8: Chemical agent detector paper.

M8A1: A chemical agent alarm detector used by the military. For nerve agents only (GA, GB, GD, or VX), it can provide an early warning of a possible vapor hazard.

M9 Chemical Agent Detector Paper: The M9 self-adhesive paper used by military personnel attaches to most surfaces, and indicates the presence of a nerve agent (G or V) or blister agent (H or L) by turning a reddish color.

Mark I & II: Nerve Agent Antidote Kit (NAAK).

Material Safety Data Sheets (MSDS): A MSDS contains descriptive information on hazardous chemicals under OSHA's Hazard Communication Standard. The data sheets also provide precautionary information on safe handling, health effects, chemical and physical properties, emergency phone numbers, and first aid procedures.

MD: Methyldichloroarsine, a blister agent.

Median lethal dosage: (LD$_{50}$): The lethal dose of a poison, when taken orally or absorbed through the skin, which is lethal to 50% of the exposed laboratory animal population.

Melting point: Temperature at which a solid changes to a liquid as it is slowly heated. Also the freezing point — temperature at which a liquid changes to a solid as it is slowly cooled.

Micron: A unit of measurement equal to one-millionth of a meter.

Mini-CAM: Miniature chemical agent monitor.

Miscible: Able to mix (but not chemically combine) in any ratio without separating into two phases.

Mission-Oriented Protective Posture (MOPP): The protective clothing used by members of the U.S. military who engage in nuclear, biological, and chemical warfare. MOPP gear provides a flexible system requiring personnel to wear only that protective clothing and equipment appropriate to the threat level, work rate imposed by the mission, temperature, and humidity.

Mists: Suspended liquid droplets generated by condensing a substance from the gaseous to the liquid state or by breaking up a liquid into a dispersed state, such as by splashing, foaming or atomizing. Mist is formed when a finely divided liquid is suspended in air.

Mitigation: An action employed to contain, reduce, or eliminate the harmful effects of a spill or release of a hazardous material.

Molecular weight: The sum of the atomic weights of all the atoms in a molecule. In regard to chemical agents, the molecular weight is a guide to persistence; an agent with a high molecular weight would tend to have a lower rate of evaporation and greater persistence.

Monitoring: To determine contamination levels and atmospheric conditions by observing and sampling using instruments and devices to identify and quantify contaminants and other factors.

Mutagen: Anything that can cause a change (mutation) in the genetic material of a living cell.

Mutual aid: An agreement between two or more agencies, jurisdictions, or political sub-divisions to supply specifically agreed-upon aid or support in an emergency situation.

Mycotoxin: A fungal toxin.

National contingency plan: Created by CERCLA to define the federal response authority and responsibility for oil and hazardous materials spills. The regulations are codified at 40 CFR 300.

National Fire Protection Association (NFPA): A voluntary membership agency to promote fire safety and allied considerations. Publishes standards of interest to hazardous materials responders such as, NFPA-471 Recommended Practice for Responding to Hazardous Materials Incidents, NFPA-472 Standard for Professional Competence of Responders to Hazardous Materials Incidents, and NFPA-473, Competencies for EMS Personnel Responding to Hazardous Materials Incidents.

National Institute for Occupational Safety and Health (NIOSH): A federal agency that performs research on occupational disease and injury, recommends limits for substances, and assists OSHA in investigations and research.

National Response Center (NRC): The national response center in Washington, D.C. is operated by the U.S. Coast Guard. The center must be informed by the spiller within 24 hours of any spill of a reportable quantity of a hazardous substance.

NBC: Nuclear, biological, and chemical.

Nerve agent: Substances that interfere with the central nervous system. Organic esters of phosphoric acid used as a chemical warfare agent because of their extreme toxicity (GA, GB, GD, GF, and VX). All are potent inhibitors of the enzyme, acetylcholinesterase, which is responsible for the degradation of the neurotransmitter, acetylcholine in neural synapses or myoneural junctions. Nerve agents are readily absorbed by inhalation and/or through intact skin.

Nerve Agent Antidote Kit (NAAK): Also called the MARK I, containing atropine and 2-PAM chloride.

Nerve Agent Pyrdostigmine Pretreatment (NAPP): This pretreatment provides a countermeasure to soman (GD) and/or tabun (GA). NAPP treatment substantially increases the effectiveness of the chemical components of the Mark I kit against soman and tabun.

Neutralize: To render chemically harmless; to bring a solution of an acid or base to a pH of 7.0.

Nonpersistent agent: An agent that upon release loses its ability to cause casualties after 10 to 15 minutes. It has a high evaporation rate, is lighter than air, and will disperse rapidly in open air.

Occupational Safety and Health Administration (OSHA): The U.S. Department of Labor through OSHA has safety and health regulatory and enforcement control over worker health in most industries, businesses, and states in the U.S.

Odor: Smell, scent, aroma, fragrance.

Off-gassing: Giving off a vapor or gas.

Organic materials: Compounds composed of carbon, hydrogen, and other elements with chain or ring structures.

Organophosphorous compound: Containing elements of phosphorous and carbon, the physiological effects of such a compound include inhibition of acetylcholinesterase. A number of pesticides including parathion and malathion, and virtually all nerve agents, are organophosphorous compounds.

Oxime: A chemical compound containing one or more oxime groups. Although blister agents, some oximes are beneficial. 2-PAM chloride restores the reactivity of cholinesterase and is used to counteract organophosphate poisoning caused by pesticides and nerve agents.

PD: Phenyldichloroarsine, a blister agent.

Permissible Exposure Limit (PEL): The maximum time-weighted average concentration mandated by OSHA to which workers may be repeatedly exposed for eight hours a day, 40 hours per week without adverse health effects.

Permeation: The passage of chemicals, on a molecular level, through intact material such as protective clothing.

Persistent agent: Chemical agents that do not hydrolyze or volatilize readily, such as VX and HD or biological agents that are highly stable, such as anthrax spores and Q-fever. At the time of release, this agent can produce casualties for an extended period of time up to several days. Usually, it has a low evaporation rate. Since its vapor is heavier than air, its vapor cloud will hug the ground and accumulate in low areas. It is an inhalation hazard, but extreme care should be taken to avoid skin contact as well.

Personal Protective Equipment (PPE): Equipment provided to shield or isolate a person from the chemical, physical, and thermal hazards that may be encountered at a hazardous materials incident. It should include protection for the respiratory system, skin, eyes, ears, face, hands, feet, head, and body.

pH: The value that represents the acidity or alkalinity of an aqueous solution. The number is a logarithm to the base 10 of the reciprocal of the hydrogen ion concentration of a solution. Pure water has a pH of 7. The pH scale is logarithmic and the intervals are exponential, so the progression of values represents far greater concentrations than one might suspect.

Phosgene: Carbonyl chloride. An extremely poisonous gas, but not immediately irritating even when fatal concentrations are inhaled.

Physical state: The (solid, liquid, or gas) of a chemical under specific conditions of temperature and pressure. Condition with respect to structure, form, phase, etc.

Physiological Action: Of or pertaining to physiology, the science dealing with the functions of living organisms or their parts.

Plug and patch: Plugging and patching refers to the use of compatible plugs and/or patches to temporarily reduce or stop the flow of materials from small holes, rips, tears, or gashes in containers.

Plume: A vapor cloud formation which has shape and buoyancy.

Poison, Class A: A D.O.T. term for extremely dangerous poisons such as poisonous gases or liquids of such a nature that a very small amount of the gas or vapor of the liquid mixed with air is dangerous to life. Examples include phosgene, cyanogen, hydrocyanic acid, and nitrogen peroxide.

Poison, Class B: A D.O.T. term for liquid, solid, paste, or semisolid substances other than Class A poison of irritating materials that are known or presumed on the basis of animal tests to be so toxic to man as to afford a hazard to health during transportation.

Polymerization: A process in which hazardous materials react in the presence of a catalyst with themselves or another material to form a polymeric system. This process oftentimes is violent.

Potentially Responsible Party (PRP): An individual or company identified by EPA as potentially liable under CERCLA for cleanup costs at a hazardous waste site. PRPs may include generators of hazardous substances, present or former owners of hazardous substances that have been disposed, site property owners, and transporters of hazardous materials to the site.

ppb: Parts per billion.

ppm: Parts per million.

ppt: Parts per trillion.

PS: Chloropicrin, a choking agent.

Public Information Officer (PIO): Person who acts as a liaison between the incident commander and the news media.

Pyridostigmine bromide: An antidote enhancer that blocks acetylcholinesterase, protecting it from nerve agents. When taken in advance of nerve agent exposure, pyridostigmine bromide increases survival provided that atropine and oxime and other measures are taken.

Reactivity: A substance's susceptibility to undergoing a chemical reaction or change that may result in dangerous side effects, such as explosion, burning, and corrosive or toxic emissions.

Reference library: Chemical text books, references, computer data programs and similar materials carried by response personnel.

Rem: Radiation Equivalent Man; the unit of dose equivalence commonly used in the United States.

Required level of protection: Chemical protective clothing/equipment required to work safely.

RETECS: Registry of Toxic Effects of Chemical Substances is a database of toxicological information compiled, maintained, and updated by NIOSH.

Rickettsia: A microorganism of the genus *Rickettsia*, made up of small rod-shaped coccoids occurring in fleas, lice, ticks, and mites by which they are transmitted to man and other animals causing diseases such as typhus, scrub typhus, and Rocky Mountain Spotted Fever in humans.

Risk assessment: The scientific process of evaluating the toxic properties of a chemical and the conditions of human exposure to it in order to ascertain the likelihood that exposed humans will be adversely affected and to characterize the nature of the effects they may experience. It may contain some or all of the following four steps: hazard identification, dose-response assessment, exposure assessment, and risk characterization.

Route of exposure: The avenue by which a chemical comes into contact with an organism (such as a person). Possible routes include inhalation, ingestion, and dermal contact.

RSCAAL: Remote Sensing Chemical Agent Alarm used by the military.

Rupture: The physical failure of a container or mechanical device, releasing or threatening to release a hazardous material.

SA: Arsine.

Sample: To take a representative portion of a material for evidence or analytical reasons.

Scenario: An outline of a natural or expected course of events.

Scene: The location impacted or potentially impacted by a hazard.

Self-Contained Breathing Apparatus (SCBA): Protective equipment consisting of an enclosed facepiece and an independent, individual supply (tank) of air; used for breathing in atmospheres containing toxic substances. A positive pressure

SCBA or a combination SCBA/supplied air breathing apparatus certified by NIOSH and the Mine Safety Health Administration or an appropriate approval agency.

Shipping papers: Term used to refer to the shipping documents that must accompany all shipments of hazardous materials and waste.

Short Term Exposure Limit (STEL): The time weighted average concentration to which workers can be exposed continuously for a short period of time, normally 15 minutes, without suffering irritation, chronic or irreversible tissue damage, etc.

SIC code: Standard Industrial Classification (SIC) codes are numerical codes that categorize industrial facilities by the type of activity in which they are engaged. All companies conducting the same type of business, regardless of their size, have the same SIC code. As an example, SIC code 2911 refers to petroleum refineries.

Solubility: The ability of one material to dissolve in or blend uniformly with another.

Specific gravity (sp gr): The ratio of the mass of a unit of volume of a substance to the mass of the same volume of a standard substance (usually water) at a standard temperature.

Spill: The release of a liquid, powder, or solid hazardous material in a manner that poses a threat to air, water, ground, or the environment.

Spores: Resistant, dormant cells of some bacteria.

Stabilization: The point where the dangerous effects or results of a hazardous materials incident have been controlled.

Staging area: The safe area established for the temporary location of available resources closer to the incident to reduce response time.

State Emergency Response Commission (SERC): State commissions required under the Superfund Amendments and Reauthorization Act (SARA) that designate emergency planning districts, appoint local emergency planning committees, and supervise and coordinate their activities.

STB: Supertropical Bleach (a decontamination agent), a mixture of calcium oxide and bleaching powder.

STEL: Short-Term Exposure Limit. The maximum concentration for a continuous, 15-minute exposure period, with four exposure periods a day with 60 minutes minimum between exposure periods.

Strict liability: The responsible party is liable even when they have exercised reasonable care.

Sump: A pit or tank that catches liquid runoff for drainage or disposal.

Symptoms: Functional evidence of disease. Information related by an individual that may indicate illness or injury.

Synapse: A site at which neurons make functional contacts with other neurons or cells.

Synergistic effect: Joint action of agents that when taken together increase each other's effectiveness.

TAP Apron: Toxicological Agent Protective Apron.

TC_{50}: Toxic Concentration 50% is the concentration in inhaled air needed to produce an observed toxic effect in 50% of the test animals in a given time period.

TD$_{50}$: Total Dosage 50% is the dosage by any other route other than inhalation that produces an observed toxic effect in 50% of the test animals in a given time period.

Thickened agent: An agent to which a polymer or plastic has been added to retard evaporation and cause it to adhere to surfaces.

Threshold: A level of chemical exposure below which there is no adverse effect and above which there is significant toxicological effect. ,

TLV: Threshold Limit Value. An estimate of the average safe airborne concentration of a substance; conditions under which it is believed that nearly all workers may be repeatedly exposed without adverse effect.

Toxic: Relating to or caused by a toxin, able to cause injury by contact or systemic action. The ability of a material to injure biological tissue; poisonous.

Toxicology: The study of nature, effects, and detection of poisons in living organisms. The basic assumption of toxicology is that there is a relationship among the dose, the concentration at the affected site, and resulting effects.

Toxin: A colloidal poisonous substance that is a specific product of the metabolic activities of a living organism and notably toxic when introduced into living tissue.

Transfer: The process of moving a liquid, gas, or some forms of solids from a leaking or damaged container to a secure container. Care must be taken to ensure the pump, transfer hoses and fittings, and the container selected are compatible with the hazardous materials. When hazardous substances are transferred, proper concern to electrical continuity (bonding/grounding) must be observed.

Triage: The sorting of and allocation of treatment to patients, particularly in warfare or disasters, according to a system of priorities designed to maximize the numbers of survivors.

TWA: Time Weighted Average. Usually, a personal, eight-hour exposure concentration to an airborne chemical hazard.

T-2: Trichothecene, one type of mycotoxin.

United Nations/North American Identification Number(s): UN/NA identification numbers are four-digit numbers assigned to identify and cross-reference a hazardous material (e.g., nitric acid, fuming = 2032; butane = 1011; white phosphorus, dry = 1381).

Upwind: In or toward the direction from which the wind blows.

V-agents: Persistent, highly toxic nerve agents absorbed primarily through the skin.

Vaccine: A preparation of killed or weakened infective or toxic agent used as an inoculation to produce active artificial immunity.

Vapor: The gaseous form of solid or liquid substances. Vapor can be changed to a solid or liquid by increasing the pressure or decreasing the temperature.

Vapor density: The weight of a given volume of vapor of gas compared to the weight of an equal volume of dry air, both measured at the same temperature and pressure. The ratio of the density of any gas or vapor to the density of air, under the same conditions of temperature and pressure. That is, a measure of how heavy the vapor is in relation to the same volume of air. Air is considered to have a molecular weight of "1." As an example, the higher the vapor density is in relation to "1," the longer it will persist in low-lying areas such as valleys, trenches, and cellar holes. If the vapor pressure is less than "1," a chemical agent would will likely be non-persistent and dissipate quickly into the atmosphere.

Vapor dispersion: Vapors from certain materials can be dispersed or moved using water spray or air movement. Reducing the concentration of the material may bring the material into its flammable range.

Vapor pressure (VP): Vapor pressure is a function of the substance and the temperature and is often used as a measure of how rapidly a liquid will evaporate. The pressure exerted when a solid or liquid is in equilibrium with its own vapor. A measure of the tendency of a liquid to become a gas at a given temperature. Chemical agents with a high vapor pressure evaporate rapidly while those with a low vapor pressure evaporate more slowly.

Vapor suppression: Vapor suppression refers to the reduction or elimination of vapors emanating from the spilled or released material through the application of specially designed agents, also called blanketing. Vapor suppression can also be accomplished by the use of solid activated material to treat hazardous materials. This process results in the formation of a solid that affords easier handing but requires proper disposal.

Vector: A carrier, or a host, that carries a pathogen from one host to another.

Venting: Venting is the process that is used to deal with liquids or liquefied compressed gases where a danger, such as an explosion or mechanical rupture of the container or vessel, is considered likely. The method of venting will depend on the nature of the hazardous material. In general, it involves the controlled release of material to reduce and contain the pressure and diminish the probability of an explosion.

Vesicant: An agent that operates on the eyes and lungs and is capable of producing blisters.

Virus: Any of various submicroscopic pathogens consisting of a core of a single nucleic acid surrounded by a protein coat, having the ability to replicate only inside a living cell.

Volatility: Passing off rapidly in the form of vapor; evaporating rapidly. It provides a measure of how much material evaporates under given conditions and varies directly with temperature. A measure of how readily a substance will vaporize. Volatility is directly related to vapor pressure.

Warm zone: The area where personnel and equipment decontamination takes place.

Weapons of Mass Detruction (WMD): Any explosive, incendiary, or poison gas, bomb, grenade, rocket having a propellant charge of more than four ounces, missile having an explosive or incendiary charge of more than one quarter ounce, mine or device similar to the above; poison gas; any weapon involving a disease organism; or any weapon that is designed to release radiation or radioactivity at a level dangerous to human life.

2-PAM Chloride: Trade names protopam chloride, or pralidoxime chloride. 2-PAM chloride can be used in the treatment of nerve agent poisoning.

Bibliography

Building a Systems Approach for Health and Medical Response to Acts of NBC Terrorism. Office of Emergency Preparedness (OEP), U.S. Department of Health and Human Services (HHS), 1996.

Chemical Accident Contamination Control. Department of the Army Field Manual FM 3-21, 1978.

Chemical/Biological Incident Handbook. Director of Central Intelligence, for the Intelligence Committee on Terrorism, and the Community Counterterrorism Board, 1995.

Code of Federal Regulations Title 49-Transportation (Parts 178-199). U.S. Government Printing Office, Washington, D.C., 1997.

Combating Terrorism (Federal Agencies' Efforts to Implement National Policy and Strategy). U.S. General Accounting Office (GAO/NSIAD-97-254), September, 1997.

Competencies for EMS Personnel Responding to Hazardous Materials Incidents - NFPA 473. National Fire Protection Association, Quincy, MA,1992.

Department of Health and Human Services Health and Medical Services Support Plan for the Federal Response Acts of Chemical/Biological (C/B) Terrorism. Office of Emergency Preparedness (OEP), of the Department of Health and Human Services (HHS), undated.

Franz, D.R., Defense Against Toxin Weapons, U.S. Army Medical Research and Materiel Command, Fort Detrick, MD, 1997.

Domestic Preparedness Program in the Defense Against Weapons of Mass Destruction (Department of Defense Report To Congress). 1997.

Emergency Response To Terrorism Self-Study. U.S. Department of Justice, Office of Justice Programs, and Federal Emergency Management Agency, United States Fire Administration, FEMA/USFA - ERT:SS, 1997.

Field Behavior of NBC Agents. Department of the Army Field Manual FM 3-6, 1986.

Baker, C.J., *The Firefighters Handbook of Hazardous Materials* (5th ed.). Maltese Enterprises, Indianapolis, IN, 1990.

Stern, K.S., *A Force Upon the Plain,* Simon & Schuster, New York, 1996.

Health Service Support in a Nuclear, Biological, and Chemical Environment. Department of the Army Field Manual FM 8-10-7, 1993.

Lewis, R.J., Sr., *Hazardous Chemicals Desk Reference* (4th ed.). Van Nostrand Reinhold, a Division of International Thompson Publishing, New York, 1997.

Hazardous Materials Emergency Response Planning Guide. The National Response Team of the National Oil and Hazardous Substances Contingency Plan, Washington, D.C., 1987.

Hazardous Materials Emergency Response Training. Boeing Commercial Airplane Group, Seattle, WA, 1994 (unpublished material).

Borak, J., M. Callan, and W. Abbott. *Hazardous Materials Exposure - Emergency Response and Patient Care.* Prentice-Hall, Englewood Cliffs, NJ, 1991.

Liquid Splash-Protective Suits for Hazardous Chemical Emergencies - NFPA 1992. National Fire Protection Association, Quincy, MA, 1994.

Sidell, F.R., *Management of Chemical Warfare Casualties: a Handbook for Emergency Medical Services.* HB Publishing, Bel Air, MD, 1995.

Medical Management of Biological Casualties Handbook (2nd ed.). U.S. Army Medical Research Institute of Infectious Diseases, Fort Detrick, MD, 1996.

Medical Management of Chemical Casualties Handbook (2nd ed.). Medical Research Institute of Chemical Defense, Aberdeen Proving Ground, MD, 1995.

Metropolitan Medical Strike Team Operational System Description. Metropolitan Washington Council of Governments, and the United States Public Health Service, Office of Emergency Preparedness, with the U.S. Department of Health and Human Services, Washington, D.C., 1996.

NATO Handbook on the Medical Aspects of NBC Defensive Operations, Part II, Biological AMedP-6(B). Department of the Army Field Manual FM 8-9, 1996.

NATO Handbook on the Medical Aspects of NBC Defensive Operations, Part III, Chemical AMedP-6(B). Department of the Army Field Manual FM 8-9, 1996.

NBC Decontamination. Department of the Army Field Manual FM 3-5, 1993.

NBC Field Handbook. Department of the Army Field Manual FM 3-7, 1994.

NBC Protection. Department of the Army Field Manual FM 3-4, 1992.

NIOSH Pocket Guide To Chemical Hazards. U.S. Department of Health and Human Services, Public Health Service, Centers for Disease Control, and the National Institute for Occupational Health and Safety. Superintendent of Documents, U.S. Government Printing Office, Washington, D.C., 1997.

2000 North American Emergency Response Guidebook (A Guidebook for First Responders During the Initial Phase of a Hazardous Materials/Dangerous Goods Incident. U.S. Department of Transportation/Transport Canada/the Secretariat of Communications and Transportation for Mexico, 1999.

Occupational Safety and Health Guidance Manual for Hazardous Waste Site Activities. National Institute for Occupational Safety and Health, Occupational and Health Administration, the U.S. Coast Guard, and the U.S. Environmental Protection Agency. Published by the U.S. Department of Health and Human Services, Public Health Service, Centers for Disease Control, and the National Institute for Occupational Safety and Control, 1985

Ronk, R., M.K. White, and H. Linn. *Personal Protective Equipment for Hazardous Materials Incidents: a Selection Guide.* U.S. Department for Health and Human Services, Public Health Service, Centers for Disease Control, and the National Institute for Occupational Safety and Health, Superintendent of Documents, U.S. Government Printing Office, Washington, D.C., 1984.

Potential Military Chemical/Biological Agents and Compounds. Department of the Army Field Manual FM 3-9, 1990.

Proceedings of the Seminar on Responding to the Consequences of Chemical and Biological Terrorism. Sponsored by the U.S. Public Health Service, Office of Emergency Preparedness, July 11–14, 1995.

Professional Competence of Responders to Hazardous Materials Incidents - NFPA 472. National Fire Protection Association, Quincy, MA, 1992.

Protective Clothing for Emergency Medical Operations - NFPA 1999. National Fire Protection Association, Quincy, MA, 1992.

Recommended Practice For Responding To Hazardous Materials Incidents - NFPA 471. National Fire Protection Association, Quincy, MA, 1992.

SALT LAKE TRIBUNE, Salt Lake, UT, January 1, 1998.

Lewis, R.J., Sr., *Sax's Dangerous Properties of Industrial Materials* (three volumes). Van Nostrand Reinhold, a Division of International Thompson Publishing, New York, 1995.

Support Function Protective Clothing for Hazardous Chemicals Operations - NFPA 1993. National Fire Protection Association, Quincy, MA, 1994.

Technical Bulletin - Assay Techniques For Detection of Exposure To Sulfur Mustard, Cholinesterase Inhibitors, Sarin, Soman, GF, and Cyanide. Department of the Army TB MED 296, 1996.

Terrorism in the United States 1995. Federal Bureau of Investigation, U.S. Department of Justice, 1996.

Terrorism in the United States 1996. Federal Bureau of Investigation, U.S. Department of Justice, 1997.

Terrorism in the United States 1997. Federal Bureau of Investigation, U.S. Department of Justice, 1998.

Treatment of Chemical Agent Casualties and Conventional Military Chemical Injuries. Department of the Army Field Manual FM8-285.

U.S. Navy Shipboard Chemical-Hazard Assessment Guide (C-HAG). 1990.

U.S. Policy on Counterterrorism (Presidential Decision Directive 39). 1995.

Vapor - Protective Suits for Hazardous Chemical Emergencies - NFPA 1991. National Fire Protection Association, Quincy, MA, 1994.

Index

VX
   antidotes for, 17
   decontamination, 22, 42
   detection of, 42
   persistence of, 20
   properties of, 41–42

W

Weapons of mass destruction,
      9–10, *see also* Biologi-
      cal agents; Chemical
      agents
Weather
   characteristics of, 1–2
   description of, 1–2
   destruction of, 2
   members of, 1
Wilmington Hazardous Materi-
      als Response Team,
      197–200
WMD, *see* Weapons of mass
      destruction

X

XM88 ACADA, 88